T0324889

Computational Methods for Geodynamics

Computational Methods for Geodynamics describes all the numerical methods typically used to solve problems related to the dynamics of the Earth and other terrestrial planets, including lithospheric deformation, mantle convection and the geodynamo.

The book starts with a discussion of the fundamental principles of mathematical and numerical modelling, which is then followed by chapters on finite difference, finite volume, finite element and spectral methods; methods for solving large systems of linear algebraic equations and ordinary differential equations; data assimilation methods in geodynamics; and the basic concepts of parallel computing. The final chapter presents a detailed discussion of specific geodynamic applications in order to highlight key differences between methods and demonstrate their respective limitations. Readers learn when and how to use a particular method in order to produce the most accurate results.

This combination of textbook and reference handbook brings together material previously available only in specialist journals and mathematical reference volumes, and presents it in an accessible manner assuming only a basic familiarity with geodynamic theory and calculus. It is an essential text for advanced courses on numerical and computational modelling in geodynamics and geophysics, and an invaluable resource for researchers looking to master cutting-edge techniques. Links to online source codes for geodynamic modelling can be found at www.cambridge.org/zadeh.

ALIK ISMAIL-ZADEH is a Senior Scientist at the Karlsruhe Institute of Technology (KIT), Chief Scientist of the Russian Academy of Sciences at Moscow (RusAS) and Professor of the Institut de Physique du Globe de Paris. He graduated from the Baku State and Lomonossov Moscow State Universities before being awarded Ph.D. and Doctor of Science degrees in geophysics from RusAS. He lectures on computational geodynamics at KIT, Abdus Salam International Center for Theoretical Physics in Trieste, and Moscow State University of Oil and Gas, while his research interests cover crust and mantle dynamics, basin evolution, salt tectonics and seismic hazards. Professor Ismail-Zadeh is the recipient of the 1995 Academia Europaea Medal and the 2009 American Geophysical Union International Award, and is Secretary-General of the International Union of Geodesy and Geophysics.

PAUL TACKLEY is Chair of the Geophysical Fluid Dynamics Group in the Institute of Geophysics, Department of Earth Sciences, Swiss Federal Institute of Technology (ETH Zürich). He received an MA from the University of Cambridge and an MS and Ph.D. from the California Institute of Technology before taking up a position in the Department of Earth and Space Sciences and Institute of Geophysics and Planetary Physics at the University of California, Los Angeles. He became a full professor there before moving to ETH Zürich in 2005, where he currently teaches courses in geodynamic modelling. Professor Tackley's research involves applying large-scale three-dimensional numerical simulations using state of the art methods and parallel supercomputers to study the structure, dynamics and evolution of the Earth and other terrestrial planets. He has served as an associate editor for various journals and is on the editorial board of *Geophysical and Astrophysical Fluid Dynamics*.

Cover illustration (front): upper images by M. Armann show numerical simulations of thermo-chemical convection in a stagnant- or episodic-lid planet such as Venus, with (left) composition ranging from basalt (red) to harzburgite (blue) and (right) potential temperature (simulations by M. Armann and P.J. Tackley); lower images by T. Nakagawa show numerical simulations of thermo-chemical convection in a mobile-lid planet such as Earth; isosurfaces show cold (blue) and hot (red) temperature anomalies and (green) basaltic composition (simulations by T. Nakagawa and P.J. Tackley).
(back): the images by I. Tsepelev show (top) the time snapshots of the thermal evolution of the descending slab (blue, dark cyan and light cyan mark the surfaces of different temperature anomalies) and pattern of mantle flow (arrows illustrate the flow's direction and magnitude) beneath the south-eastern Carpathians. The model evolution is restored numerically using the quasi-reversibility method for data assimilation (Ismail-Zadeh *et al.*, 2008).

Computational Methods for Geodynamics

Alik Ismail-Zadeh

Karlsruhe Institute of Technology (KIT)
Moscow Institute of Mathematical Geophysics, Russian Academy of Sciences (MITPAN)
Institute de Physique du Globe de Paris (IPGP)

Paul J. Tackley

Swiss Federal Institute of Technology Zurich (ETH)

CAMBRIDGE
UNIVERSITY PRESS

Shaftesbury Road, Cambridge CB2 8EA, United Kingdom

One Liberty Plaza, 20th Floor, New York, NY 10006, USA

477 Williamstown Road, Port Melbourne, VIC 3207, Australia

314–321, 3rd Floor, Plot 3, Splendor Forum, Jasola District Centre, New Delhi – 110025, India

103 Penang Road, #05–06/07, Visioncrest Commercial, Singapore 238467

Cambridge University Press is part of Cambridge University Press & Assessment,
a department of the University of Cambridge.

We share the University's mission to contribute to society through the pursuit of education, learning and research at the highest international levels of excellence.

www.cambridge.org
Information on this title: www.cambridge.org/9780521867672

First published 2010

A catalogue record for this publication is available from the British Library

Library of Congress Cataloging-in-Publication data
Ismail-Zadeh, Alik.
Computational methods for geodynamics / Alik Ismail-Zadeh, Paul J. Tackley.
p. cm.
Includes bibliographical references and index.
ISBN 978-0-521-86767-2
1. Geodynamics–Data processing. I. Tackley, Paul J. II. Title.
QE501.3.I86 2010
551–dc22 2010017823

ISBN 978-0-521-86767-2 Hardback

Additional resources for this publication at www.cambridge.org/zadeh

To David, Junko, and Sonya
as a sign of deep affection

Contents

Colour plate section between pages 238 *and* 239.

Foreword

Geodynamics is the application of the basic principles of physics, chemistry and mathematics to understanding how the internal activity of the Earth results in all the geological phenomena and structures apparent at the surface, including seafloor speading and continental drift, mountain building, volcanoes, earthquakes, sedimentary basins, faulting, folding, and more. Geodynamics also deals with how the Earth's internal activity and structure reveals itself externally in ways both geophysical, its gravitational and magnetic fields, and geochemical, the mineralogy of its rocks and the isotopic composition of its rocks, atmosphere, and ocean. The discipline of geodynamics did not exist until about the early 1970s. The plate tectonics revolution was the impetus for the birth of the subject. Today, geodynamics goes beyond the Earth to consider the interiors and surfaces of other planets and moons in our solar system. While this aspect of the science could be termed planetary dynamics, it involves the same geodynamical processes that shape the Earth, though often with intriguingly different outcomes for the other bodies.

Mathematical modeling, which attempts to understand a phenomenon quantitatively, lies at the heart of geodynamics. In the early years of the subject analytic and semi-analytic methods were sufficient to gain insights into the workings of the Earth's interior. After four decades of progress in the subject it is, generally speaking, no longer possible to address the remaining questions with such simple models. Indeed, it has been necessary for some time now, to employ sophisticated numerical computational models to achieve further understanding of the complex dynamics of the Earth. Accordingly, researchers now entering the field of geodynamics need to acquire the skills to understand the numerical methods upon which computational geodynamics codes are based. An understanding of the methods is required not only for intelligent use of existing codes but also to enable adaptations of the codes and future improvements in them. The present book responds to this need by thoroughly discussing the many different numerical schemes designed to provide approximate solutions to the ordinary and partial differential equations encountered in geodynamical problems and by emphasizing the fundamental principles behind the numerical approaches. It is a book that goes far beyond the black box utilization of numerical tools by providing the student with a deep understanding of the numerical approaches upon which the codes are based.

The book begins with an introductory chapter discussing the basic equations and the boundary and initial conditions of geodynamics followed by general remarks on the numerical approach to their solution. The succeeding chapters discuss all the widely used numerical schemes in detail, finite differences, finite volumes, finite elements, and spectral decomposition. The succeeding two chapters apply these methods to solving linear algebraic equations and ordinary differential equations. The concluding chapters deal with data assimilation, parallel computing, and applications in geodynamics. A number of appendices provide mathematical background.

The authors are distinguished geodynamicists with decades of experience in numerical modeling. They are at the forefront of geodynamical modeling and are responsible for the initial development and continued improvement of state-of-the-art codes. They have written a clear and comprehensive book that everyone working in the field of geodynamics would be well advised to read and keep handy for future reference.

Gerald Schubert
Distinguished Professor of Geophysics and Planetary Physics
University of California
Los Angeles, California
USA

Preface

The book of nature is written in the language of mathematics
(Galileo Galilei, 1564–1642)

All the mathematical sciences are founded on relations between physical laws and laws of numbers, so that the aim of exact science is to reduce the problems of nature to the determination of quantities by operations with numbers
(James Clerk Maxwell, 1831–1879)

It is impossible to explain honestly the beauties of the laws of nature in a way that people can feel, without their having some deep understanding of mathematics and its methods
(Richard Feynman, 1918–1988)

Great advances in understanding of the planet Earth and in computational tools permitting accurate numerical modelling are transforming the geosciences in general and geodynamics particularly. Research on dynamical processes in the Earth and planets relies increasingly on sophisticated quantitative models. Improved understanding of fundamental physical processes such as mantle convection, lithospheric deformation, and core dynamos in the Earth and terrestrial planets depends heavily on better numerical modelling. Characteristic of this new intellectual landscape is the need for strong interaction across traditional disciplinary boundaries: Earth sciences, applied mathematics, and computer science.

Solid Earth scientists, with few exceptions, rarely achieve mathematical competence beyond elementary calculus and a few statistical formulae. Meanwhile, in some sense it has become a fashion nowadays, when scientists dealing with geodynamics make numerical modelling as their primary research tool. Most of these scientists employ standard commercial software or the codes developed by representatives of the geodynamic community and do not take care of numerical methods and their limitations behind the software and codes. Sometimes numerical results of complicated models, being wrong from a mathematical point of view, can feature the Earth dynamics in a 'realistic' way and can hence lead to wrong physical interpretations. To distinguish between wrong and true solutions, geodynamicists should know more about numerical techniques and their applicability.

Our motivation to write the book grew steadily from about two decades of experience with students and young scientists in geodynamics (both geologists or geophysicists), who were sometimes disarmed when it comes to understanding of essential features of mathematical and numerical modelling, like how a numerical code works to produce accurate results, which computational methods are behind the codes employed and what is the difference between, let us say, finite differences and finite elements methods, etc. To understand mathematical and computer limitations of numerical modelling, every user should firmly know how the employed numerical methods work for the problems under study. Even though most geoscience students take several semesters of prerequisite courses in maths and/or computer sciences, these students have sometimes little education and experience in quantitative thinking and computation to prepare them to participate in the new world of quantitative geosciences. In order to participate fully in the research of the future, it will be essential for geoscientists to be conversant not only with the language of geology and geophysics but also with the languages of applied mathematics and computers. If all areas of the geosciences are to assimilate into the world of quantitative science, students in geosciences will need a different kind of education than we provide today.

The book *Computational Methods for Geodynamics* bridges two cultures within geosciences (quantitative and qualitative) and assists in solving the problems related to dynamics of the Earth using computational methods. We did not consider filling the gap between geodynamics on the one hand and applied mathematics and computer science on the other hand, but rather to contribute to the understanding of computational geodynamics via basics of numerical modelling, computational methods and challenges in numerical simulations. We believe that this book will complement several excellent textbooks on geodynamics, like *Geodynamics* by Turcotte and Schubert, *Mantle Convection in the Earth and Planets* by Schubert, Turcotte and Olson and *The Solid Earth* by Fowler, in terms of quantitative understanding of geodynamical problems. It will assist students in choosing appropriate numerical methods and algorithms to analyse their research problems (e.g. mantle and lithospheric dynamics, thermal or thermo-chemical convection, geodynamo). This book offers readers the possibility of finding efficient computational methods to be employed in geodynamic modelling (not spending a lot of time in search for the methods in a vast amount of research papers and specialised mathematical books) and seeing examples of how a particular method works for a specific problem. The book can also be of interest to researchers dealing with computational geodynamics as well as to other quantitative geoscientists.

We tried to make the mathematical language of the book not too complicated, and the maths formulations are kept at the level necessary to understand the computational methods. The book is organised into the following parts: fundamental formulations; basic numerical approaches and essential numerical methods; and applications. The first chapter defines the discipline of computational geodynamics and describes the main principles of mathematical and numerical modelling of geodynamic problems. Chapters 2 to 5 describe the finite difference, finite volume, finite element and spectral methods. Chapters 6 and 7 present the basic numerical methods for solving the systems of linear algebraic equations and ordinary differential equations. Chapter 8 is devoted to the methods for data assimilation in geodynamics and presents some applications. The basic concepts of parallel computing are presented in Chapter 9. We discuss how different numerical methods have been used in modelling of various geodynamics problems in Chapter 10.

We have to apologise that the book does not contain all computational methods for geodynamic modelling. We omitted the mesh-free methods, like the discrete element method

(DEM) or the element-free Galerkin method, because these methods are not employed often in modelling of dynamics of the Earth interior and are used mostly to simulate the bulk behaviour of granular material, strain localisation and shear band formations. The reader is referred to some classical works on this topic (e.g. Cundall and Strack, 1979; Liu, 2003) as well as application of these methods to geodynamics (e.g. Hansen, 2003; Egholm, 2007; and references therein).

We have tried to show how mathematical and numerical methods contribute to understanding dynamics of the Earth interior, and how the boundaries between the disciplines are becoming arbitrary and irrelevant. We hope that the book will allow students to learn the languages of the different disciplines in context. Scientists educated in this way, regardless of their ultimate professional speciality, would share a common scientific language, facilitating both cross-disciplinary understanding and collaboration. Mathematics and computational methods provide the best way of understanding complex natural systems, and a good mathematical education for geoscientists is the best route for enabling the most able people to address really important problems in Earth sciences.

Acknowledgements

The idea of writing a book on computational methods for geodynamics emerged during several conversations of Alik Ismail-Zadeh with Simon Mitton of the University of Cambridge. We are grateful to Simon for his kind encouragement and assistance in producing the book's proposal to Cambridge University Press. This idea was further developed during the 2003 sabbatical leave of Ismail-Zadeh at the University of California at Los Angeles engineered by Gerald Schubert. We are very thankful to him for his fruitful discussions on computational geodynamics that helped in the selection of topics for this book. We are very grateful to several anonymous reviewers for constructive comments on the content of this book, which improved our original book proposal and resulted in the volume that you hold in your hands.

We thank our colleagues for in-depth and fruitful discussions on numerical methods, computational geodynamics and mathematical approaches in Earth Sciences and/or for their review of parts of the book's manuscript: Grigory Barenblatt, Klaus-Jürgen Bathe, Dave Bercovici, Uli Christensen, Taras Gerya, Mike Gurnis, Uli Hansen, Satoru Honda, Wolf Jacoby, Boris Kaus, Vladimir Keilis-Borok, Alex Korotkii, Jahja Mamedov, Dave May, Boris Naimark, Michael Navon, Neil Ribe, Harro Schmeling, Gerald Schubert, Alexander Soloviev, Chris Talbot, Andrei Tikhonov, Valery Trubitsyn, Igor Tsepelev and Dave Yuen. We are very thankful to Fedor Winberg, who helped in a search of the relevant literature. We acknowledge with great pleasure the support from the institutions where the book's chapters were written: Swiss Federal Institute of Technology in Zurich, Institut de Physique du Globe de Paris, Karlsruhe Institute of Technology, Moscow Institute of Mathematical Geophysics of the Russian Academy of Sciences, University of California at Los Angeles and the University of Tokyo.

We will be very grateful for comments, inquiries and complaints on the content of the book.

1 Basic concepts of computational geodynamics

1.1 Introduction to scientific computing and computational geodynamics

Present life without computers is almost impossible: industry and agriculture, government and media, transportation and insurance are major users of computational power. The earliest and still principal users of computers are researchers who solve problems in science and engineering or more specifically, who obtain solutions of mathematical models that represent some physical situation. The methods, tools and theories required to obtain such solutions are together called *scientific computing*, and the use of these methods, tools and theories to resolve scientific problems is referred to as *computational science*. A majority of these methods, tools, and theories were developed in mathematics well before the advent of computers. This set of mathematical theories and methods is an essential part of numerical mathematics and constitutes a major part of scientific computing. The development of computers signalled a new era in the approach to the solution of scientific problems. Many of the numerical methods initially developed for the purpose of hand calculation had to be revised; new techniques for solving scientific problems using electronic computers were intensively developed. Programming languages, operating systems, management of large quantities of data, correctness of numerical codes and many other considerations relevant to the efficient and accurate solution of the problems using a large computer system became subjects of the new discipline of computer science, on which scientific computing now depends heavily. Mathematics itself continues to play a major role in scientific computing: it provides the information about the suitability of a model and the theoretical foundation for the numerical methods.

There is now almost no area of science that does not use computers for modelling. In geosciences, meteorologists use parallel supercomputers to forecast weather and to predict the change of the Earth's climate; oceanographers use the power of computers to model oceanic tsunamis and to estimate harmful effects of the hazards on coastal regions; solid Earth physicists employ computers to study the Earth's deep interior and its dynamics. The planet Earth is a complex dynamical system. To gain a better understanding of the evolution of our planet, several concepts from various scientific fields and from mathematics should be combined in computational models. Great advances in understanding the Earth as well as in experimental techniques and in computational tools are transforming geoscience in general and geodynamics particularly.

Modern geodynamics was born in the late 1960s with the general acceptance of the plate tectonics paradigm. At the beginning, simple analytical models were developed to explain plate tectonics and its associated geological structures. These models were highly successful in accounting for many of the first order behaviours of the Earth. The necessity to go beyond these basic models to make them more realistic and to understand better the Earth shifted the emphasis to numerical simulations. These numerical models have grown increasingly complex and capable over time with improvements in computational power and numerical algorithms. This has resulted in the development of a new branch of geoscience: *computational geodynamics*.

Characteristic of this new intellectual landscape is the need for strong interaction across traditional disciplinary boundaries: *geodynamics*, *mathematics* and *computer science*. Computational geodynamics can be defined as a blending of these three areas to obtain a better understanding of some phenomena through a match between the problem, computer architecture and algorithms. The computational approach to geodynamics is inherently *multi-disciplinary*. Mathematics provides the means to establish the credibility of numerical methods and algorithms, such as error analysis, exact solutions, uniqueness and stability analysis. Computer science provides the tools, ranging from networking and visualisation tools to algorithms matching modern computer architectures.

1.2 Mathematical models of geodynamic problems

Many geodynamic problems can be described by mathematical models, i.e. by a set of partial differential equations and boundary and/or initial conditions defined in a specific domain. Models in computational geodynamics predict quantitatively what will happen when the crust and the mantle deform slowly over geological time, often with the complications of simultaneous heat transport (e.g. thermal convection in the mantle), phase changes in the deep interior of the Earth, complex rheology (e.g. non-Newtonian flow, elasticity and plasticity), melting and melt migration, chemical reactions (e.g. thermo-chemical convection), solid body motion (e.g. idealised continent over the mantle), lateral forces, etc.

A mathematical model links the causal characteristics of a geodynamic process with its effects. The causal characteristics of the modelled process include, for example, parameters of the initial and boundary conditions, coefficients of the differential equations, and geometrical parameters of a model domain. The aim of the *direct* (sometimes called *forward*) mathematical problem is to determine the relationship between the causes and effects of the geophysical process and hence to find a solution to the mathematical problem for a given set of parameters and coefficients.

An *inverse* mathematical problem is the opposite of a direct problem. An inverse problem is considered when there is a lack of information on the causal characteristics (but information on the effects of the geophysical process exists). Inverse problems can be subdivided into time-reverse problems (e.g. to restore the development of a geodynamic process), coefficient problems (e.g. to determine the coefficients of the model equations and/or boundary conditions), geometrical problems (e.g. to determine the location of heat sources in a model domain or the geometry of the model boundary), and some others.

Inverse problems are often ill-posed. Jacques Hadamard, a French mathematician, introduced the idea of well- (and ill-) posed problems in the theory of partial differential equations (Hadamard, 1902). A mathematical model for a geophysical problem has to be *well-posed* in the sense that it has to have the properties of (1) existence, (2) uniqueness and (3) stability of a solution to the problem. Problems for which at least one of these properties does not hold are called *ill-posed*. The requirement of stability is the most important one. If a problem lacks the property of stability then its solution is almost impossible to compute because computations are polluted by unavoidable errors. If the solution of a problem does not depend continuously on the initial data, then, in general, the computed solution may have nothing to do with the true solution. We should note that despite the fact that many inverse problems are ill-posed, there are methods for solving the problems (see, for example, Tikhonov and Arsenin, 1977). While most geodynamic models are concerned with direct (forward) problems, there is increasing interest in the inverse problem (or data assimilation), as discussed in Chapter 8.

1.3 Governing equations

In this section we present the basic equations that govern geodynamic processes. The equations are partial differential equations (PDEs), involving more than one independent variable. PDEs can be distinguished by the following property. Consider a partial differential equation in the following form: $A\Psi_{xx} + B\Psi_{xy} + C\Psi_{yy} = f(x, y, \Psi, \Psi_x, \Psi_y)$, where A, B and C are constants. Depending on $D = B^2 - 4AC$, a PDE is called *elliptic* if $D < 0$, *parabolic* if $D = 0$ or *hyperbolic* if $D > 0$. Examples of these in solid Earth dynamics are the solution of gravitational potential (elliptic), thermal diffusion (parabolic) and seismic wave propagation (hyperbolic).

Because the mantle behaves basically as a viscous fluid for the geological time scale, the governing equations describe the flow of highly viscous fluid. The basic conservation laws used to derive these equations are only briefly summarised (see Chandrasekhar, 1961, and Schubert *et al.*, 2001, for details).

1.3.1 The equation of continuity

Consider a fluid in which the density ρ is a function of position x_j ($j = 1, 2, 3$ hereinafter). Let u_j denote the components of the velocity. We shall use the notation of Cartesian tensors with the usual summation convention. Consider the physical law of the *conservation of mass*: the rate of change of the mass contained in a fixed volume V of the fluid is given by the rate at which the fluid flows out of it across the boundary S of the volume. Mathematically it is expressed as

$$\frac{\partial}{\partial t}\int_V \rho d\tau = -\int_S \rho u_j dS_j, \tag{1.1}$$

where τ is the volume element. The use of the Gauss–Ostrogradsky (divergence) theorem transforms the law of mass conservation into the following equation

$$\frac{\partial}{\partial t}\int_V \rho d\tau = -\int_V \frac{\partial}{\partial x_j}(\rho u_j)d\tau. \tag{1.2}$$

An alternative form of the equation, which is useful for numerical analysis is the Lagrangian continuity equation:

$$\frac{D\rho}{Dt} \equiv \frac{\partial \rho}{\partial t} + u_j\frac{\partial \rho}{\partial x_j} = -\rho\frac{\partial u_j}{\partial x_j}, \tag{1.3}$$

which can also be written in Eulerian form:

$$\frac{\partial \rho}{\partial t} = -\frac{\partial}{\partial x_j}(\rho u_j). \tag{1.4}$$

For an incompressible fluid, the equation of continuity reduces to:

$$\frac{\partial u_j}{\partial x_j} = 0, \text{ because } \frac{\partial \rho}{\partial t} + u_j\frac{\partial \rho}{\partial x_j} = \frac{D\rho}{Dt} = 0. \tag{1.5}$$

1.3.2 The equation of motion

Consider the physical law of the *conservation of momentum*: the rate of change of the momentum contained in a fixed volume V of the fluid is equal to the volume integral of the external body forces acting on the elements of the fluid plus the surface integral of normal and shear stresses acting on the bounding surface S of the volume V minus the rate at which momentum flows out of the volume across the boundaries of V by the motions prevailing on the surface S. Mathematically it is expressed as

$$\frac{\partial}{\partial t}\int_V \rho u_i d\tau = \int_V \rho F_i d\tau + \int_S \sigma_{ij} dS_j - \int_S \rho u_i u_j dS_j, \tag{1.6}$$

where F_i $(= g_i)$ is the ith component of external (usually gravity) force per unit of mass; and σ_{ij} is the stress tensor. We note that

$$\frac{\partial}{\partial t}(\rho u_i) = \rho\frac{\partial u_i}{\partial t} + u_i\frac{\partial \rho}{\partial t} = \rho\frac{\partial u_i}{\partial t} - u_i\frac{\partial}{\partial x_j}(\rho u_j). \tag{1.7}$$

If we substitute now expression (1.7) into (1.6), we obtain

$$\int_V \left(\rho\frac{\partial u_i}{\partial t} - u_i\frac{\partial}{\partial x_j}(\rho u_j)\right) d\tau + \int_S \rho u_i u_j dS_j = \int_V \rho F_i d\tau + \int_S \sigma_{ij} dS_j. \tag{1.8}$$

Integrating by parts the second term of the first volume integral, we obtain

$$-\int_V u_i \frac{\partial}{\partial x_j}(\rho u_j) d\tau + \int_S \rho u_i u_j dS_j = \int_V \rho u_j \frac{\partial u_i}{\partial x_j} d\tau. \tag{1.9}$$

Application of the Gauss–Ostrogradsky theorem to the last term in (1.8) gives:

$$\int_S \sigma_{ij} dS_j = \int_V \frac{\partial \sigma_{ij}}{\partial x_j} d\tau. \tag{1.10}$$

Substituting Eqs. (1.9) and (1.10) in (1.8) we obtain the *equation of motion* which is valid for any arbitrary volume V

$$\rho \frac{\partial u_i}{\partial t} + \rho u_j \frac{\partial u_i}{\partial x_j} = \rho F_i + \frac{\partial \sigma_{ij}}{\partial x_j}. \tag{1.11}$$

For linear viscous creep, the stress is related to the rate of increase of strain (strain rate) as

$$\begin{aligned} \sigma_{ij} &= -P\delta_{ij} + 2\eta\dot{\varepsilon}_{ij} + \left(\eta_B - \frac{2}{3}\eta\right)\delta_{ij}\frac{\partial u_k}{\partial x_k} \\ &= -P\delta_{ij} + \eta\left(\frac{\partial u_i}{\partial x_j} + \frac{\partial u_j}{\partial x_i} - \frac{2}{3}\delta_{ij}\frac{\partial u_k}{\partial x_k}\right) + \eta_B\delta_{ij}\frac{\partial u_k}{\partial x_k}, \end{aligned} \tag{1.12}$$

where P is the pressure, δ_{ij} is the Kronecker delta, η is the viscosity, η_B is the bulk viscosity, and $\dot{\varepsilon}_{ij}$ is the strain rate tensor. As compaction or dilation is normally accommodated elastically, η_B is usually assumed to be zero. By substituting the relationship (1.12) into the equation of motion (1.11) and assuming $\eta_B = 0$, we obtain

$$\rho \frac{\partial u_i}{\partial t} + \rho u_j \frac{\partial u_i}{\partial x_j} = \rho F_i - \frac{\partial P}{\partial x_i} + \frac{\partial}{\partial x_j}\left\{\eta\left(\frac{\partial u_i}{\partial x_j} + \frac{\partial u_j}{\partial x_i} - \frac{2}{3}\delta_{ij}\frac{\partial u_k}{\partial x_k}\right)\right\}. \tag{1.13}$$

For an incompressible, constant-viscosity fluid, equation (1.13) simplifies to

$$\rho \frac{\partial u_i}{\partial t} + \rho u_j \frac{\partial u_i}{\partial x_j} = \rho F_i - \frac{\partial P}{\partial x_i} + \eta \nabla^2 u_i. \tag{1.14}$$

Equation (1.14) represents the original form of the *Navier–Stokes* equations.

Now we show that in geodynamical applications the Navier–Stokes equations (1.14) are transformed into the Stokes equations. Let us define new dimensionless variables and parameters (denoted by a tilde) as $t = \tilde{t}l_*/\kappa_*$, $\mathbf{x} = \tilde{\mathbf{x}}l_*$, $\mathbf{u} = \tilde{\mathbf{u}}\kappa_*/l_*$, $P = \tilde{P}\eta_*\kappa_*/l_*^2$, $\rho = \tilde{\rho}\rho_*$, and $\eta = \tilde{\eta}\eta_*$, where $\rho_* = 4 \times 10^3$ kg m^{-3}, $\eta_* = 10^{21}$ Pa s, $l_* = 3 \times 10^6$ m, and $\kappa_* = 10^{-6}$ m^2 s^{-1} are typical values of the density, viscosity, length and thermal diffusivity for the Earth's mantle, respectively. We assume that $F_i = (0, 0, g)$, where $g = 9.8$ m s^{-2} is the acceleration due to gravity. After the replacement of the variables by their dimensionless form (and omitting tildes), we obtain:

$$\frac{1}{Pr}\rho\left(\frac{\partial u_i}{\partial t} + u_j\frac{\partial u_i}{\partial x_j}\right) = -\frac{\partial P}{\partial x_i} + \frac{\partial}{\partial x_j}\left\{\eta\left(\frac{\partial u_i}{\partial x_j} + \frac{\partial u_j}{\partial x_i} - \frac{2}{3}\delta_{ij}\frac{\partial u_k}{\partial x_k}\right)\right\} + La\,\rho\delta_{i3}, \tag{1.15}$$

where the dimensionless parameter $Pr = \frac{\eta_*}{\rho_* \kappa_*} = 2.5 \times 10^{23}$ is the Prandtl number; and the dimensionless parameter $La = \frac{\rho_* g l_*^3}{\eta_* \kappa_*} \sim 10^9$ is the Laplace number. Note that $La = Ra/(\alpha \Delta T)$, where Ra is the Rayleigh number controlling the vigour of thermal convection, α is the thermal expansivity and ΔT is the typical temperature variation. Therefore, (1.15) are reduced to the following elliptic equations called the *Stokes equations*:

$$0 = -\frac{\partial P}{\partial x_i} + \frac{\partial}{\partial x_j}\left\{\eta\left(\frac{\partial u_i}{\partial x_j} + \frac{\partial u_j}{\partial x_i} - \frac{2}{3}\delta_{ij}\frac{\partial u_k}{\partial x_k}\right)\right\} + \frac{Ra}{\alpha \Delta T}\rho\delta_{i3} \tag{1.16}$$

or, in dimensional units,

$$0 = -\frac{\partial P}{\partial x_i} + \frac{\partial}{\partial x_j}\left\{\eta\left(\frac{\partial u_i}{\partial x_j} + \frac{\partial u_j}{\partial x_i} - \frac{2}{3}\delta_{ij}\frac{\partial u_k}{\partial x_k}\right)\right\} + \rho F_i. \tag{1.17}$$

For incompressible flow the $-\frac{2}{3}\delta_{ij}\frac{\partial u_k}{\partial x_k}$ term is omitted. For constant viscosity and incompressible flow the second term reduces to $\eta \nabla^2 u_i$ as in Eq. (1.14).

1.3.3 The heat equation

Consider the physical law of the *conservation of energy*. Counting the gains and losses of energy that occur in a volume V of the fluid, per unit time, we have

$$\frac{\partial}{\partial t}\int_V \rho E d\tau = \int_S u_i \sigma_{ij} dS_j + \int_V \rho u_i F_i d\tau - \int_S k\frac{\partial T}{\partial x_j} dS_j - \int_S \rho E u_j dS_j + \int_V \rho H d\tau. \tag{1.18}$$

Here the first term of the right-hand side of the Eq. (1.18) is the rate at which work is done on the boundary; the second term represents the rate at which work is done on each element of the fluid inside V by the external forces; the third term is the rate at which energy in the form of heat is conducted across S; the fourth term is the rate at which energy is convected across S by the prevailing mass motion (k is the coefficient of heat conduction); and the fifth term is the rate at which energy is added by internal heat sources. The first and third terms of Eq. (1.18) can be represented as follows:

$$\int_S u_i \sigma_{ij} dS_j = \frac{1}{2}\frac{\partial}{\partial t}\int_V \rho u_i^2 d\tau + \frac{1}{2}\int_S \rho u_i^2 u_j dS_j - \int_V \rho u_i F_i d\tau + \int_V \Phi d\tau, \tag{1.19}$$

where $\Phi = \frac{\partial u_i}{\partial x_j}\sigma_{ij}$ is the viscous dissipation function, and

$$\int_S k\frac{\partial T}{\partial x_j} dS_j = \int_V \frac{\partial}{\partial x_j}\left(k\frac{\partial T}{\partial x_j}\right) d\tau. \tag{1.20}$$

The energy E per unit mass of the fluid can be written as

$$E = \frac{1}{2}u_i^2 + c_V T, \tag{1.21}$$

where c_V is the specific heat at constant volume and T is the temperature. This allows the fourth term of (1.16) to be rewritten as:

$$-\int_S \rho E u_j dS_j = -\int_S \rho\left[\frac{1}{2}u_i^2 + c_V T\right] u_j dS_j = -\frac{1}{2}\int_S \rho u_i^2 u_j dS_j - \int_V \frac{\partial}{\partial x_j}(\rho u_j c_V T)d\tau. \tag{1.22}$$

Substituting Eqs. (1.19)–(1.22) into (1.18), we obtain

$$\int_V \frac{\partial}{\partial t}(\rho c_V T)d\tau = \int_V \frac{\partial}{\partial x_j}\left(k\frac{\partial T}{\partial x_j}\right)d\tau + \int_V \Phi d\tau - \int_V \frac{\partial}{\partial x_j}(\rho c_V T u_j)d\tau + \int_V \rho H d\tau. \tag{1.23}$$

Since Eq. (1.23) is valid for any arbitrary volume V, we must have

$$\frac{\partial}{\partial t}(\rho c_V T) + \frac{\partial}{\partial x_j}(\rho c_V T u_j) = \frac{\partial}{\partial x_j}\left(k\frac{\partial T}{\partial x_j}\right) + \Phi + \rho H. \tag{1.24}$$

Noting that the left-hand side of the equation is the Lagrangian time derivative D/Dt, and applying the derivative separately to T and ρ results in:

$$\rho\frac{D}{Dt}(c_V T) + c_V T\frac{D\rho}{Dt} = \frac{\partial}{\partial x_j}\left(k\frac{\partial T}{\partial x_j}\right) + \Phi + \rho H, \tag{1.25}$$

which after some manipulation using thermodynamic expressions leads to the form

$$\rho c_p\frac{DT}{Dt} - \alpha T\frac{DP}{Dt} = \frac{\partial}{\partial x_i}\left(k\frac{\partial T}{\partial x_i}\right) + \Phi + \rho H. \tag{1.26}$$

This is a general form, valid for compressible flow. Various other forms exist. For example for incompressible flow, applying the incompressible continuity equation (1.5) to equation (1.24) results in the simplified form:

$$\rho\frac{\partial}{\partial t}(c_V T) + \rho u_j\frac{\partial}{\partial x_j}(c_V T) = \frac{\partial}{\partial x_j}\left(k\frac{\partial T}{\partial x_j}\right) + \Phi + \rho H. \tag{1.27}$$

We note that Eq. (1.27) is a parabolic equation. Equation (1.26) is often written using the ∇ operator as:

$$\rho c_p\left(\frac{\partial T}{\partial t} + \mathbf{u}\cdot\nabla T\right) - \alpha T\left(\frac{\partial P}{\partial t} + \mathbf{u}\cdot\nabla P\right) = \nabla\cdot(k\nabla T) + \Phi + \rho H. \tag{1.28}$$

1.3.4 The rheological law

In the mid twentieth century, E. C. Bingham introduced the term of ‘rheology’ in colloid chemistry, which has a meaning of ‘everything flows’ (in Greek $\pi\alpha\nu\tau\alpha\ \rho\varepsilon\iota$), the motto of the subject from Heraclitus (Reiner, 1964). A rheological law describes a relationship between stress and strain (strain rate) in a material. We often hear that the Earth's mantle exhibits the rheological properties of a fluid or a solid. The Deborah number, a dimensionless number expressing the ratio between the time of relaxation and time of observation, can assist in the understanding of the behaviour of geomaterials. If the time of observation is very large (or the time of relaxation of the geomaterial under observation is very small), the mantle is considered to be a fluid and hence it flows. On the other hand, if the time of relaxation of the geomaterial is larger than the time of observation, the mantle is considered to be a solid. Therefore, the greater the Deborah number, the more solid the geomaterial (and vice versa, the smaller the Deborah number, the more fluid it is). In nature, geomaterials (e.g. rocks comprising the crust, lithosphere and mantle) exhibit more complicated rheological behaviour than fluid or solid materials. We consider here a few principal rheological relationships. For detailed information on rock rheology, the reader is referred to Ranalli (1995) and Karato (2008).

In geodynamic modelling a viscous rheology is extensively used, because the mantle behaves as a highly viscous fluid at geological time scales. The equation describing the relationship between the viscous stress and strain rate can be presented in the following form:

$$\tau_{ij} = C^{\frac{1}{n}} \dot{\varepsilon}_{ij} \dot{\varepsilon}^{\frac{1-n}{n}}, \tag{1.29}$$

where τ_{ij} is the deviatoric stress tensor, C is a proportionality factor defined from the thermodynamic conditions, $\dot{\varepsilon} = (0.5\dot{\varepsilon}_{kl}\dot{\varepsilon}_{kl})^{1/2}$ is the second invariant of the strain rate tensor, and n is a power-law exponent. If $n = 1$, Eq. (1.29) describes a *Newtonian fluid* with $C/2$ as the fluid's viscosity, which depends on temperature and pressure as discussed below. For $n > 1$, Eq. (1.29) represents a *non-Newtonian (non-linear) fluid.*

At high temperatures (that are a significant fraction of the melt temperature) the atoms and dislocations in a crystalline solid become sufficiently mobile to result in creep when the solid is subject to deviatoric stresses. At very low stresses diffusion processes dominate, and the crystalline solid behaves as a Newtonian fluid with a viscosity that depends exponentially on pressure and the inverse absolute temperature. The proportionality factor C in (1.29) can be then represented as:

$$C(T,P) = C^* d^m \exp\left(\frac{E+PV}{RT}\right), \tag{1.30}$$

where T is the absolute temperature, P is pressure, C^* is the proportionality factor that does not depend on temperature and pressure, E is the activation energy, V is the activation volume, R is the universal gas constant, and d is the grain size. For dislocation creep, grain size is unimportant and $m = 0$, but for diffusion creep m is between 2 and 3. At higher stresses the motion of dislocations becomes the dominant creep process resulting in a non-Newtonian fluid behaviour described by Eqs. (1.29)–(1.30), with typically $n = 3.5$. Thermal convection in the mantle and some aspects of lithosphere dynamics are attributed to these thermally activated creep processes. The temperature–pressure dependence of the rheology of geomaterials is important in understanding the role of convection in transporting heat.

During dislocation creep as mentioned above, diffusion-controlled climb of edge dislocations is the limiting process. At low temperatures this is extremely slow, but can be bypassed at stresses high enough to force dislocations through obstacles, a process known as low-temperature (Peierls) plasticity. In this case, the exponential proportionality factor C becomes stress-dependent. A commonly assumed form of the strain rate dependence on stress is:

$$\dot{\varepsilon} = A \exp\left[-\frac{H_0}{RT}\left(1 - \frac{\sigma}{\sigma_P}\right)^2\right], \tag{1.31}$$

where σ_P is the Peierls stress, which is of order 2–9 GPa, and σ is the second invariant of the stress tensor.

Creep processes can relax elastic stresses in the lower lithosphere. Such behaviour can be modelled with a rheological law that combines linear elasticity and linear or non-linear viscosity. A material that behaves elastically on short time scales and viscously on long time scales, is referred to as a viscoelastic material. The most commonly employed rheology to simulate numerically lithosphere dynamics is the *viscoelastic* (*Maxwell*) rheology. According to the Hooke law of elasticity, the elastic strain ε_{ij} and the deviatoric stress τ_{ij} are related as

$$\tau_{ij} = \mu \varepsilon_{ij}, \tag{1.32}$$

where μ is the shear modulus. For the fluid we assume a linear Newtonian relation between viscous strain rate and the stress (consider Eq. (1.29) with $n = 1$ and $C = 2\eta$)

$$\tau_{ij} = 2\eta \frac{\partial \varepsilon_{ij}}{\partial t}, \tag{1.33}$$

where η is the fluid viscosity. The Maxwell model for a viscoelastic geomaterial assumes that the strain rate of the geomaterial is a superposition of the elastic and viscous strain rates, namely,

$$\frac{\partial \varepsilon_{ij}}{\partial t} = \frac{\tau_{ij}}{2\eta} + \frac{1}{\mu}\frac{\partial \tau_{ij}}{\partial t} \quad \text{or} \quad \left(1 + 2t_r \frac{\partial}{\partial t}\right)\tau_{ij} = 2\eta \frac{\partial \varepsilon_{ij}}{\partial t}, \tag{1.34}$$

where $t_r = \eta/\mu$ is the viscoelastic relaxation (or Maxwell relaxation) time. We see that on time scales short compared with the time of relaxation t_r the geomaterial behaves elastically, and on time scales long compared with t_r the material behaves as a Newtonian fluid.

Because the effective viscosity of the shallow lithosphere is very high, its deformation is no longer controlled by dislocation creep; instead it is determined by (at lower pressures) the movement of blocks of the lithosphere along pre-existing faults of various orientations and (at higher pressures) deformation accommodated by distributed microcracking. The dynamic friction along such faults depends only weakly upon the strain rate, and is often idealised using the rheological model of a *perfectly plastic material*, which does not exhibit work-hardening but flows plastically under constant stress. Hence the stress–strain relationship for the lithosphere obeys the von Mises equations (Prager and Hodge, 1951)

$$\tau_{ij} = \kappa \dot{\varepsilon}_{ij}/\dot{\varepsilon}, \tag{1.35}$$

where κ is the yield limit. The second invariant of the stress, $\tau = (0.5\tau_{kl}\tau_{kl})^{1/2}$, equals the yield limit for any non-zero strain rate. When $\tau < \kappa$, there is no plastic deformation and hence no motion along the faults. A comparison of Eqs. (1.29) and (1.35) shows that the perfectly plastic rheology can be considered as the limit of non-Newtonian power-law rheology as $n \to \infty$ (and $C = \kappa$). In rocks, the yield stress κ depends on pressure. If κ increases linearly with pressure, as is commonly assumed, then this gives the *Drucker–Prager yield criterion*, $\kappa = a + bP$, where a and b are constants and P is the pressure.

Brittle failure may be treated by the *Mohr–Coulomb failure criterion*, which expresses a linear relationship between the shear stress and the normal stress resolved on the failure plane, which is oriented at a particular angle,

$$\tau_f = \sigma_f \tan\phi + c, \tag{1.36}$$

where τ_f and σ_f are the shear stress and normal stress acting on the failure plane, ϕ is the *angle of internal friction* and c is the *cohesion*. It is often more convenient to express this in terms of the maximum shear stress $\tau_{\max}$ and $\bar{\sigma}$, the average of the maximum and minimum principle stresses:

$$\tau_{\max} = \bar{\sigma}\sin\phi + c\cos\phi. \tag{1.37}$$

In numerical models the Mohr–Coulomb criterion is often approximated by the Drucker–Prager criterion, with $\tau_{\max}$ equal to the second stress invariant and pressure used in place of $\bar{\sigma}$.

Thus, a fluid behaviour of geomaterials is described by Eqs. (1.29)–(1.31), and (1.33), elastic behaviour by Eq. (1.32), viscoelastic by Eq. (1.34), perfectly plastic by Eq. (1.35) and brittle by Eq. (1.36)–(1.37). These relationships are used frequently in geodynamic modelling.

1.3.5 Other equations

The equations of continuity, motion and heat balance compose the basic equations governing models of mantle and lithosphere dynamics. Together with the basic equations, additional equations are necessary to describe the behaviour of mantle rocks, namely, equations of state, rheological law (or equation for viscosity), equation for phase transformations, etc. In many practical applications, a linear dependence of density on temperature (equation of state) is assumed:

$$\rho = \rho_0[1 - \alpha(T - T_0)], \tag{1.38}$$

where ρ_0 is a reference density, α is the coefficient of thermal expansivity and T_0 is a reference temperature. If phase transformations of mantle rocks are considered the state equation is modified. The viscosity of mantle rocks is the least well-known parameter used in numerical modelling of geodynamic problems. The mantle viscosity can depend on temperature, pressure, grain size, content of water or melt, stress, etc. We shall use various representations of viscosity in our geodynamic model examples (see Chapter 10).

1.3.6 Boussinesq approximation

Mantle dynamics is controlled by heat transfer, and the mantle properties are normally functions of temperature. The variations in density due to temperature variations are generally small and yet are the cause of the mantle motion. If the density variation is not large, one may treat the density as constant in the continuity equation (i.e. the fluid is assumed to be *incompressible* (Eq. 1.5)) and in the energy equation (e.g. in the unsteady and advection terms) and treat it as variable only in the gravitational (buoyancy) term of the momentum equation.

Consider the Stokes equation (1.17) and split the term $\rho F_i = \rho g_i$ into two parts: $\rho_0 g_i + (\rho - \rho_0) g_i$. The first part can be included with pressure and the density variation is retained in the gravitational term. The remaining term can be expressed as:

$$(\rho - \rho_0) g_i = -\rho_0 g_i \alpha (T - T_0). \tag{1.39}$$

Such simplification of the model is called the Boussinesq approximation. In the strict form of this, all physical properties except viscosity are constant. The dimensionless mass and energy conservation equations then become

$$-\frac{\partial P}{\partial x_i} + \frac{\partial}{\partial x_j}\left\{\eta\left(\frac{\partial u_i}{\partial x_j} + \frac{\partial u_j}{\partial x_i}\right)\right\} = RaT\delta_{i3}, \quad \frac{\partial T}{\partial t} + u_j\frac{\partial T}{\partial x_j} = \frac{\partial^2 T}{\partial x_j^2} + H. \tag{1.40}$$

If fluid is compressible, compressibility is incorporated in a model using either the *extended Boussinesq approximation*, in which the density is still assumed constant in the continuity equation but the extra terms are included in the energy equation, or the *anelastic approximation*, in which the density is assumed to vary with position but not with time. Both approximations are discussed in detail in Section 10.3.

1.3.7 Stream function formulation

The stream function formulation is a way of eliminating pressure and reducing two velocity components to a single scalar, in two-dimensional geometry. The velocity field $\mathbf{v} = (u_1, u_3)$ in two dimensions (x_1, x_3) is related to derivatives of a scalar stream function ψ:

$$\mathbf{v} = \left(\frac{\partial \psi}{\partial x_3}, -\frac{\partial \psi}{\partial x_1}\right). \tag{1.41}$$

Often, the opposite sign is used. It is easily verified that this satisfies the incompressible continuity equation (Eq. 1.5). Substituting this into the constant-viscosity Boussinesq Stokes equation for thermally driven flow,

$$-\nabla P + \nabla^2 \mathbf{v} = RaT\mathbf{e}, \tag{1.42}$$

where $\mathbf{e}$ is the unit vector, and taking the out of plane (y) component of the curl of this equation yields:

$$\nabla^4 \psi = -Ra\frac{\partial T}{\partial x_1}. \tag{1.43}$$

Hence, three variables (two velocity components and pressure) have been reduced to one scalar. This can be solved as a fourth-order differential equation, or split into two second-order equations:

$$\nabla^2 \omega = -Ra \frac{\partial T}{\partial x_1}, \quad \nabla^2 \psi = \omega, \tag{1.44}$$

where ω is the vorticity, which is the out of plane (y) component of $\nabla \times \mathbf{v}$. This formulation can also be used to re-express the variable-viscosity Stokes equation, but a more complicated expression results (see Malevsky and Yuen, 1992).

1.3.8 Poloidal and toroidal decomposition

A way of simplifying the Stokes and continuity equations in three dimensions is to express the velocity field in terms of poloidal and toroidal mass flux potentials:

$$\rho \mathbf{v} = \nabla \times \nabla \times (W\mathbf{e}) + \nabla \times (Z\mathbf{e}), \tag{1.45}$$

where $\mathbf{v} = (u_1, u_2, u_3)$ is the velocity field, W is the poloidal potential, and Z is the toroidal potential. This automatically satisfies the continuity equation, which is therefore eliminated, and reduces the three velocity components to two scalars. If the flow is incompressible, then W and Z become velocity potentials:

$$\mathbf{v} = \nabla \times \nabla \times (W\mathbf{e}) + \nabla \times (Z\mathbf{e}). \tag{1.46}$$

In the case of homogeneous boundary conditions and viscosity that does not vary in the horizontal directions, there is no source for the toroidal term (see Ricard and Vigny, 1989), so this further reduces to

$$\mathbf{v} = \nabla \times \nabla \times (W\mathbf{e}). \tag{1.47}$$

Assuming constant properties and the Boussinesq approximation, by taking the x_3-component of the double curl of the momentum equation (1.42), substituting Eq. (1.47), and using identities such as $\nabla \times \nabla \times \mathbf{a} = \nabla(\nabla \cdot \mathbf{a}) - \nabla^2 \mathbf{a}$, the Stokes equation can be reduced to the simple form:

$$\nabla^4 W = Ra\, T. \tag{1.48}$$

The pressure has been eliminated, so the number of variables has been reduced from four (pressure and three velocity components) to one. A poloidal–toroidal decomposition can also be used for flow in which viscosity varies (see Christensen and Harder, 1991) and/or the boundary conditions are not homogeneous (see Hager and O'Connell, 1981), but then the toroidal component must be retained and the resulting equations become much more complex.

1.4 Boundary and initial conditions

The equations given above govern the slow movements of the Earth's mantle and lithosphere. They are the same equations whether the movement is, for example, a thermal plume rising beneath a particular region, subduction of the lithosphere, a mid-ocean ridge, convective flow in the upper mantle or whole mantle convection. However, the movements are different for these cases, although the governing equations are the same. Why? If all parameters entering the governing equations are the same, the answer is because of the boundary and initial conditions, which are different for each of the above examples.

For example, rising mantle plumes require mainly free-slip conditions at the boundaries of a model domain. Meanwhile spreading at a mid-ocean ridge is driven partly by forces due to distant subduction, so for a local model of a mid-ocean ridge a velocity field should be imposed at the upper boundary of a model domain. The boundary and initial conditions dictate the particular solutions to be obtained from the governing equations. Therefore, once we have the governing equations, then the real driver for any particular solution is the boundary conditions.

Let us review the proper physical boundary conditions. When the condition on a surface of the Earth assumes zero relative velocity between the surface and the air immediately at the surface, we refer to the condition as the *no-slip* (or *rigid*) condition. If the surface is stationary, then

$$u_1 = u_2 = u_3 = 0. \tag{1.49}$$

When the velocity at the boundary is a finite, non-zero value and there is no mass flow in to or out of the model domain, the velocity vector immediately adjacent to the boundary must be tangential to this boundary. If $\mathbf{n}$ is a unit normal vector at a point on the boundary and $\mathbf{u}_\tau$ is the projection of the velocity vector onto the tangent plane at the same point on the boundary, the condition at this boundary can be given as

$$\mathbf{u} \cdot \mathbf{n} = 0, \quad \partial \mathbf{u}_\tau / \partial \mathbf{n} = 0. \tag{1.50}$$

These conditions are called *free-slip* conditions. The actual surface of the Earth can move upwards and downwards. The above conditions, in which the upper boundary of the model domain represents the Earth's surface and there is no vertical motion at the boundary, are idealisations made to simplify the model. Modelling an actual free surface that deflects vertically is more complicated but methods exist, as discussed in Chapter 10.

There is an analogous 'no-slip' condition associated with the temperature at the surface. If the temperature at the surface is denoted by T_u, then the temperature immediately in contact with the surface is also T_u. If in a given problem the temperature is known, then the proper condition on the temperature at the upper boundary of the model domain is

$$T = T_u. \tag{1.51}$$

On the other hand, if the temperature at the surface is not known, e.g. if it is changing with time due to heat transfer to the surface, then the Fourier law of heat conduction provides the boundary condition at the surface. If we let $\dot{q}_u$ denote the instantaneous heat flux to the surface, then from the Fourier law

$$\dot{q}_u = -\left(k\frac{\partial T}{\partial n}\right)_u, \tag{1.52}$$

where n denotes the direction normal to the surface. The surface rocks are responding to the heat transfer to the surface, $\dot{q}_u$, hence changing T_u, which in turn affects $\dot{q}_u$. This general, unsteady heat transfer problem must be solved by treating the viscous flow and the thermal response of the surface rocks simultaneously. This type of boundary condition is a boundary condition on the temperature gradient at the surface, in contrast to stipulating the surface temperature itself as the boundary condition. That is, from Eq. (1.52),

$$\left(\frac{\partial T}{\partial n}\right)_u = -\frac{\dot{q}_u}{k}. \tag{1.53}$$

While the above discussion refers to the top boundary of the domain, similar conditions also apply to the lower boundary, which in global models is the core–mantle boundary. At the sides, no-slip or free-slip conditions are sometimes assumed, but if the model is intended to represent the entire mantle then periodic boundaries are most realistic. In local or regional models, which are often applied to model the crust and/or lithosphere, it is quite common for material to flow in or out of the domain, either with a prescribed velocity and temperature or with some other conditions such as prescribed normal stress, but we do not give mathematical details here.

The boundary conditions discussed above are physical boundary conditions imposed by nature. Meanwhile in numerical modelling we should sometimes introduce additional conditions to properly define the mathematical problem under question.

In general, when the value of the variable is given at a boundary of the model domain, the condition is referred to as a Dirichlet boundary condition. When the gradient of the variable in a particular direction (usually normal to the boundary) is prescribed to the model boundary, the condition is called a Neumann boundary condition. Sometimes a linear combination of the two quantities is given, and in this case the boundary condition is referred to as a mixed boundary condition.

1.5 Analytical and numerical solutions

Mathematical models of geodynamic processes can be solved analytically or numerically. Analytical solutions are those that a researcher can obtain by solving mathematical models by using a pencil, a piece of paper, and his or her own brain activity. Simple mathematical models allow analytical solutions, which have been (and still are) of great importance because of their power: the solutions are precise and can be presented by exact formulas. However, the usefulness of this power is limited as many mathematical models of geodynamics are too complicated to be solved analytically.

Numerical solutions are those that researchers can obtain by solving numerical models using computational methods and computers. Numerical models allow the solution of complex problems of geodynamic processes, although the solutions are not exact. In some geodynamic applications an analytical solution to part of the complex problem can be implemented into the numerical model to make the model much more effective.

An analytical solution to a specified mathematical problem can be used to verify a numerical solution to the problem; in fact, it is the simplest way to benchmark a numerical code. Unfortunately, many two- and three-dimensional mathematical problems in geodynamics have no analytical solutions. But when analytical solutions to such problems are obtained in some cases, it is like finding water in a desert. For example, an analytical solution to a three-dimensional model of viscous flow (e.g. describing movements of salt diapirs in sedimentary basins) was recently obtained by Trushkov (2002). Considering the equations of slow viscous incompressible flow coupled with the equation for density advection, Trushkov (2002) found an exact solution to this set of partial differential equations. This solution can be used to verify numerical solutions to the problem of gravitational instability.

1.6 Rationale of numerical modelling

Only a few of the differential and partial differential equations describing geodynamical models can be solved exactly, and hence the equations are transformed into discrete equations to be solved numerically. Although the widespread access to high-performance computers has resulted in an over-reliance on numerical answers when there are other possibilities, and a corresponding false sense of security about the possibilities of serious numerical problems or errors, it is now possible without too much trouble to find solutions to most equations that are routinely encountered.

The rationale of the numerical modelling is described graphically in Fig. 1.1. The initial stage of numerical modelling is to describe geodynamic complex reality by a simplification of the reality; namely, to introduce the concept of the geodynamic problem, forces acting on the system (lithosphere, crust, mantle), physical parameters to be used in the modelling, etc.

A physical model is then developed to which the physical laws can be applied. The next step in the numerical modelling is to describe the physical model by means of mathematical equations. The comparison with observations allows the model to be tested (validated). If the mathematical model is found to be inadequate, it must be changed: the assumed process is not the correct one, or some significant factors have been missed. The mathematical model should be properly determined, at least after the numerical values of some still unknown parameters have been determined (that is, the model is tuned).

Once the mathematical model is developed, proper numerical tools and methods have to be determined, and relevant numerical codes (software) should be constructed (or otherwise obtained). The mathematical model should be transformed into the computational model containing discrete equations to be solved by using computers. An important element of numerical modelling is *verification* of the model, namely, the assessment of the accuracy

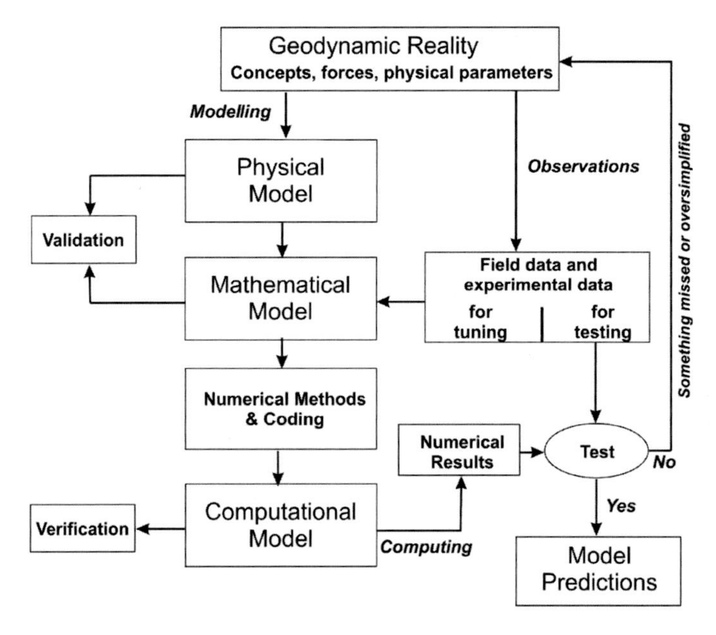

Fig. 1.1. **Flowchart of numerical modelling.**

of the solution to the computational model by comparison with known solutions (analytic or numerical). Once the computational model is verified, the model can be computed and numerical results obtained can be tested against observations. If there is good agreement between the numerical results and observed (field or experimental) data, the model results can be considered as the model predictions. Sometimes researchers dealing with numerical modelling make a serious mistake, when all available data have been used to tune the model and no data have been left to test its validity or, even worse, when the data used for the model tuning are employed to test model results.

1.7 Numerical methods: possibilities and limitations

By a numerical method we mean a procedure that permits us to obtain the solution to a mathematical problem with an arbitrary precision in a finite number of steps that can be performed rationally. The number of steps depends on the desired accuracy. A numerical method usually consists of a set of directions for the performance of certain arithmetical or logical operations in predetermined order. This set of directions must be complete and unambiguous. A set of directions to perform mathematical operations designed to lead to the solution of a given problem is called an *algorithm*.

Numerical methods came with the birth of electronic computers. Although many of the key ideas for numerical solution methods were established several centuries ago, they were of little use before computers appeared. Interest in numerical methods increased dramatically with the development of computer power. Computer solution of the equations describing geodynamic processes has become so important that it occupies the attention of many researchers in geodynamics.

Numerical methods provide possibilities to obtain accurate solutions to geodynamic problems. However, the numerical results are always approximate. There are reasons for differences between computed results and observations. Errors arise from each part of the process used to produce numerical solution (we discuss sources of the errors in Section 1.9): (i) the physical model is too simplified compared with geodynamic reality; (ii) the equations (mathematical model) may contain approximations or idealisations; (iii) approximations are made in the discretisation process; and (iv) in solving the discrete equations, iterative methods are used and insufficient iterations are taken. Additionally, uncertainty in physical parameters can lead to differences between computed results and observations.

1.8 Components of numerical modelling

Numerical simulations in geodynamics enable one to analyse and to predict the dynamics of the Earth's interior. Computers are employed to solve numerically models of geodynamic processes. The basic elements of the numerical modelling are as follows: (i) a mathematical model describing geodynamics; (ii) a discretisation method to convert the mathematical equations into discrete equations to be solved numerically; (iii) numerical method(s) to solve the discretised equations; (iv) computer code(s) (i.e. software) to be developed or to be used, if already developed, that solve numerically the discrete equations; (v) computer hardware, which performs the calculations; (vi) results of numerical modelling to be visualised, analysed and interpreted by (vii) geoscientist(s).

Models of geodynamical processes described by partial differential (or integro-differential) equations cannot be solved analytically except in special cases. To obtain an approximate solution numerically, we have to use the discretisation method, which approximates the differential equations by a set of algebraic equations, which can then be solved on a computer. The approximations are applied to small domains in space and/or time so the numerical solution provides results at discrete locations in space and time. Much as the accuracy of observations depends on the quality of the tools used, the accuracy of numerical solutions depends on the quality of the discretisations used.

When the governing equations are known accurately, solutions of any desired accuracy can be achieved. However, for many geodynamic processes (e.g. thermo-chemical convection, mantle flow in the presence of phase transformations and complex rheology) the exact equations governing the processes are either not available or numerical solution of the full equations is not feasible. This requires the introduction of models. Even if we solve the equations exactly, the solution would not be a correct representation of reality. In order to validate the models, we have to rely on observations. Even when the exact treatment is possible, models are often needed to reduce the cost.

Discretisation errors can be reduced by using more accurate interpolation or approximations or by applying the approximations to smaller regions, but this usually increases the time and cost of obtaining the solution. Compromise is usually needed. We shall present some schemes in detail but shall also point out ways of creating more accurate approximations. Compromises are also needed in solving the discretised equations. Direct solvers, which obtain accurate solutions, are seldom used in new codes, because they are too costly. Iterative methods are more common but the errors due to stopping the iteration process too soon need to be taken into account. The need to analyse and estimate numerical errors cannot be overemphasised.

Visualisation of numerical solutions using vectors, contours, other kinds of plots or movies (videos) is essential to interpret numerical results. However, there is the danger that an erroneous solution may look good but may not correspond to the true solution of a mathematical problem. It is especially important in the case of geodynamic problems because of the complex dynamics of the Earth components (crust, mantle and core). Sometimes incorrect numerical results are interpreted as physical phenomena. Users of commercial software should be especially careful, as the optimism of salesmen is legendary. Colour figures of results of numerical experiments sometimes make a great impression but are of no value if they are not quantitatively correct. Results must be examined critically before they are believed.

We follow Ferziger and Peric (2002) in the description of the components of numerical modelling.

Mathematical model. The starting point of numerical modelling is a mathematical model, i.e. the set of partial differential or integro-differential equations and boundary conditions. The equations governing a thermo-convective viscous flow in the Earth's mantle have been presented in Section 1.4. An appropriate model should be chosen for a geodynamic application (e.g. incompressible, viscous, two- or three-dimensional, etc.). As already mentioned, this model may include simplifications of the exact conservation laws. A solution method is usually designed for a particular set of equations.

Coordinate systems. The conservation equations can be written in many different forms, depending on the coordinate system. For example, one can select Cartesian, cylindrical, spherical and some others. The choice depends on the target problem, and may influence the discretisation method and grid type to be used.

Discretisation method. After selecting the mathematical model, one has to choose a suitable discretisation method, i.e. a method of approximating the differential equations by a set of algebraic equations for the variables at some set of discrete locations in space and time. There are many approaches, but the most popular at this time are finite difference, finite element and finite volume methods. Spectral methods were popular in the past, particularly for three-dimensional modelling, but their use is decreasing due to limitations. Other methods, like boundary element and discrete element methods are also used in geodynamic modelling, but less often.

Each type of method yields the same solution if the grid is very fine. However, some methods are more suitable to some classes of problems than others. The preference is often determined by the attitude of the developer. We shall discuss the pros and cons of the various methods later.

Numerical grid. This defines the discrete locations at which the unknowns are to be calculated. The grid is essentially a discrete representation of the geometric domain, on which the problem is to be solved. It divides the solution domain into a finite number of sub-domains (e.g. elements, control volumes, points, etc.)

Structured (regular) grids consist of families of grid lines with the property that members of a single family do not cross each other and cross each member of the other families only once. The position of any grid point within the domain is uniquely identified by a set of two (in two-dimensional spaces) and three (in three-dimensional spaces) indices, e.g. (i, j, k). This is the simplest grid structure, since it is logically equivalent to a Cartesian grid. Each point has four nearest neighbours in two dimensions (2-D) and six in three dimensions (3-D). An example of a structured two-dimensional grid is illustrated in Fig. 1.2a. The simplest example of a numerical grid is an orthogonal grid.

For complex model domain geometries, *unstructured grids* are most appropriate. Such grids are best adapted to the finite element or finite volume approaches. The elements may have any shape, and there is no restriction on the number of neighbour elements or nodes. In practice, grids made of triangles or quadrilaterals in 2-D (see Fig. 1.2b), and tetrahedral or hexahedral in 3-D are most often used. Such grids can be generated automatically by existing algorithms (see Section 4.8).

Finite approximations. Following the choice of grid type, approximations should be selected to be used in the discretisation process. In a finite difference method, approximations for the derivatives at the grid points have to be selected. In the finite element method, one has to choose the shape functions (elements) and weight functions. The choice

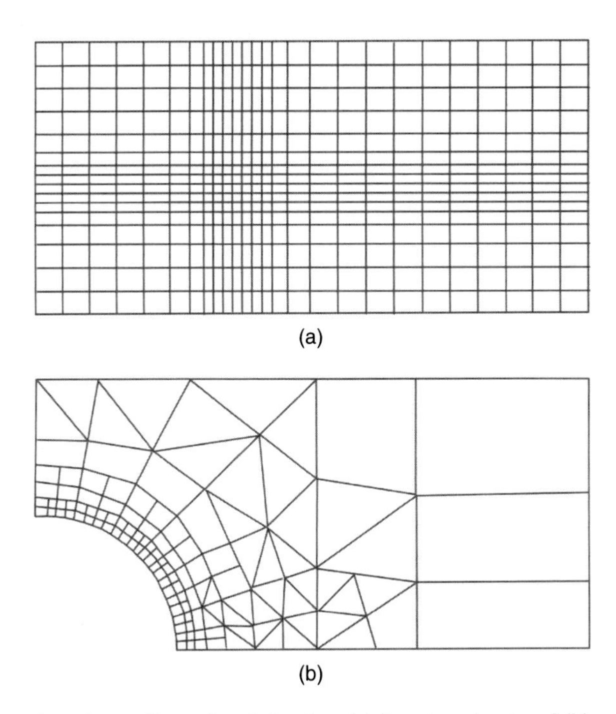

Fig. 1.2. Examples of two-dimensional structured (a) and unstructured (b) grids.

of the discretisation process influences the accuracy of the approximation. It also affects the difficulty of developing the solution method, coding it, debugging it, and the speed of the code. More accurate approximations involve more nodes and give fuller coefficient matrices. A compromise between simplicity, ease of implementation, accuracy and computational efficiency has to be made.

Solution method. Discretisation yields a large set of equations, and the method of solution depends on the problem. For non-stationary geodynamic processes, numerical methods for solving initial value problems for ordinary differential equations should be employed. At each time step a set of algebraic equations has to be solved. When the equations are non-linear, an iteration scheme is used to solve them. We present some solvers in Chapters 6 and 7.

Convergence criteria. When iterative methods are employed to solve discrete equations, convergence criteria should be established. Usually, there are two levels of iterations: *inner iterations*, within which the linear equations are solved, and *outer iterations*, that deal with the non-linearity and coupling of the equations. Deciding when to stop the iterative process on each level is important from the accuracy and efficiency points of view.

1.9 Properties of numerical methods

Numerical solution methods have certain important properties; they are summarised below following Ferziger and Peric (2002).

Consistency. The difference between the discretised and exact equations is called the *truncation error*. For a method to be consistent, the truncation error must become zero when the mesh spacing tends to zero. The truncation error is usually proportional to a power of the grid spacing Δx and/or the time step Δt. If the principal term of an equation is proportional to $(\Delta x)^n$ or $(\Delta t)^n$ we call the method an nth-order approximation; $n > 0$ is required for consistency. Even if the approximations are consistent, it does not necessarily mean that the solution of the set of discrete equations will become the exact solution to the differential equation in the limit of small step size. For this to happen, the solution method has to be *stable*.

Stability. A numerical solution method is stable if it does not magnify the errors that appear in the course of numerical solution process. For unsteady problems, stability guarantees that the method produces a bounded solution whenever the solution of the exact equation is bounded. For iterative methods, a stable method is one that does not diverge. Stability can be difficult to analyse, especially when solving non-linear and coupled equations with prescribed boundary conditions. There are few stability results for complicated discrete problems, so we should rely on experience and intuition. It is common to estimate the stability of a method for linear problems with constant coefficients without boundary conditions. The results obtained in this way can often be applied to more complex problems.

Convergence. A numerical method is said to be convergent if the solution of the discretised equations tends to the exact solution of the differential equation as the grid spacing tends to zero. For many non-linear problems in geodynamics, which are strongly influenced by boundary conditions, the convergence (as well as *stability*) of a method is difficult to demonstrate. Therefore, convergence is usually checked using numerical experiments, i.e. repeating the calculation on a series of successively refined grids. If the method is stable and if all approximations used in the discretisation process are consistent, it is usually found that the solution converges to a grid-independent solution. For sufficiently small grid sizes, the rate of convergence is governed by the order of the principal truncation error component. This allows one to estimate the error in the solution.

Conservation. Since the equations to be solved are conservation laws, the numerical scheme should also respect these laws. This means that, at steady state and in the absence of sources, the amount of a conserved quantity leaving a closed volume is equal to the amount entering that volume. Conservation is an important property of the solution method, since it imposes a constraint on the solution error. If conservation of mass, momentum and energy are insured, the error can only improperly distribute these quantities over the solution domain. Non-conservative schemes can produce artificial sources, changing the balance both locally and globally. However, non-conservative schemes can be consistent and stable and therefore lead to correct solutions in the limit of very fine grids. The errors due to non-conservation are in most cases significant only on relatively coarse grids. Meanwhile it is difficult to estimate the size of the grid at which these errors are small enough, and hence conservative schemes are preferred.

Boundedness. Numerical solution should lie within proper *bounds*. Physically non-negative quantities (like density and viscosity) must always be positive. In the absence of sources, some equations (e.g. the heat equation for the temperature when no heat sources are present) require that the minimum and maximum values of the variable be found on the boundaries of the domain. These conditions should be inherited by the numerical approximation.

Accuracy. This is the most important property of numerical modelling. Numerical solutions of geodynamic problems are only *approximate solutions*. In addition to the errors that might be introduced in the course of the development of the solution algorithm, in programming or setting up the boundary conditions, numerical solutions always include three kinds of systematic error.

- *Modelling errors*, which are defined as the difference between the actual process and the exact solution of the mathematical model (modelling errors are introduced by simplifying the model equations, the geometry of the model domain, the boundary conditions, etc.).
- *Discretisation errors*, defined as the difference between the exact solution of the conservation equations and the exact solution of the algebraic system of equations obtained by discretising these equations.
- *Iteration errors*, defined as the difference between the iterative and exact solutions of the algebraic system of equations.

It is important to be aware of the existence of these errors, and even more to try to distinguish one from another. Various errors may cancel each other, so that sometimes a solution obtained on a coarse grid may agree better with the experiment than a solution on a finer grid – which, by definition, should be more accurate.

1.10 Concluding remarks

The success in numerical modelling of geodynamical processes is based on the following basic, but simple, rules.

(i) '*People need simplicity most, but they understand intricacies best*' (B. Pasternak, writer). Start from a simple mathematical model, which describes basic physical laws by a set of equations, and then develop to more complex models. Never start from a complex model, because in this case you cannot understand the contribution of each term of the equations to the solution of the model.

(ii) Use analytical methods at first (if possible) to solve the mathematical problem. If it is impossible to derive an analytical solution, transform the mathematical problem into a discrete problem.

(iii) Study the numerical methods behind your computer code. Otherwise it becomes difficult to distinguish true and erroneous solutions to the discrete problem, especially when your problem is complex enough.

(iv) Test your model against analytical and/or asymptotic solutions, and simple model examples. Develop benchmark analysis of different numerical codes and compare numerical results with laboratory experiments. Remember that the numerical tool you employ is not perfect, and there are small bugs in every computer code. Therefore the testing is the most important part of your numerical modelling.

(v) Learn relevant statements concerning the existence, uniqueness and stability of the solution to the mathematical and discrete problems. Otherwise you can solve an improperly posed problem, and the results of the modelling will be far from the true solution of your model problem.

(vi) Try to analyse numerical models of a geophysical phenomenon using as little as possible tuning of model parameters. Two tuning parameters already give enough possibility to constrain a model well with respect to observations. Data fitting is sometimes quite attractive and can take one far from the principal aim of numerical modelling in geodynamics: to understand geophysical phenomena and to simulate their dynamics. If the number of tuning model parameters are greater than two, test carefully the effect of each of the parameters on the modelled phenomenon. Remember: '*With four exponents I can fit an elephant*' (E. Fermi, physicist).

(vii) Make your numerical model as accurate as possible, but never put the aim to reach a great accuracy. '*Undue precision of computations is the first symptom of mathematical illiteracy*' (N. Krylov, mathematician).

How complex should a numerical model be? '*A model which images any detail of the reality is as useful as a map of scale 1:1*' (J. Robinson, economist). This message is quite important for geoscientists who study numerical models of complex geodynamical processes. Geoscientists will never create a model that represents the Earth dynamics in full complexity, but we should try to model the dynamics in such a way as to 'simulate' basic geophysical processes and phenomena.

Does a particular model have a predictive power? Each numerical model has a predictive power, otherwise the model is useless. The predictability of the model varies with its complexity. Remember that a solution to the numerical model is an approximate solution to the equations, which have been chosen in the belief that they describe dynamic processes of the Earth. Therefore, a numerical model predicts dynamics of the Earth as well as the mathematical equations describe this dynamics.

2 Finite difference method

2.1 Introduction: basic concepts

Finite difference (FD) approximations for derivatives were already in use by Euler (1768). The simplest FD procedure for dealing with the problem $dx/dt = f(t,x)$, $x(0) = x_0$ is obtained by replacing $(dx/dt)_{n-1}$ with the crude approximation $(x_n - x_{n-1})/\Delta t$, $\Delta t = t_n - t_{n-1}$. This leads to the recurrence relation $x_n = x_{n-1} + \Delta t f(t_{n-1}, x_{n-1})$ for $n > 0$. This procedure is known as the Euler method (see Section 7.2 for more detail). Therefore, we see that for one-dimensional (1-D) problems the FD approach has been deeply ingrained in computational algorithms for quite some time. For two-dimensional (2-D) problems the first computational application of FD methods was most probably carried out by Runge (1908). He studied the numerical solution of Poisson's equation $\Delta u = \partial^2 u/\partial x^2 + \partial^2 u/\partial y^2 = c$, where c is a constant. A few years later Richardson (1910) published his work on the application of iterative methods to the solution of continuous equilibrium problems by FDs. The celebrated paper by Courant *et al.* (1928) is often considered as the birth date of the modern theory of numerical methods for partial differential equations.

The goal of the FD method is to reduce the ordinary or partial differential equations to discrete equations approximating the differential equations and making then suitable for computer implementation. The following two actions are required before using the FD method: (i) a continuous domain of unknown functions (e.g. velocity, temperature, pressure) should be represented by a computational domain, and (ii) each differential operator entering in the governing equations, as well as boundary and initial conditions, should be replaced by their discrete analogues. One can initially be deceived by the seeming elementary nature of the FD method. A little knowledge is dangerous since these approximations raise many serious and difficult mathematical questions of adequacy, accuracy, convergence and stability.

The basic approximation involves the replacement of a continuous domain Ω by a mesh of discrete points within the domain. Consider a 2-D model domain $\Omega = \{0 \le x \le A, 0 \le y \le B\}$ (Fig. 2.1a) and a uniform mesh or grid (Fig. 2.1b). To obtain the uniform mesh, the line segments [0, A] and [0, B] are divided into N_x and N_y parts, respectively ($h_x = 1/N_x$ and $h_y = 1/N_y$), and lines parallel to the axes are drawn through the dividing points. The points of the line intersections generate the uniform mesh. Instead of developing a solution defined everywhere in Ω, approximations are obtained only at the isolated nodes x_{ij}. Intermediate values of differential operators entering in the governing equations may be obtained from this discrete solution by an interpolation.

Another example of a computational grid is a non-uniform mesh (Fig. 2.1c). It can be generated in the following way. Introduce arbitrary points at the line segments [0, A] and

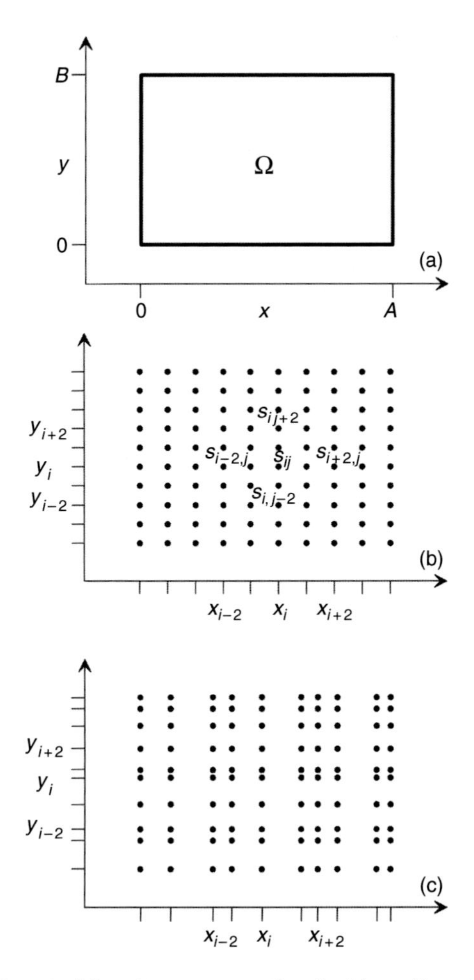

Fig. 2.1. **Examples of the model domain (a) and two computational grids, uniform (b) and non-uniform (c).**

$[0, B]$, i.e. $0 < x_1 < x_2 < \cdots < x_{N_x-1} < A$ and $0 < y_1 < y_2 < \cdots < y_{N_x-1} < B$. The set of points $\{x_i, y_j\}$ defines the non-uniform mesh. The distance between neighbouring mesh points $h_x^i = x_i - x_{i-1}$ and $h_y^j = y_j - y_{j-1}$ $\left(\sum_{i=1}^{N_x} h_x^i = 1 \text{ and } \sum_{j=1}^{N_y} h_y^j = 1\right)$ depends on the number (i, j) of the mesh point and hence is a function of the mesh.

Discretisation of the continuous problem (governing equations together with boundary and initial conditions) is based on a transformation of the problem to a discrete one replacing derivatives by finite difference approximations. Partial derivatives can be approximated by finite differences in many ways. All such approximations introduce truncation errors (see Section 1.9). Several simple approximations will be presented here. More detailed approximations can be found in Samarskii (1977) and Morton and Mayers (2005).

Consider the 2-D boundary value problem

$$Lu = f, \tag{2.1}$$

in a domain Ω (Fig. 2.1a) subject to certain boundary conditions; here L is the differential operator, $u = u(x,y)$ is the unknown variable, and f is the given function. The points s_{ij} (Fig. 2.1b) form a discrete approximation for Ω with uniform spacing $h_x = x_i + _1 - x_i$ and $h_y = y_j + _1 - y_j$. Now we consider several examples of the operator L and its approximations by finite differences.

Example 1: $Lu = \dfrac{\partial u(x,y)}{\partial x}$
We expand $u(x + h_x, y)$ in the Taylor series about the point (x,y):

$$u(x + h_x, y) = u(x,y) + h_x \frac{\partial u(x,y)}{\partial x} + \frac{h_x^2}{2}\frac{\partial^2 u(x,y)}{\partial x^2} + \frac{h_x^3}{6}\frac{\partial^3 u(x,y)}{\partial x^3} + O(h_x^4), \tag{2.2}$$

where $O(h)$ represents the asymptotic notation for the truncation error. Now dividing (2.2) by h_x, we obtain:

$$\frac{\partial u(x,y)}{\partial x} = [u(x + h_x, y) - u(x,y)]/h_x + O(h_x). \tag{2.3}$$

The *forward difference* of (2.3) provides the first-order approximation

$$\frac{\partial u}{\partial x} \approx [u(x + h_x, y) - u(x,y)]/h_x \tag{2.4}$$

for $\partial u/\partial x$ evaluated at point (x,y). The order of an approximation is defined as follows. If $w(x)$ is an approximation of the function $W(x)$, the approximation is of order m with respect to some quantity h, if n is the largest possible positive real number such that $|W - w| = O(h^n)$ as $h \to 0$. Equation (2.4) can be rewritten as

$$(\partial u/\partial x)_{ij} = (u_{i+1,j} - u_{ij})/h_x + O(h_x), \tag{2.5}$$

where $u_{ij} = u(ih_x, jh_y)$ is the exact solution to (2.1).

As an alternative to the forward difference approximation of (2.5), a *backward difference* is obtained in a similar fashion. The Taylor series for $u(x - h_x, y)$ about (x,y) can be represented as

$$u(x - h_x, y) = u(x,y) - h_x \frac{\partial u(x,y)}{\partial x} + \frac{h_x^2}{2}\frac{\partial^2 u(x,y)}{\partial x^2} - \frac{h_x^3}{6}\frac{\partial^3 u(x,y)}{\partial x^3} + O(h_x^4). \tag{2.6}$$

Similarly, we obtain the following expression

$$(\partial u/\partial x)_{ij} = (u_{ij} - u_{i-1j})/h_x + O(h_x), \tag{2.7}$$

which results in a first-order backward difference approximation upon suppression of the truncation error.

A higher-order approximation to $\partial u/\partial x$ can be obtained by subtraction of (2.6) from (2.2). The result is represented as

$$u(x + h_x, y) - u(x - h_x, y) = 2h_x \frac{\partial u(x,y)}{\partial x} + \frac{h_x^3}{3}\frac{\partial^3 u(x,y)}{\partial x^3} + O(h_x^5). \tag{2.8}$$

Dividing both sides of (2.8) by $2h_x$ we obtain the second-order approximation

$$(\partial u/\partial x)_{ij} = (u_{i+1j} - u_{i-1j})/2h_x + O(h_x^2). \tag{2.9}$$

This is known as a *central difference* approximation. While it is true that (2.9) has a second-order truncation error this does not mean that its application will always give rise to a more useful numerical technique than (2.5).

The third- and higher-order approximations of the first derivative of the function can be obtained by expanding the function in the Taylor series about four and more points. Namely, consider the following Taylor expansions:

$$\begin{aligned}
&u(x+2h_x,y)\\
&\quad = u(x,y) + 2h_x\frac{\partial u(x,y)}{\partial x} + \frac{4h_x^2}{2}\frac{\partial^2 u(x,y)}{\partial x^2} + \frac{8h_x^3}{6}\frac{\partial^3 u(x,y)}{\partial x^3} + \frac{16h_x^3}{24}\frac{\partial^4 u(x,y)}{\partial x^4} + O(h_x^5),\\
&u(x+h_x,y)\\
&\quad = u(x,y) + h_x\frac{\partial u(x,y)}{\partial x} + \frac{h_x^2}{2}\frac{\partial^2 u(x,y)}{\partial x^2} + \frac{h_x^3}{6}\frac{\partial^3 u(x,y)}{\partial x^3} + \frac{h_x^3}{24}\frac{\partial^4 u(x,y)}{\partial x^4} + O(h_x^5),\\
&u(x-h_x,y)\\
&\quad = u(x,y) - h_x\frac{\partial u(x,y)}{\partial x} + \frac{h_x^2}{2}\frac{\partial^2 u(x,y)}{\partial x^2} - \frac{h_x^3}{6}\frac{\partial^3 u(x,y)}{\partial x^3} + \frac{h_x^3}{24}\frac{\partial^4 u(x,y)}{\partial x^4} + O(h_x^5),\\
&u(x-2h_x,y)\\
&\quad = u(x,y) - 2h_x\frac{\partial u(x,y)}{\partial x} + \frac{4h_x^2}{2}\frac{\partial^2 u(x,y)}{\partial x^2} - \frac{8h_x^3}{6}\frac{\partial^3 u(x,y)}{\partial x^3} + \frac{16h_x^3}{24}\frac{\partial^4 u(x,y)}{\partial x^4} + O(h_x^5).
\end{aligned} \tag{2.10}$$

Combining the equations such a way to eliminate the terms with the second, third and fourth derivatives, we obtain two high-order approximations for the first derivative:

$$(\partial u/\partial x)_{ij} = (2u_{i+1j} + 3u_{ij} - 6u_{i-1j} + u_{i-2j})/6h + O(h_x^3) \tag{2.11}$$

and

$$(\partial u/\partial x)_{ij} = (-u_{i+2j} + 8u_{i+1j} - 8u_{i-1j} + u_{i-2j})/12h + O(h_x^4). \tag{2.12}$$

Example 2: $Lu = \dfrac{\partial^2 u(x,y)}{\partial x^2}$

Elementary approximations for second partial derivatives are obtained from the Taylor series of (2.2) and (2.6). For example, on addition of those two equations we find

$$\frac{1}{h_x^2}\left[u(x+h_x,y) - 2u(x,y) + u(x-h_x,y)\right] = \frac{\partial^2 u}{\partial x^2} + O(h_x^2) \tag{2.13}$$

and therefore,

$$\left.\frac{\partial^2 u}{\partial x^2}\right|_{ij} \approx \frac{u_{i+1j} - 2u_{ij} + u_{i-1j}}{h_x^2}. \tag{2.14}$$

In general, higher-order FD approximations for the second (and higher) derivatives of the function can be obtained by involving more points as in the case of (2.11) and (2.12). Other differential operators are presented in Table 2.1, where u_{ij} denotes the discrete approximation of the solution. In what we presented above, the finite difference approximations at a point have been considered. It should be made clear that sometimes the order of approximation at a point does not coincide with that at an entire mesh (Samarskii, 1977).

Mathematical statements of geodynamical problems require additional (boundary and initial) conditions to be introduced along with the governing equations. Only these conditions can define the unique solution to the problem from a set of possible solutions. Therefore, additional conditions should be approximated by finite differences as well. A set of difference equations, approximating differential equations and boundary and initial conditions, is referred to as *a finite difference scheme*. As an example, we consider the boundary-value problem for the 1-D heat equation:

$$\begin{aligned} &\frac{\partial u}{\partial t} = \frac{\partial^2 u}{\partial x^2} + f(x),\ 0 < x < 1,\ 0 < t \le t_0, \\ &u(0,t) = g_0(t),\ u(1,t) = g_1(t), \\ &u(x,0) = u_0(x). \end{aligned} \tag{2.15}$$

We choose a uniform mesh $s(h,\tau) = \{x_i = ih_x,\ t_j = j\tau,\ i = 0, 1, \ldots, N_x,\ j = 0, 1, \ldots, N_t\}$ and replace the problem (2.15) by the discrete problem:

$$\begin{aligned} &\frac{U_i^{j+1} - U_i^j}{\tau} = \frac{U_{i+1}^j - 2U_i^j + U_{i-1}^j}{h^2} + f(x_i), \quad 1 \le i \le N_x - 1,\ 0 \le j \le N_t - 1, \\ &U_0^j = g_0(t_j),\ U_{N_x}^j = g_1(t_j), U_i^0 = u_0(x_i). \end{aligned} \tag{2.16}$$

Table 2.1. Difference formula for partial derivatives and simple differential operators.

Differential operator	Difference formula
$(\partial u/\partial y)_{ij}$	$(u_{ij+1} - u_{ij})/h_y$, $(u_{ij} - u_{ij-1})/h_y$, $(u_{ij+1} - u_{ij-1})/2h_y$
$(\partial^2 u/\partial y^2)_{ij}$	$(u_{ij+1} - 2u_{ij} + u_{ij-1})/h_y^2$
$(\partial^2 u/\partial x \partial y)_{ij}$	$(u_{i+1j+1} - u_{i+1j-1} - u_{i-1j+1} + u_{i-1j-1})/4h^2$ $(h = h_x = h_y)$
$(\nabla^2 u)_{ij}$	$(u_{i+1j} + u_{ij+1} - 4u_{ij} + u_{ij-1} + u_{i-1j})/h^2$ $(h = h_x = h_y)$
$(\nabla^4 u)_{ij}$	$(u_{i+2j} + 2u_{i+1j+1} - 8u_{i+1j} + 2u_{i+1j-1} + u_{ij+2} - 8u_{ij+1} + 20u_{ij} - 8u_{ij-1} + u_{ij-2} + 2u_{i-1j+1} - 8u_{i-1j} + 2u_{i-1j-1} + u_{i-2j})/h^4$ $(h = h_x = h_y)$

The finite-difference problem (2.16) represents an example of *explicit* schemes, that is, the solution to the problem at the $(j+1)$ time step U^{j+1} is determined entirely from the solution at the preceding time step as

$$U_i^{j+1} = U_i^j + (\tau/h^2)[U_{i+1}^j - 2U_i^j + U_{i-1}^j] + \tau f(x_i). \quad (2.17)$$

Another discrete representation of the problem (2.15)

$$\frac{U_i^{j+1} - U_i^j}{\tau} = \frac{U_{i+1}^{j+1} - 2U_i^{j+1} + U_{i-1}^{j+1}}{h^2} + f(x_i), \quad 1 \le i \le N_x - 1,\ 0 \le j \le N_t - 1,$$

$$U_0^j = g_0(t_j),\ U_{N_x}^j = g_1(t_j), U_i^0 = u_0(x_i) \quad (2.18)$$

leads to an *implicit* finite difference scheme. To determine the solution U^{j+1} in this case, one should solve a set of linear algebraic equations with a tridiagonal matrix using, e.g. the sweep method (see Section 2.3):

$$-(1/h^2)U_{i+1}^{j+1} + (1/\tau + 2/h^2)U_i^{j+1} - (1/h^2)U_{i-1}^{j+1} = (1/\tau)U_i^j + f(x_i). \quad (2.19)$$

2.2 Convergence, accuracy and stability

When the mathematical problem is solved using a numerical method, it is important to know the accuracy of the numerical solution (or how close is the solution, obtained using the numerical method, to the exact solution to the problem). In the domain Ξ with the boundary Γ we consider the following problem:

$$Lu = f(x),\ x \in \Xi, \quad lu = g(x),\ x \in \Gamma, \quad (2.20)$$

where L and l are linear differential operators, and $f(x)$ and $g(x)$ are known functions. We assume that the solution u to (2.20) exists and it is unique. The domain $\Xi \cup \Gamma$ is replaced by a set of the discrete mesh points. The problem (2.20) can be then transformed into a finite difference problem:

$$L_h U_h = f(x_h),\ x_h \in \varpi_h, \quad l_h U_h = g(x_h),\ x_h \in \gamma_h, \quad (2.21)$$

where the parameter h characterises the density of the mesh points, ϖ_h is the set of internal mesh points, and γ_h is the set of boundary mesh points. Comparing U_h and $u(x)$ provides an estimate of the accuracy of the solution U_h to the problem (2.21) with respect to the choice of spatial step h. To estimate the accuracy, we consider the residual $\delta u_h = U_h - u_h$ between the exact solution at a mesh point u_h and the finite difference solution U_h. Using (2.21) we find that the residual satisfies the similar problem, namely:

$$L_h \delta u_h = \xi_h,\ x_h \in \varpi_h, \quad l_h \delta u_h = \zeta_h,\ x_h \in \gamma_h, \quad (2.22)$$

where $\xi_h = f(x_h) - L_h u_h$ and $\zeta_h = g(x_h) - l_h u_h$ are approximation errors between equations and between boundary conditions in Eqs. (2.20) and (2.21), respectively.

The finite difference solution (2.21) converges to the solution of the problem (2.20) (or simply the FD scheme (2.21) converges), if $\|\delta u_h\| = \|U_h - u_h\| \to 0$ at $h \to 0$, where the norm is defined in the relevant space. The scheme converges with the rate $O(h^n)$ (or the FD scheme is the nth-order scheme), if the inequality $\|\delta u_h\| = \|U_h - u_h\| \leq Mh^n$ holds at $h \leq h^*$, where M is a positive constant independent of h and $n > 0$ (Samarskii, 1977). The FD scheme is the nth-order approximation if $\|\xi_h\| = O(h^n)$ and $\|\zeta_h\| = O(h^n)$ (note that these norms can belong to different spaces). Therefore, when finite differences are used to approximate the problem, one should pay attention to the fact that the accuracy of the FD scheme depends on the order of the approximation of the equations as well as the boundary conditions of the problem.

Application of the FD method to the mathematical problem allows the solution of continuous differential equations with boundary and initial conditions to be reduced to the solution of a set of linear algebraic equations. To obtain the approximate solution, we need to solve the set of equations. Techniques for this will be discussed in Chapter 6.

Input data (right-hand sides of the equations, boundary and initial conditions) are introduced with certain errors. During computations round-off errors are inevitably obtained. Therefore, it is essential to use FD schemes that prohibit the rapid (e.g. exponential) growth of small errors during computations. Such FD schemes are referred to as *stable schemes*. Otherwise, they are unstable and cannot be used in the modelling. The solution U_h of the discrete problem depends continuously on the input data ψ_h as

$$\left\| U_h^* - U_h \right\| \leq C \left\| \psi_h^* - \psi_h \right\|, \tag{2.23}$$

where U_h^* is the solution of the discrete problem (FD scheme) with the input data ψ_h^*, and C is a positive constant independent of h. The FD scheme is stable if the inequality (2.23) holds at sufficiently small $h \leq h^*$.

2.3 Finite difference sweep method

There are several approaches to solving the set of FD equations, including a direct matrix solver and iterative relaxation methods, both of which are discussed later. One of the simplest approaches to solve the FD equations is the sweep method, which was introduced in the middle of the twentieth century (e.g. see Marchuk, 1958; Richtmayer and Morton, 1967; and references therein). Consider the following FD equation

$$A_i U_{i-1} - B_i U_i + C_i U_{i+1} = -D_i, i = 1, 2, \ldots, N-1, \tag{2.24}$$

with boundary conditions:

$$U_0 = \lambda_1 U_1 + \chi_1, \; U_N = \lambda_2 U_{N-1} + \chi_2, \tag{2.25}$$

where $A_i, B_i, C_i, D_i, \lambda_1, \lambda_2, \chi_1$, and χ_2 are known parameters. We search for the solution of (2.24) in the form

$$U_i = \alpha_{i+1} U_{i+1} + \beta_{i+1}, i = 0, 1, \dots, N-1, \tag{2.26}$$

where α_i and β_i are yet unknown coefficients. Inserting (2.26) in (2.24) for U_i and U_{i-1}, we obtain

$$(\alpha_{i+1}(\alpha_i A_i - B_i) + C_i)\, U_{i+1} + ((\alpha_i A_i - B_i)\beta_{i+1} + \beta_i A_i + D_i) = 0, \tag{2.27}$$

and therefore, both terms of the equations should equal to zero. Doing so, we obtain the recurrence formulas for the coefficients α_i and β_i: $\alpha_{i+1} = C_i/(B_i - \alpha_i A_i)$, $\beta_{i+1} = (A_i \beta_i + D_i)/(B_i - \alpha_i A_i)$, $i = 1, 2, \dots, N-1$.

The parameters α_1, β_1 and U_N are determined from Eqs. (2.25) and (2.26) at $i = 0$ and $i = N-1$, respectively: $\alpha_1 = \lambda_1$, $\beta_1 = \chi_1$, and $U_N = (\chi_2 + \lambda_2 \beta_N)/(1 - \lambda_2 \alpha_N)$. Thus, we obtain the exact solution of the problem (2.24)–(2.25) in the following form:

$$\begin{aligned}
U_i &= \alpha_{i+1} U_{i+1} + \beta_{i+1}, i = 0, 1, \dots, N-1, \\
U_N &= (\chi_2 + \lambda_2 \beta_N)/(1 - \lambda_2 \alpha_N), \\
\alpha_{i+1} &= C_i/(B_i - \alpha_i A_i),\ \beta_{i+1} = (A_i \beta_i + D_i)/(B_i - \alpha_i A_i), i = 1, 2, \dots, N-1, \\
\alpha_1 &= \lambda_1,\ \beta_1 = \chi_1.
\end{aligned} \tag{2.28}$$

This approach is referred to as the sweep method. As the values U_i are determined recurrently starting from the right boundary, the method (2.28) is called the right-side sweep method. The left-side sweep method can be defined as

$$\begin{aligned}
U_{i+1} &= \xi_{i+1} U_i + \varsigma_{i+1}, i = 0, 1, \dots, N-1, \\
U_0 &= (\chi_1 + \lambda_1 \varsigma_1)/(1 - \lambda_1 \xi_1), \\
\xi_i &= A_i/(B_i - \xi_{i+1} C_i),\ \varsigma_i = (C_i \varsigma_{i+1} + D_i)/(B_i - \xi_{i+1} C_i), i = 1, 2, \dots, N-1, \\
\xi_N &= \lambda_2,\ \varsigma_N = \chi_2.
\end{aligned} \tag{2.29}$$

The sweep method (2.28) is stable if $|\alpha_i| \le 1$. The following restrictions provide stability to the solution: $A_i > 0$, $C_i > 0$, $B_i \ge A_i + C_i$, $0 \le \lambda_{1,2} < 1$.

2.4 Principle of the maximum

In this section we introduce the principle of the maximum, which can assist in estimations of the dependence of the solution to a FD problem on boundary conditions. Details of the principle and the proofs of the following statements can be found in the book by Samarskii (1977). Consider the following operator:

$$(\Lambda U)_i = A_i U_{i-1} - B_i U_i + C_i U_{i+1},\ i = 1,\ 2, \dots,\ N-1, \tag{2.30}$$

where

$$A_i > 0, \ C_i > 0, \text{ and } B_i \geq A_i + C_i. \tag{2.31}$$

Statement 1. *The principle of the maximum (minimum).* If $(\Lambda U)_i \geq 0$ (or $(\Lambda U)_i \leq 0$) for all indices i, than the non-constant function U_i cannot take maximum (or minimum) value at mesh points $i = 1, 2, \ldots, N-1$.

As a consequence of this statement, we conclude that (i) if $(\Lambda U)_i \geq 0$, $U_0 \leq 0$, and $U_N \leq 0$, then the function U_i is non-positive for $i = 1, 2, \ldots, N-1$; and (ii) if $(\Lambda U)_i \leq 0$, $U_0 \geq 0$, and $U_N \geq 0$, then the function U_i is non-negative for $i = 1, 2, \ldots, N-1$.

Consider now Eq. (2.24) with the following boundary conditions:

$$U_0 = \theta_1, \ U_N = \theta_2. \tag{2.32}$$

Statement 2. The FD problem (2.24) and (2.32) is referred to as a *monotonic* FD scheme, if inequalities (2.31) are held.

It can be shown that a monotonic scheme converges to the unique solution.

Statement 3. If the right-hand side of (2.24) is null function $D_i \equiv 0$, then $\|U_i\| \leq \max\{|\theta_1|, \ |\theta_2|\}$.

2.5 Application of a finite difference method to a two-dimensional heat equation

2.5.1 Statement of the problem

In the model domain $\Omega = \{0 \leq x \leq H_x, 0 \leq y \leq H_y,\}$ we consider the 2-D heat problem

$$\rho\frac{\partial}{\partial t}(cT) + \rho u\frac{\partial}{\partial x}(cT) + \rho v\frac{\partial}{\partial y}(cT) = \frac{\partial}{\partial x}\left(k\frac{\partial T}{\partial x}\right) + \frac{\partial}{\partial y}\left(k\frac{\partial T}{\partial y}\right) + Q, \tag{2.33}$$

$$T(x, y, t = 0) = T_*(x, y), \tag{2.34}$$

$$k\frac{\partial T}{\partial \mathbf{n}} + \alpha(T - T_b) = q, \tag{2.35}$$

where $\mathbf{x} = (x, y)$ are the Cartesian coordinates; T, t and $\mathbf{u} = (u, v)$ are temperature, time and velocity, respectively; ρ is the density; c is the thermal capacity; k is the thermal conductivity; Q is the heat source; $T_*(x, y)$ is the initial temperature; $[t = 0, t = \vartheta]$ is the model time interval; $\mathbf{n}$ is the outward unit normal vector at a point on the model boundary $\partial\Omega$; T_b is the background (pre-defined) temperature; q is the heat flux; and α is a numerical parameter controlling the type of boundary condition. The condition (2.35) is the mixed boundary condition, and one can prescribe temperature (by assuming that $\alpha \to \infty$) or heat flux (by assuming that $\alpha \to 0$) at the model boundary.

We introduce the following dimensionless variables:

$$T = \tilde{T}T_0,\ c = \tilde{c}c_0,\ k = \tilde{k}k_0,\ t = \frac{\tilde{t}b^2\rho_0 c_0}{k_0},\ (u,v) = \frac{(\tilde{u},\tilde{v})k_0}{b\rho_0 c_0},$$

$$\rho = \tilde{\rho}\rho_0,\ x = \tilde{x}bAsp, y = \tilde{y}b,\ \ Q = \tilde{Q}\frac{k_0 T_0}{b^2}, \tag{2.36}$$

where T_0, ρ_0 and c_0 are the typical temperature, density and thermal capacity, respectively; and Asp is the aspect ratio (H_x/H_y). Using the dimensionless variable, Eq. (2.33) can be represented in the following form (we omit the sign 'tilde' from the variables):

$$\frac{\partial T}{\partial t} + \frac{1}{Asp}u(x,y)\frac{\partial T}{\partial x} + v(x,y)\frac{\partial T}{\partial y} = \frac{1}{Asp^2}\frac{\partial}{\partial x}\left(\kappa(x,y)\frac{\partial T}{\partial x}\right) + \frac{\partial}{\partial y}\left(\kappa(x,y)\frac{\partial T}{\partial y}\right) + q(x,y). \tag{2.37}$$

Here we assume that the thermal capacity and density are constant in the model domain and $q(x,y) = \dfrac{Q(x,y)}{c\rho}$, $\kappa(x,y) = \dfrac{k(x,y)}{c\rho}$.

We present the dimensionless boundary conditions in the following form:

$$a_{11}\frac{\partial T}{\partial x} + a_{12}\frac{\partial T}{\partial y} + a_2 T = a_3, \tag{2.38}$$

where $a_{11} = 0.0,\ a_{12} = 0.0,\ a_2 = 1.0,\ a_3 = T_+/T_0$ in the case of the Dirichlet condition (where T_+ is the given temperature), $a_{11} = \cos(\gamma)/Asp,\ a_{12} = \sin(\gamma),\ a_2 = 0.0,\ a_3 = qb/(kk_0T_0)$ in the case of the von Neumann condition, and $a_{11} = \cos(\gamma)/Asp,\ a_{12} = \sin(\gamma),\ a_2 = \frac{\alpha}{k}\frac{b}{k_0},\ a_3 = \frac{q}{k}\frac{b}{k_0T_0} + \frac{\alpha T_*}{k}\frac{b}{k_0T_0}$ in the case of the mixed boundary condition (2.35) (where γ is the angle between the axis x and the vector $\mathbf{n}$, and is normally 0 or $\pi/2$ for a rectangular grid).

In this section we define the following boundary conditions. Temperature $T = T_1$ is prescribed at the upper boundary of the model domain. Temperature $T = T_2$ (Problem 1) or heat flux $\left.\frac{\partial T}{\partial y}\right|_{y=0} = g(x,t)$ (Problem 2) is given at its lower boundary. A zero heat flux is set at the horizontal boundaries of the model domain. Thus, the mathematical problem reduces to the determination of the solution to Eq. (2.37) with the relevant boundary and initial conditions.

2.5.2 Finite difference discretisation

To solve the problem numerically, the finite difference method is employed. Initially we rearrange the terms in Eq. (2.33) in the following way:

$$\frac{\partial T}{\partial t} + \frac{1}{Asp}\tilde{u}(x,y)\frac{\partial T}{\partial x} + \tilde{v}(x,y)\frac{\partial T}{\partial y} = \kappa(x,y)\left(\frac{1}{Asp^2}\frac{\partial^2 T}{\partial x^2} + \frac{\partial^2 T}{\partial y^2}\right) + q(x,y),$$

$$\tilde{u}(x,y) = u(x,y) - \frac{1}{Asp}\frac{\partial\kappa(x,y)}{\partial x},\ \tilde{v}(x,y) = v(x,y) - \frac{\partial\kappa(x,y)}{\partial y}. \tag{2.39}$$

Now consider a regular spatial mesh $(N+2)\times(M+2)$ with the spacing $\Delta x = x_i - x_{i-1}$ and $\Delta y = y_i - y_{i-1}$. A difference operator of spatial approximation is constructed on this mesh. The first and second derivatives are approximated by central difference derivatives and standard second-order difference derivatives, respectively

$$\frac{\partial T}{\partial t} + u_{ij}\left[\frac{T_{i+1,j} - T_{i-1,j}}{2Asp\Delta x}\right] + v_{ij}\left[\frac{T_{i,j+1} - T_{i,j-1}}{2\Delta y}\right]$$
$$= \kappa_{ij}\left(\frac{T_{i+1,j} - 2T_{i,j} + T_{i-1,j}}{Asp^2\Delta x^2} + \frac{T_{i,j+1} - 2T_{i,j} + T_{i,j-1}}{\Delta y^2}\right)$$
$$+ q_{ij} + O(\Delta x^2 + \Delta y^2), \qquad (2.40)$$

where $u_{ij} = \tilde{u}(x_i, y_j)$ and $v_{ij} = \tilde{v}(x_i, y_j)$. The approximation is accurate to $O(\Delta x^2 + \Delta y^2)$. Note that only spatial derivatives of temperature are discretised in Eq. (2.40).

The thermal conductivity in the Earth's interior varies with temperature and depth. We do not consider here its dependence on temperature. To solve Eq. (2.40), thermal diffusivity must be determined mid-way between the grid points at which the temperature is defined. This can be done by using an appropriate average of the values defined at the two adjacent temperature points; a harmonic average gives the most accurate approximation of the heat flux, if it is assumed that κ varies in a stepwise manner. Another method is to smooth the coefficients of thermal diffusivity in the vicinity of the interface between two materials, where the diffusivity has a large gradient:

$$\hat{\kappa} = \frac{\sum \kappa_i w_i}{\sum w_i},$$
$$w_i(x,y) = \begin{cases} 1 - \sin\left(\dfrac{\pi\sqrt{(x-x_i)^2 + (y-y_i)^2}}{2\max(a,b)}\right), & \text{if } \left(\dfrac{x-x_i}{a}\right)^2 + \left(\dfrac{y-y_i}{b}\right)^2 \le 1, \\ 0, \text{ otherwise.} \end{cases} \qquad (2.41)$$

The parameters a and b depend on the number of the points adjacent to point (x, y), which are used in the smoothing of the function $\kappa(x, y)$.

The standard approximation of Neumann boundary conditions by simple forward and backward differences yields an accuracy of the first order. To obtain approximations to accuracy of the second order, we use relations based on the Taylor expansion of the function T and its first and second derivatives. In the case of forward differences, we have

$$T_{k+2} - T_{k+1} = \Delta x \left.\frac{\partial T}{\partial x}\right|_{k+1} + \frac{\Delta x^2}{2}\left.\frac{\partial^2 T}{\partial x^2}\right|_{k+1} + \frac{\Delta x^3}{6}\left.\frac{\partial^3 T}{\partial x^3}\right|_{k+1} + O(\Delta x^4)$$
$$= \Delta x \left.\frac{\partial T}{\partial x}\right|_{k} + \left[\Delta x^2 + \frac{\Delta x^2}{2}\right]\left.\frac{\partial^2 T}{\partial x^2}\right|_{k}$$

$$+\left[\frac{\Delta x^3}{2}+\frac{\Delta x^3}{2}+\frac{\Delta x^3}{6}\right]\left.\frac{\partial^3 T}{\partial x^3}\right|_k + O(\Delta x^4), \tag{2.42}$$

$$3\,[T_{k+1}-T_k] = 3\Delta x\left.\frac{\partial T}{\partial x}\right|_k + \frac{3\Delta x^2}{2}\left.\frac{\partial^2 T}{\partial x^2}\right|_k + \frac{3\Delta x^3}{6}\left.\frac{\partial^3 T}{\partial x^3}\right|_k + O(\Delta x^4). \tag{2.43}$$

Subtracting (2.43) from (2.42) we obtain

$$\left.\frac{\partial T}{\partial x}\right|_k = \frac{-T_{i+2}+4T_{i+1}-3T_i}{2\Delta x} + \frac{\Delta x^2}{3}\left.\frac{\partial^3 T}{\partial x^3}\right|_k + O(\Delta x^3)$$

or

$$T_k = -\frac{1}{3}T_{k+2} + \frac{4}{3}T_{k+1} - \frac{2\Delta x}{3}\left.\frac{\partial T}{\partial x}\right|_k + \frac{2\Delta x^3}{9}\left.\frac{\partial^3 T}{\partial x^3}\right|_k + O(\Delta x^4). \tag{2.44}$$

The formula for backward differences is obtained in a similar way:

$$\left.\frac{\partial T}{\partial x}\right|_k = \frac{T_{i-2}-4T_{i-1}+3T_i}{2\Delta x} + \frac{\Delta x^2}{3}\left.\frac{\partial^3 T}{\partial x^3}\right|_k + O(\Delta x^3) \tag{2.45}$$

or

$$T_k = -\frac{1}{3}T_{k-2} + \frac{4}{3}T_{k-1} + \frac{2\Delta x}{3}\left.\frac{\partial T}{\partial x}\right|_k - \frac{2\Delta x^3}{9}\left.\frac{\partial^3 T}{\partial x^3}\right|_k + O(\Delta x^4). \tag{2.46}$$

Therefore, the boundary conditions can be presented by the following FD equations:

$$T_{0,j} = -\frac{1}{3}T_{2,j} + \frac{4}{3}T_{1,j},\ T_{N+1,j} = -\frac{1}{3}T_{N-1,j} + \frac{4}{3}T_{N,j}$$

$$T_{i,M+1} = T_1$$

$T_{i,0} = T_2$ (in the case of Problem 1) or

$$T_{i,0} = -\frac{1}{3}T_{i,2} + \frac{4}{3}T_{i,1} - \frac{2\Delta y}{3}\left.\frac{\partial T}{\partial x}\right|_i \quad \text{(in the case of Problem 2).} \tag{2.47}$$

2.5.3 Monotonic finite difference scheme

Relations (2.40) and (2.47) can be written in the operator form as

$$\frac{\partial T}{\partial t} + AT = F_A. \tag{2.48}$$

The operator A can be represented in the canonical form:

$$AT = \gamma T_{ij} - \varphi_1 T_{i+1j} - \phi_1 T_{i-1j} - \varphi_2 T_{ij+1} - \phi_2 T_{ij-1},\ \gamma = \kappa_{ij}\left(\frac{2}{Asp^2\Delta x^2} + \frac{2}{\Delta y^2}\right),$$

$$\varphi_1 = u_{ij}\frac{-1}{2Asp\Delta x} + \kappa_{ij}\frac{1}{Asp^2\Delta x^2},\ \phi_1 = u_{ij}\frac{1}{2Asp\Delta x} + \kappa_{ij}\frac{1}{Asp^2\Delta x^2},$$

$$\varphi_2 = v_{ij}\frac{-1}{2\Delta y} + \kappa_{ij}\frac{1}{\Delta y^2},\ \phi_2 = v_{ij}\frac{1}{2\Delta y} + \kappa_{ij}\frac{1}{\Delta y^2}. \tag{2.49}$$

The following conditions provide the scheme (2.49) to be monotonic (see Statement 2 in Section 2.4):

$$\gamma \geq \varphi_1 + \phi_1 + \varphi_2 + \phi_2,\ \varphi_m > 0,\ \phi_m > 0,\ m = 1, 2. \tag{2.50}$$

The first condition is always fulfilled, and other two conditions constrain the discrete Péclet number θ_k^{ij}:

$$|\theta_k^{ij}| < 1\ (k = 1, 2),\ \theta_1^{ij} = \frac{u_{ij}Asp\Delta x}{2\kappa_{ij}},\ \theta_2^{ij} = \frac{v_{ij}\Delta y}{2\kappa_{ij}}. \tag{2.51}$$

For Péclet numbers higher than this critical value, i.e. when advection dominates over diffusion, using central differences for the advective derivatives gives an unconditionally unstable scheme, in which spurious oscillations appear and grow exponentially with time. In order to obtain a stable advection scheme in situations where diffusion is slow compared to advection or even zero, it is necessary to bias the advective derivatives to the upwind direction, i.e. the direction that material is locally coming from, as discussed later (see Chapter 6), or to refine the mesh as the discrete Péclet number depends on the grid size. An alternative way to obtain a stable solution is to use a regularisation function $g(x, y)$ (Samarskii, 1967; 1977) replacing κ_{ij} by $\kappa_{ij}(1+g_{ij})$ in Eq. (2.51). The regularisation function g is chosen to be zero at the grid points, where the scheme is monotonic (i.e. the conditions (2.50) are satisfied), and to be $\eta\left(\theta_k^{ij}\right)^2$ otherwise. The scheme becomes monotonic at $\eta > 0.25$.

2.5.4 Solution method

Several methods exist to solve the discrete problem (Eq. (2.48)) including explicit time-stepping as in the 1-D diffusion example (Eq. (2.17)) or implicit time-stepping with the coefficients (Eq. (2.49)) placed into a single matrix. Here, to solve the problem (2.48), we use a stabilisation method belonging to the class of splitting methods. This method is absolutely stable provided that the split operators are positively semi-definite (Marchuk, 1989), and its accuracy of approximation in time is of the order of $O(\tau^2)$, where τ is the time step. Moreover, since the operator A can be represented as the sum of two operators of banded structure (see Eq. (2.52)), at each time step the use of the stabilisation method reduces computations to the solution of a series of banded equations.

Representing the operator A as the sum of two operators (splitting it in coordinates), we transform (2.48) into the form

$$\frac{\partial T}{\partial t} + (\Lambda_1 + \Lambda_2)T = F_A, \tag{2.52}$$

where the FD operators A, Λ_1 and Λ_2 are determined to an accuracy of $O(\Delta x^2) + O(\Delta y^2)$ as

$$\begin{aligned} AT &= T_{i-1j}\left(-\alpha_{ij} - \mu_{ij}\right) + T_{ij-1}\left(-\beta_{ij} - \sigma_{ij}\right) + T_{ij}\left(2\mu_{ij} + 2\sigma_{ij}\right) \\ &\quad + T_{ij+1}\left(\beta_{ij} - \sigma_{ij}\right) + T_{i+1j}\left(\alpha_{ij} - \mu_{ij}\right), \\ \Lambda_1 T &= T_{i-1j}\left(-\alpha_{ij} - \mu_{ij}\right) + T_{ij}\left(2\mu_{ij}\right) + T_{i+1j}\left(\alpha_{ij} - \mu_{ij}\right), \\ \Lambda_2 T &= T_{ij-1}\left(-\beta_{ij} - \sigma_{ij}\right) + T_{ij}\left(2\sigma_{ij}\right) + T_{ij+1}\left(\beta_{ij} - \sigma_{ij}\right), \\ \alpha_{ij} &= \frac{u_{ij}}{2Asp\Delta x}, \beta_{ij} = \frac{v_{ij}}{2\Delta y}, \mu_{ij} = \frac{\kappa_{ij}}{Asp^2\Delta x^2}, \sigma_{ij} = \frac{\kappa_{ij}}{\Delta y^2}. \end{aligned} \tag{2.53}$$

Since the matrices of the operators Λ_1 and Λ_2 are constructed from components of the vector T of different orders, simultaneous operations with both matrices require that they be brought into correspondence with one of the types of the vector T. With the matrices constructed in this way, they can be divided into independent blocks of banded structure. The solution is sought at inner points of the model mesh. Let the vector T be of the form $T = (T_{11}, T_{12}, \ldots, T_{1M}, T_{21}, T_{22}, \ldots, T_{2M}, \ldots, T_{N1}, T_{N2}, \ldots, T_{NM})^T$. Then the matrix of the operator A has the form:

$\frac{4}{3}A_1 + B_1$	$-\frac{1}{3}A_1 + C_1$	0	$\ldots$	0
A_2	B_2	C_2	$\ldots$	$\ldots$
$\ldots$	$\ldots$	$\ldots$	$\ldots$	$\ldots$
$\ldots$	$\ldots$	A_{N-1}	B_{N-1}	C_{N-1}
0	$\ldots$	0	$A_N - \frac{1}{3}C_N$	$\frac{4}{3}A_N + B_N$

, (2.54)

where the diagonal matrix A_i is written as

a_{i1}	0	0	$\ldots$	0
0	a_{i2}	0	$\ldots$	0
0	$\ldots$	$\ldots$	$\ldots$	0
0	$\ldots$	0	a_{iM-1}	0
0	$\ldots$	0	0	a_{iM}

(2.55)

and $a_{ij} = -\alpha_{ij} - \mu_{ij}$. The matrix B_i is tridiagonal and has the following form for Problem 1:

r_{i1}	s_{i1}	0	$\ldots$	0
q_{i2}	r_{i2}	s_{i2}	0	$\ldots$
$\ldots$	$\ldots$	$\ldots$	$\ldots$	$\ldots$
$\ldots$	0	q_{iM-1}	r_{iM-1}	s_{iM-1}
0	$\ldots$	0	q_{iM}	r_{iM}

. (2.56)

For Problem 2, the matrix B_i is written as

$$\begin{array}{|c|c|c|c|c|}\hline \frac{4}{3}q_{i1}+r_{i1} & -\frac{1}{3}q_{i1}+s_{i1} & 0 & \cdots & 0 \\ \hline q_{i2} & r_{i2} & s_{i2} & 0 & \cdots \\ \hline \cdots & \cdots & \cdots & \cdots & \cdots \\ \hline \cdots & 0 & q_{iM-1} & r_{iM-1} & s_{iM-1} \\ \hline 0 & \cdots & 0 & q_{iM} & r_{iM} \\ \hline \end{array}, \tag{2.57}$$

where $q_{ij} = -\beta_{ij} - \sigma_{ij}$, $r_{ij} = 2(\mu_{ij} + \sigma_{ij})$, and $s_{ij} = \beta_{ij} - \sigma_{ij}$. The tridiagonal matrix C_i has the form

$$\begin{array}{|c|c|c|c|c|}\hline c_{i1} & 0 & 0 & \cdots & 0 \\ \hline 0 & c_{i2} & 0 & \cdots & 0 \\ \hline 0 & \cdots & \cdots & \cdots & 0 \\ \hline 0 & \cdots & 0 & c_{iM-1} & 0 \\ \hline 0 & \cdots & 0 & 0 & c_{iM} \\ \hline \end{array}, \tag{2.58}$$

where $c_{ij} = \alpha_{ij} - \mu_{ij}$. The vector F_A can be written as

$$F_A = \left(\tilde{F}_1, \tilde{F}_2, \ldots, \tilde{F}_N\right)^T, \; \tilde{F}_i = ((\beta_{i1} + \sigma_{i1})T_2, \; 0, \; \ldots, \; 0, \; (-\beta_{iM} + \sigma_{iM})T_1)^T \tag{2.59}$$

for Problem 1 and as

$$F_A = \left(\tilde{F}_1, \tilde{F}_2, \ldots, \tilde{F}_N\right)^T, \; \tilde{F}_i = \left(-(\beta_{i1} + \sigma_{i1})\frac{2\Delta y}{3}\left.\frac{\partial T}{\partial x}\right|_i, \; 0, \; \ldots, \; 0, \; (-\beta_{iM} + \sigma_{iM})T_1\right)^T \tag{2.60}$$

for Problem 2. The matrix of the operator Λ_2 has the form:

$$\begin{array}{|c|c|c|c|c|}\hline B_1 & 0 & \cdots & \cdots & 0 \\ \hline 0 & B_2 & 0 & \cdots & \cdots \\ \hline \cdots & \cdots & \cdots & \cdots & \cdots \\ \hline \cdots & \cdots & 0 & B_{N-1} & 0 \\ \hline 0 & \cdots & \cdots & 0 & B_N \\ \hline \end{array}. \tag{2.61}$$

For Problem 1, the matrix B_i is

$$\begin{array}{|c|c|c|c|c|c|c|}\hline r_{i1} & s_{i1} & 0 & \cdots & \cdots & \cdots & 0 \\ \hline q_{i2} & r_{i2} & s_{i2} & 0 & \cdots & \cdots & \cdots \\ \hline 0 & \cdots & \cdots & \cdots & \cdots & \cdots & 0 \\ \hline \cdots & \cdots & \cdots & 0 & q_{iM-1} & r_{iM-1} & s_{iM-1} \\ \hline 0 & \cdots & \cdots & \cdots & 0 & q_{iM} & r_{iM} \\ \hline \end{array}. \tag{2.62}$$

For Problem 2, the matrix is

$\frac{4}{3}q_{i1} + r_{i1}$	$-\frac{1}{3}q_{i1} + s_{i1}$	0	$\dots$	$\dots$	$\dots$	0
q_{i2}	r_{i2}	s_{i2}	0	$\dots$	$\dots$	$\dots$
0	$\dots$	$\dots$	$\dots$	$\dots$	$\dots$	0
$\dots$	$\dots$	$\dots$	0	q_{iM-1}	r_{iM-1}	s_{iM-1}
0	$\dots$	$\dots$	$\dots$	0	q_{iM}	r_{iM}

, (2.63)

where $q_{ij} = -\beta_{ij} - \sigma_{ij}$, $r_{ij} = 2\sigma_{ij}$ and $s_{ij} = \beta_{ij} - \sigma_{ij}$.

The matrix Λ_1 is constructed from the condition that the vector T has the following structure: $T = (T_{11}, T_{21}, \dots, T_{N1}, T_{12}, T_{22}, \dots, T_{N2}, \dots, T_{1M}, T_{2M}, \dots, T_{NM})^T$. Then Λ_1 has the form

A_1	0	$\dots$	$\dots$	0
0	A_2	$\dots$	$\dots$	0
$\dots$	$\dots$	$\dots$	$\dots$	$\dots$
0	$\dots$	$\dots$	A_{M-1}	0
0	$\dots$	$\dots$	0	A_M

, (2.64)

where the matrix A_j is

$\frac{4}{3}a_{1j} + b_{1j}$	$-\frac{1}{3}a_{1j} + c_{1j}$	0	$\dots$	$\dots$	$\dots$	0
a_{2j}	b_{2j}	c_{2j}	0	$\dots$	$\dots$	$\dots$
$\dots$	$\dots$	$\dots$	$\dots$	$\dots$	$\dots$	$\dots$
$\dots$	$\dots$	$\dots$	0	a_{N-1j}	b_{N-1j}	c_{N-1j}
0	$\dots$	$\dots$	$\dots$	0	$-\frac{1}{3}c_{Nj} + a_{Nj}$	$\frac{4}{3}c_{Nj} + b_{Nj}$

. (2.65)

Here $a_{ij} = -\alpha_{ij} - \mu_{ij}$, $b_{ij} = 2\mu_{ij}$ and $c_{ij} = \alpha_{ij} - \mu_{ij}$.

The temperature at a new time step is calculated by the following FD scheme:

$$\left(E + \frac{\tau}{2}\Lambda_1\right)\xi^{j+1/2} = -AT^j + \frac{F^{j+1} + F^j}{2} \tag{2.66}$$

$$\left(E + \frac{\tau}{2}\Lambda_2\right)\xi^{j+1} = \xi^{j+1/2} \tag{2.67}$$

$$T^{j+1} = T^j + \tau\xi^{j+1} \tag{2.68}$$

where E is the unit matrix, and ξ is the auxiliary variable. Systems (2.66) and (2.67) are solved by the scheme of single division (a modification of the sweep method). System (2.68) is solved explicitly.

2.5.5 Verification of the finite difference scheme

To verify the correctness of the algorithm, we use the trial function

$$T = e^{-2\pi^2\kappa t}\cos(\pi x)\sin(\pi y) + uvx^2y^2 + y(T_1 - T_2) + T_2 \tag{2.69}$$

and the following parameters: $N = 200$, $M = 200$, $\tau = 10^{-4}$, $T_1 = 0.0$, $T_2 = 1.0$, and $Asp = 1.0$. The boundary and initial conditions in this case are

$$\begin{aligned}
T_0 &= \cos(\pi x)\sin(\pi y) + uvx^2y^2 + y(T_1 - T_2) + T_2 \\
T|_{y=0} &= T_2 \quad T|_{y=1} = uvx^2 + T_1 \text{ (in Problem 1) and} \\
\left.\frac{\partial T}{\partial y}\right|_{y=0} &= \pi e^{-2\pi^2\kappa t}\cos(\pi x) + T_1 - T_2 \text{ (in Problem 2)} \\
\left.\frac{\partial T}{\partial x}\right|_{x=0} &= 0 \quad \left.\frac{\partial T}{\partial x}\right|_{x=1} = 2uvy^2.
\end{aligned} \tag{2.70}$$

According to the boundary conditions, the right-hand side for the solution of the system has the form

$$\begin{aligned}
F^j &= F_\Pi + F^j_\Gamma, \\
F_\Pi &= -2\kappa uv(x^2 + \frac{y^2}{Asp^2}) + 2uvxy(vx + \frac{u}{Asp}y) + v(T_1 - T_2) \\
&\quad + \exp(-2\pi^2\kappa t)\left(v\cos(\pi x)\cos(\pi y) - \frac{u}{Asp}\sin(\pi x)\sin(\pi y)\right. \\
&\quad \left. +\kappa\left[\frac{1}{Asp^2} - 1\right]\pi\cos(\pi x)\sin(\pi y)\right), \\
F^j_\Gamma &= \left(F^j_1, F^j_2, \ldots, F^j_{N-1}, F^j_N + \tilde{F}^j_N\right)^T,
\end{aligned} \tag{2.71}$$

in the case of Problem 1

$$F^j_i = \left((\beta_{i1} + \rho_{i1})T_2, 0, \ldots, 0, \nu_{iM}(uv(\Delta x \cdot i)^2 + T_1)\right)^T \tag{2.72}$$

and in the case of Problem 2

$$\begin{aligned}
F^j_i = \Big(&-\frac{2}{3}\Delta y(\beta_{i1} + \rho_{i1})(\pi\exp(-2j\pi^2\kappa\tau)\cos(\pi x) + T_1 - T_2), \\
&0, \ldots, 0, \rho_{iM}(uv(\Delta x \cdot i)^2 + T_1)\Big)^T,
\end{aligned}$$

and

$$\tilde{F}^j_N = \left(\varpi_{N1}\frac{4\Delta x}{3}uv(1\cdot\Delta y)^2, \ldots, \varpi_{NM}\frac{4\Delta x}{3}uv(M\cdot\Delta y)^2\right)^T, \tag{2.73}$$

Table 2.2. Misfit between the exact and the FD solutions.

Number of	Case 1		Case 2		Case 3		Case 4		Case 5	
j-iterations	Problem 1	Problem 2	Problem 1	Problem 2	Problem 1	Problem 2	Problem 1	Problem 2	Problem 1	Problem 2
1	−4.3E−8	2.0E−6	−1.2E−8	−2.1E−8	−3.9E−8	−2.0E−6	−4.1E−8	2.0E−6	−4.1E−8	2.0E−6
2	−8.7E−8	3.0E−6	−2.5E−8	−4.2E−8	7.8E−8	−2.9E−6	−8.3E−8	2.9E−6	8.2E−8	2.9E−6
3	−1.3E−7	3.9E−6	−3.8E−8	−4.2E−8	7.8E−8	−3.8E−6	−1.2E−7	3.8E−6	1.2E−7	3.8E−6
5	−2.1E−7	5.3E−6	−6.5E−8	−1.0E−7	−1.9E−7	5.2E−6	−1.2E−7	5.2E−6	−2.0E−7	5.2E−6
10	−4.2E−7	8.0E−6	−1.3E−7	−2.1E−7	−3.7E−7	−7.8E−6	−4.1E−7	7.8E−6	−3.9E−7	7.8E−6
20	−8.3E−7	1.1E−5	−2.8E−7	−4.4E−7	7.2E−7	1.1E−5	−8.0E−7	1.1E−5	−3.9E−7	1.1E−5
50	−1.9E−6	−1.9E−5	−7.9E−7	−1.2E−6	1.6E−6	1.7E−5	−1.8E−6	−1.7E−5	1.7E−6	−1.7E−5
100	−3.5E−6	−2.6E−5	−1.8E−6	2.7E−6	3.0E−6	−2.4E−5	−3.4E−6	−2.4E−5	−3.2E−6	−2.4E−5

where the superscript T means transposition; $\beta_{ij} = b_{ij}/2\Delta y$, $\rho_{ij} = \kappa_{ij}/\Delta y^2$ and $\varpi_{ij} = \frac{\kappa_{ij}}{Asp^2\,\Delta x^2}$.

Equation (2.48) with the given right-hand side F^j and boundary conditions are solved by the FD scheme (2.66)–(2.68). We consider five cases of the problem parameters: (1) $u = 1.0$, $v = 1.0$, $\kappa = 1.0$; (2) $u = 1.0$, $v = 1.0$, $\kappa = 0.0$; (3) $u = 0.0$, $v = 0.0$, $\kappa = 1.0$; (4) $u = 1.0$, $v = 0.0$, $\kappa = 1.0$; and (5) $u = 0.0$, $v = 1.0$, $\kappa = 1.0$. Table 2.2 presents the difference between the exact and the finite difference solutions.

3 Finite volume method

3.1 Introduction

The finite volume (FV) method is commonly used in computational fluid dynamics and offers an intuitive and conservative way of discretising the governing equations in a manner that combines some of the advantages of finite difference and finite element methods. The general discretisation approach is to divide the domain into control volumes and integrate the equations over each control volume, with the divergence theorem used to turn some of the volume integrals into surface integrals. The resulting discretised equations equate fluxes across control volume faces (e.g. heat fluxes) to sources and sinks inside the volume (e.g. changes in temperature), and can be solved with standard direct or iterative methods (Chapter 6). The finite volume formulation is conservative because the flux flowing across a shared volume face is the same for each adjoining volume, and this is an important property in some applications. The method can be used with unstructured grids, although this chapter focuses mainly on rectangular grids, on which the discretised equations become very similar to finite difference equations. For implementation details related to using unstructured grids the reader is referred to Versteeg and Malalasekera (2007).

3.2 Grids and control volumes: structured and unstructured grids

Each control volume contains a node on which scalar quantities are defined. For a simple, rectangular structured grid the node is straightforwardly located in the control volume centre, as indicated by the example in Fig. 3.1.

In the more general case of an unstructured grid, there are two possible relationships between the nodes and the control volume faces: either (i) the nodes can be defined first, then the faces constructed to be mid-way between the nodes, or (ii) the control volumes can be defined first, then the nodes are chosen to be located at the centroids of the volumes. The first approach is more commonly used and thus is further described here. Figure 3.2 illustrates an unstructured collection of grid points. One method for constructing the control volumes is to first draw lines connecting the points together (Fig. 3.2, faint lines), then bisect these lines with perpendicular control volume boundaries (thick lines). The control volumes are polygons. Triangular or rectangular meshes are common, leading to hexagonal or rectangular control volumes. Next, as illustrated later, the PDE(s) are integrated over a

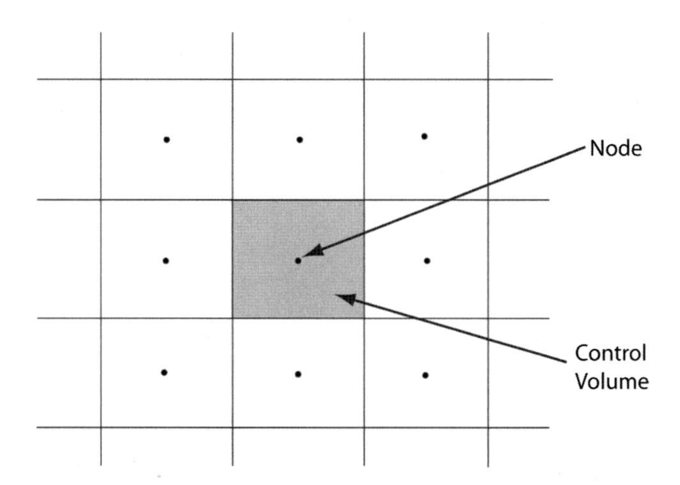

Fig. 3.1. Nodes (dots) and control volumes for a rectangular structured grid.

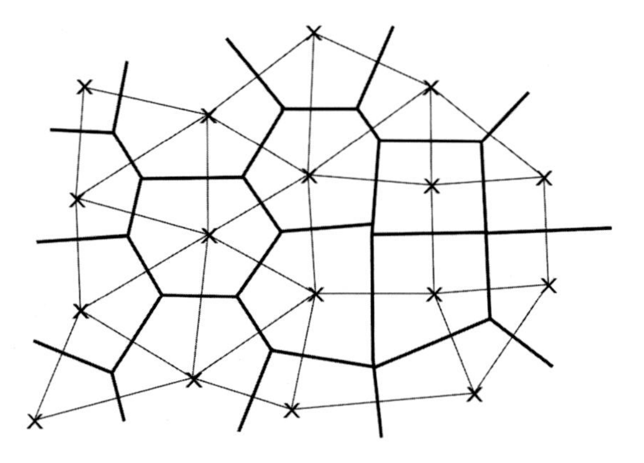

Fig. 3.2. Nodes (X) and control volumes (thick lines) for an unstructured grid. The control volume faces have been constructed such that they bisect and are perpendicular to the (thin) lines connecting the nodes.

control volume and discretised in terms of fluxes across control volume boundaries and body sources or sinks.

3.3 Comparison to finite difference and finite element methods

The finite element (FE) method also uses an integrated form of the equations, but in the FE method the equations are integrated over a weight function, whereas in the finite volume (FV) method they are integrated over a control volume. The FV method is thus similar to the FE method using a weight function that is unity inside the element (control volume) and zero outside. The second aspect to compare is the calculation of interpolated quantities and derivatives, which are needed to calculate fluxes at control volume faces. In the FV method linear interpolation is typically used to interpolate quantities to control volume faces and finite differences (FD) used to obtain derivatives. In the FE method, the variation of properties and variables between nodes is instead typically defined by Lagrange polynomials (i.e. shape functions) of any order. Therefore, the FV method is similar to the FE method using linear shape functions.

On a rectangular grid the discretisation is greatly simplified. If quantities on control volume faces are calculated by linear interpolation and finite differencing, then the discretised equations end up being equivalent to FD equations; on a staggered grid in the case of the momentum (velocity–pressure) equations. A major difference, however, is that FV schemes are always conservative, whereas it is possible to derive FD schemes (e.g. for advection) that are not conservative.

The finite volume method can therefore be seen as combining some advantages of the finite element method, such as being able to handle arbitrary geometries and unstructured grids, with some advantages of finite differences, in that the discretised equations still resemble the physical equations and thus have a clear intuitive physical meaning. A disadvantage is that, unlike with FE or FD methods, how to introduce more accurate, higher order versions of the method is not well defined.

3.4 Treatment of advection–diffusion problems

3.4.1 Diffusion

The finite volume approach can be illustrated with the simple diffusion equation:

$$\frac{\partial T}{\partial t} = \nabla \cdot (\kappa \nabla T), \tag{3.1}$$

where T is temperature, t is time and κ is the coefficient of thermal diffusivity, which may be spatially varying. Integrating this over a control volume CV, one obtains:

$$\int_{CV} \frac{\partial T}{\partial t} dV = \int_{CV} \nabla \cdot (\kappa \nabla T) dV. \tag{3.2}$$

Next, using the divergence theorem, i.e.

$$\int_{volume} \nabla \cdot \mathbf{A} dV = \int_{surface} \mathbf{A} \cdot \mathbf{n} dS, \tag{3.3}$$

we can convert the integrated divergence terms in Eq. (3.2) into fluxes over the control volume faces:

$$\int_{CV} \frac{\partial T}{\partial t} dV = \int_{surface} (\kappa \nabla T) \cdot \mathbf{n} dS, \tag{3.4}$$

which basically states that the rate of change of energy is equal to the net flux of heat into the control volume. To discretise this, assume that dT/dt is constant over the volume and then approximate the heat fluxes using first-order finite differences (which become second-order accurate if the face is mid way between the nodes)

$$\frac{\partial T_i}{\partial t} V_i = \sum_{j=1}^{n_i} F_j L_j = \sum_{j=1}^{n_i} \kappa_j \frac{(T_j - T_i)}{|\mathbf{x}_j - \mathbf{x}_i|} L_j, \tag{3.5}$$

where the sum is over the n_i adjoining cells, F_j is the heat flux through side j, L_j is the length of the side and $\mathbf{x}_i$ is the position of the ith grid point.

In a Cartesian grid with constant grid spacing in each of the three directions the equation simplifies to:

$$\begin{aligned}\frac{\partial T_{ijk}}{\partial t} &= \frac{1}{h_x^2}\left(\kappa_{i+\frac{1}{2}jk}\left(T_{i+1jk} - T_{ijk}\right) - \kappa_{i-\frac{1}{2}jk}\left(T_{ijk} - T_{i-1jk}\right)\right) \\ &+ \frac{1}{h_y^2}\left(\kappa_{ij+\frac{1}{2}k}\left(T_{ij+1k} - T_{ijk}\right) - \kappa_{ij-\frac{1}{2}k}\left(T_{ijk} - T_{ij-1k}\right)\right) \\ &+ \frac{1}{h_z^2}\left(\kappa_{ijk+\frac{1}{2}}\left(T_{ijk+1} - T_{ijk}\right) - \kappa_{ijk-\frac{1}{2}}\left(T_{ijk} - T_{ijk-1}\right)\right),\end{aligned} \tag{3.6}$$

where T_{ijk} is the temperature at node (i,j,k) and h_x, h_y and h_z are the grid spacing in the x, y and z directions respectively. This can be recognised as a finite difference approximation of the original equation, with diffusivity defined halfway between temperature points.

The thermal conductivity or diffusivity needs to be known at the centres of the cell faces. Normally, they are first defined at the nodal points then linearly interpolated to the control volume edges. If, however, the thermal conductivity varies strongly from one control volume to the next, a different interpolation scheme might give more physically accurate results. If, for example, it is assumed that the thermal conductivity is constant in each control volume and that the heat flux is continuous from one cell to the next, then for a one-dimensional problem the appropriate thermal conductivity at the control volume edge is a harmonic average of the nodal values. Similar considerations will apply later to viscosity interpolation for solving the Stokes equation.

In the case of a steady state, the time-derivative can be dropped, and Eq. (3.6) can be written as:

$$0 = a_{ijk}^C T_{ijk} + a_{ijk}^E T_{i+1jk} + a_{ijk}^W T_{i-1jk} + a_{ijk}^N T_{ij+1k} + a_{ijk}^S T_{ij-1k} + a_{ijk}^U T_{ijk+1} + a_{ijk}^D T_{ijk-1}, \tag{3.7}$$

where the a coefficients are combinations of κ and h identified from Eq. (3.6) and the superscripts refer to the position relative to point (i,j,k) (Centre, East, West, North, South, Up, Down). These a coefficients will be different for each (i,j,k) if the thermal conductivity varies spatially, but the same if it is constant. To complete the system, values of T or its gradient must be specified at the domain boundary. The resulting set of linear discretised equations for this boundary value problem can then be solved by using standard direct or iterative methods (Chapter 6).

For time-dependent diffusion problems, the different time-integration techniques discussed in Chapter 7 can be applied to the time derivative. In the case of first-order implicit time integration, the discretised version of Eq. (3.6) becomes

$$\frac{T_{ijk}^{n+1}-T_{ijk}^{n}}{\Delta t}=\frac{1}{h_x^2}\left(\kappa_{i+\frac{1}{2}jk}\left(T_{i+1jk}^{n+1}-T_{ijk}^{n+1}\right)-\kappa_{i-\frac{1}{2}jk}\left(T_{ijk}^{n+1}-T_{i-1jk}^{n+1}\right)\right)$$
$$+\frac{1}{h_y^2}\left(\kappa_{ij+\frac{1}{2}k}\left(T_{ij+1k}^{n+1}-T_{ijk}^{n+1}\right)-\kappa_{ij-\frac{1}{2}k}\left(T_{ijk}^{n+1}-T_{ij-1k}^{n+1}\right)\right)$$
$$+\frac{1}{h_z^2}\left(\kappa_{ijk+\frac{1}{2}}\left(T_{ijk+1}^{n+1}-T_{ijk}^{n+1}\right)-\kappa_{ijk-\frac{1}{2}}\left(T_{ijk}^{n+1}-T_{ijk-1}^{n+1}\right)\right), \quad (3.8)$$

where the superscript refers to the time ($n+1$ being advanced by Δt from n), leading to the following equation for temperature at the new step,

$$a_{ijk}^{C}T_{ijk}^{n+1}+a_{ijk}^{E}T_{ijk}^{n+1}+a_{ijk}^{W}T_{ijk}^{n+1}+a_{ijk}^{N}T_{ijk}^{n+1}+a_{ijk}^{S}T_{ijk}^{n+1}+a_{ijk}^{U}T_{ijk}^{n+1}+a_{ijk}^{D}T_{ijk}^{n+1}=a^{T}T_{ijk}^{n}, \quad (3.9)$$

where the a coefficients are the same as in (3.7) except that a_{ijk}^{C} now includes an extra term involving Δt. This is a coupled set of linear equations that must be solved simultaneously using, for example, one of the methods in Chapter 6.

Alternatively, the time derivative may be treated explicitly, leading to the form below in which there is no coupling between grid points at time $(n+1)$, so that the new temperature at each grid point can be calculated very quickly and simply:

$$T_{ijk}^{n+1}=T_{ijk}^{n}+\Delta t\left(\frac{1}{h_x^2}\left(\kappa_{i+\frac{1}{2}jk}\left(T_{i+1jk}^{n}-T_{ijk}^{n}\right)-\kappa_{i-\frac{1}{2}jk}\left(T_{ijk}^{n}-T_{i-1jk}^{n}\right)\right)\right.$$
$$+\frac{1}{h_y^2}\left(\kappa_{ij+\frac{1}{2}k}\left(T_{ij+1k}^{n}-T_{ijk}^{n}\right)-\kappa_{ij-\frac{1}{2}k}\left(T_{ijk}^{n}-T_{ij-1k}^{n}\right)\right)$$
$$\left.+\frac{1}{h_z^2}\left(\kappa_{ijk+\frac{1}{2}}\left(T_{ijk+1}^{n}-T_{ijk}^{n}\right)-\kappa_{ijk-\frac{1}{2}}\left(T_{ijk}^{n}-T_{ijk-1}^{n}\right)\right)\right). \quad (3.10)$$

3.4.2 Advection

Consider the following Eulerian advection–diffusion equation:

$$\frac{\partial T}{\partial t}=-\mathbf{v}\cdot\nabla T+\nabla\cdot(\kappa\nabla T). \quad (3.11)$$

For incompressible flow, $\mathbf{v}\cdot\nabla T=\nabla\cdot(\mathbf{v}T)$ so Eq. (3.11) can be rewritten as:

$$\frac{\partial T}{\partial t}=\nabla\cdot(-\mathbf{v}T+\kappa\nabla T). \quad (3.12)$$

Integrating over a control volume and using the divergence theorem leads to:

$$\int_{CV}\frac{\partial T}{\partial t}dV=\int_{surface}(-\mathbf{v}T+\kappa\nabla T)\cdot\mathbf{n}dS, \quad (3.13)$$

which can be discretised as:

$$\frac{\partial T_i}{\partial t} = \frac{1}{\Delta V_i} \sum_{side=1}^{n_i} L_{side} \left(-\mathbf{v}_{side} T_{side} + \kappa_{side} (\nabla T \cdot \mathbf{n})_{side}\right), \tag{3.14}$$

where $\mathbf{v}_{side}$ is the velocity perpendicular to the side, and T_{side} is the interpolated temperature at the side.

The choice of interpolation method to obtain T_{side} is critical to both the accuracy and stability of the scheme, and hence is the topic of many papers in the numerical literature. The choice of interpolant is influenced by whether advection and diffusion are treated together as implied by the above equation, or whether diffusion and advection are split into separate steps (operator splitting). If they are treated together, then an important quantity is the local Péclet number, i.e. ratio of advection to diffusion ($Pe = vh/\kappa$, where h is the grid spacing). For low Péclet numbers (at which diffusion is important), linear interpolation gives accurate and stable results, but for higher Pe this leads to an unstable scheme in which oscillations or 'wiggles' grow exponentially with time. The more advanced advection schemes are developed to treat pure advection (infinite Pe), so the remainder of the discussion here will focus on this case.

The simplest scheme to give stability at all Péclet numbers is the *upwind* (or *donor cell*) method, in which T_{side} is taken to be the temperature of the node from which material is flowing, i.e.

$$T_{side} = \begin{cases} T_{LEFT}, & \mathbf{v}_{side} > 0, \\ T_{RIGHT}, & \mathbf{v}_{side} < 0, \end{cases} \tag{3.15}$$

where T_{LEFT} is the temperature at the node to the left (i.e. negative coordinate direction) of the side, and T_{RIGHT} is the temperature at the node to the right of the side. This scheme is stable for any Pe (subject to the usual Courant–Friedrichs–Lewy condition if explicit time integration is used) and gives results with no artificial overshoots. It is, however, extremely diffusive, with an initially sharp gradient or pulse becoming rapidly smeared out as it is advected. For real applications, more sophisticated methods are used to reduce numerical diffusion. Owing to the large community that uses finite volume advection schemes, there are numerous improved advection schemes; an overview of the main ones is given here.

In the Multidimensional Positive Definite Advection Transport Algorithm (MPDATA) (Smolarkiewicz, 1984) an expression for the numerical diffusion inherent in the donor cell scheme is derived, and written in the form of artificial 'diffusive velocities'. The numerical diffusion is then partially reversed by using the negative of these 'diffusive velocities', which are termed 'anti-diffusive velocities', in an additional upstream advection step. This corrective approach can be repeated iteratively to gain more accurate results, although significant numerical diffusion is still present even after several iterations.

Another approach is to use quadratic interpolation to obtain T_{side}, which is the basis of the so-called QUICK (Quadratic Upstream Interpolation of Convective Kinetics) scheme (Leonard, 1979). The three points required for quadratic interpolation are the two points straddling the side, plus a third one in the upstream direction. Unfortunately the QUICK scheme can give overshoots and wiggles, and is unstable under some conditions. It is typically used to solve the combined advection–diffusion problem, with the diffusion giving some stability at low *Pe*. Modified versions have been proposed to remedy such problems; see Versteeg and Malalasekera (2007) for more discussion.

The best class of finite volume schemes to use at present is probably Total Variation Diminishing (TVD) schemes, which developed from Flux Corrected Transport (FCT) schemes; these are further discussed in Section 7.9, but a brief conceptual overview is given here. TVD schemes can be regarded as a generalisation of upwind schemes in which the interpolation of T_{side} is such that the total variation of the field (defined in Section 7.9) does not increase (suppressing the formation of wiggles), and the order of interpolation for T_{side} adapts according to the local smoothness of the field: from high order for smooth fields, to upwind donor cell for very rough fields (for which the gradient changes rapidly). The smoothness of the field is measured by the ratio of the upwind-side gradient to the downwind-side gradient (ξ_i in Eq. (7.54)). A *limiter function*, which uses this as the argument, then defines the amount of high-order correction that is applied to the donor cell interpolation ($\gamma(\xi)$ in Eqs. (7.52)–(7.54)). The requirements that the total variation of the field does not increase, and that the scheme must be at least second order, constrain the possible forms of this limiter function; however, several choices are still available. The *superbee* limiter (Equation (7.54)) (Roe, 1985) is the most aggressive choice, giving the maximum correction possible without increasing the total variation, while the Min-Mod limiter (Roe, 1985) is the least aggressive, giving the minimum correction necessary to give second-order accuracy. Intermediate limiters have been proposed by several authors including Sweby (1984) and van Leer (1974).

The choice of limiter must depend on the physical problem and should be chosen on the basis of testing. The superbee limiter is excellent for advecting discontinuities, but is not ideal for smoothly varying fields because it progressively sharpens gradients, eventually turning smooth variations into discontinuities. On the other hand, the Min-Mod limiter retains some numerical diffusion, so sharp gradients are progressively smeared out. The appropriate balance between numerical diffusion and artificial sharpening must be sought.

These TVD schemes have limitations, in particular degenerating to first-order at local minima and maxima (Osher and Chakravarthy, 1984). ENO (essentially non-oscillatory) and WENO (weighted essentially non-oscillatory) schemes have been introduced to overcome this limitation and provide a better scheme that can handle both sharp interfaces and smooth gradients (Shu, 1997). The reader is referred to the relevant literature for further details.

3.5 Treatment of momentum–continuity equations

3.5.1 Discretisation

The simplest version of the momentum and continuity equations, i.e. using the Boussinesq approximation and the infinite Prandtl number approximation, is chosen here to illustrate the basics of FV discretisation:

$$\nabla \cdot \sigma - \nabla P = -RaT\mathbf{e}, \quad \nabla \cdot \mathbf{v} = 0, \tag{3.16}$$

where σ is the deviatoric stress tensor, P is pressure, Ra is the Rayleigh number, T is temperature, $\mathbf{v}$ is velocity and $\mathbf{e} = (0, 0, 1)$. When defining control volumes and nodes for this system, it is advantageous and common practice to stagger the locations of the nodes on which the different variables are defined, which is shown in Fig. 3.3 for a two-dimensional grid and Fig. 3.4 for a three-dimensional grid. This staggered grid avoids the possibility of checkerboard oscillations in the pressure field that can occur with a collocated grid (velocity and pressure defined at the same location), and also means that all first derivatives are calculated using adjacent points, which gives greater accuracy than using a collocated arrangement. Control volumes are different for each equation (three momentum components and pressure). For the momentum equations the control volume is centred on the velocity component being calculated, whereas for the continuity equation the control volume is centred on the pressure point, as illustrated in Fig. 3.5.

In practice, stress is eliminated from Eq. (3.16) and the equations are expressed in terms of the three velocity components and pressure. The discretisation of the stress components in Cartesian geometry using finite differences is given below, and assumes that the grid

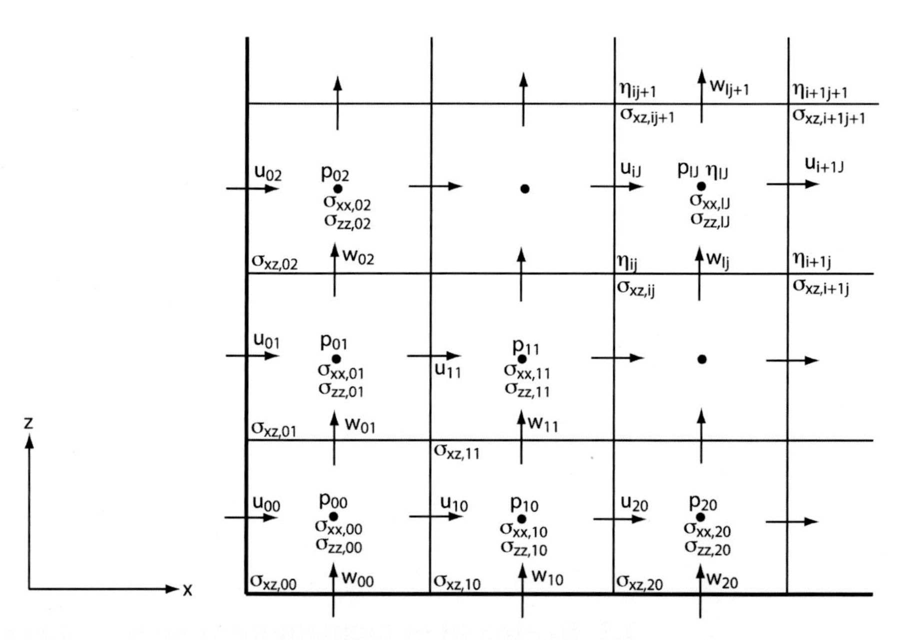

Fig. 3.3. The locations of pressure p, velocity components (u, w), stresses σ_{xx}, σ_{zz} and σ_{xz}, and viscosity components η_{ij} and η_{IJ} on a two-dimensional staggered grid. Other scalar variables such as temperature and composition are normally defined at the same point as the pressure. The boundary runs along the bottom and the left side, with the points numbered accordingly, while the general indexing scheme is illustrated by the points in the top right. The lines show the edges of 'cells', which correspond to control volumes for the continuity and energy equations, but not for the momentum equations, as shown in Fig. 3.5.

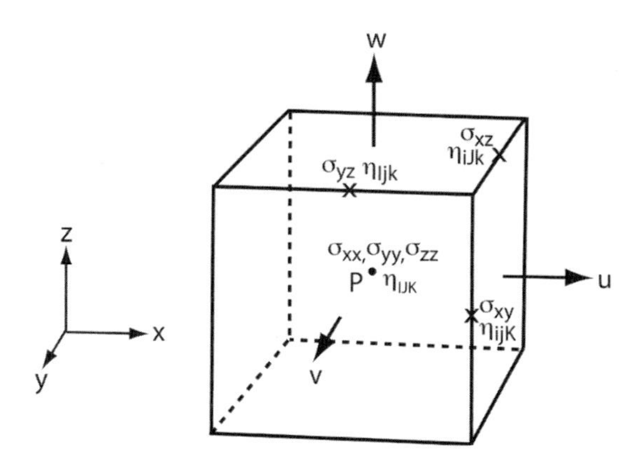

Fig. 3.4. The locations of pressure p, the three velocity components (u, v, w), the six stress components (σ_{xx}, σ_{yy}, σ_{zz}, σ_{xy}, σ_{yz} and σ_{xz}), and the four viscosity points ($\eta_{ijK}\eta_{iJk}\eta_{Ijk}\eta_{IJK}$) in one cell of a three-dimensional staggered grid. Other scalar variables such as temperature and composition are normally defined at the same point as the pressure.

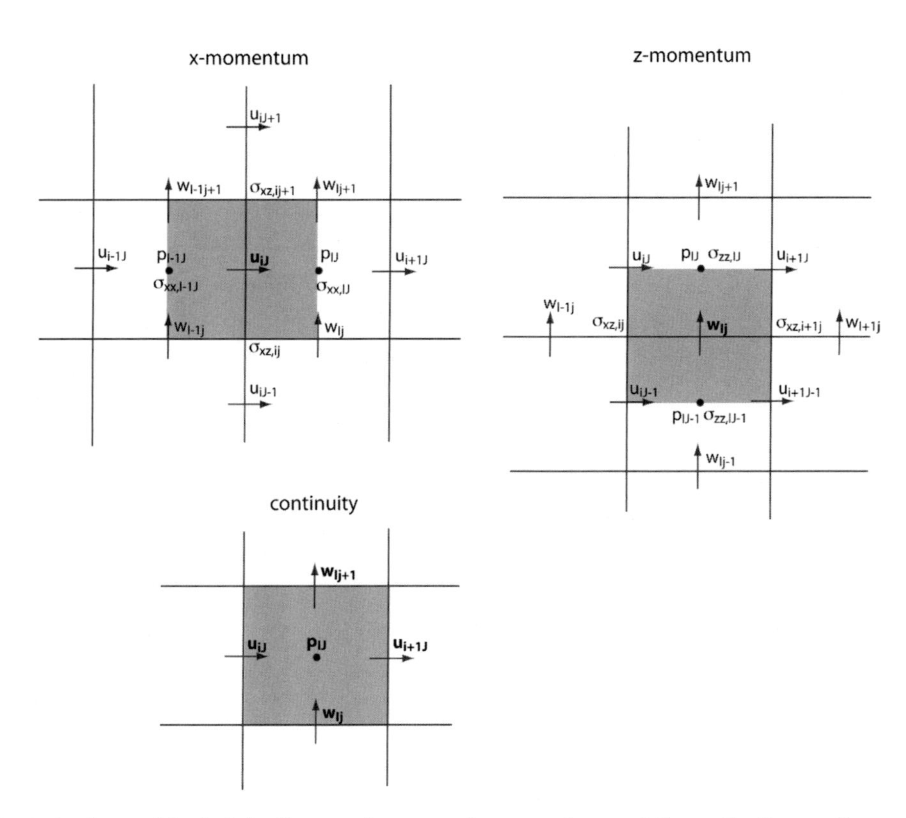

Fig. 3.5. Control volumes (shaded) for the x- and z-momentum equations and the continuity equation on the two-dimensional staggered grid shown in Fig. 3.3. The variable in the centre of the control volume (bold) is calculated (or adjusted, if an iterative scheme) using the variables shown. The definitions of variables are as in Fig. 3.3 and the text. For the continuity equation, the velocities surrounding each point are also adjusted based on the pressure correction (see text). The energy equation uses the same control volume as the continuity equation.

spacings are constant in each direction. The grid indexing scheme is the same as that in (Versteeg and Malalasekera, 2007), with upper case indices referring to coordinates that intersect the scalar (pressure) point, and lower case indices referring to coordinates that are half a grid spacing lower than that. This is a backward-staggered grid, in which the velocity components at a lower coordinate (by half a grid spacing) share the same index as the pressure point. Using u, v and w to denote the x, y and z components of velocity, the variables in one unit cell are thus $p_{IJK}, u_{iJK}, v_{IjK}, w_{IJk}$ the normal stress components are $\sigma_{xx,IJK}, \sigma_{yy,IJK}, \sigma_{zz,IJK}$ and the shear stress components are $\sigma_{xy,ijK}, \sigma_{xz,iJk}, \sigma_{yz,Ijk}$. So:

$$\begin{aligned}
\sigma_{xx,IJK} &= 2\eta_{IJK}\frac{u_{i+1JK} - u_{iJK}}{\Delta x}, \\
\sigma_{yy,IJK} &= 2\eta_{IJK}\frac{v_{Ij+1K} - v_{IjK}}{\Delta y}, \\
\sigma_{zz,IJK} &= 2\eta_{IJK}\frac{w_{IJk+1} - w_{IJk}}{\Delta z}, \\
\sigma_{xy,ijK} = \sigma_{yx,ijK} &= \eta_{ijK}\left(\frac{v_{IjK} - v_{I-1jK}}{\Delta x} + \frac{u_{iJK} - u_{iJ-1K}}{\Delta y}\right), \\
\sigma_{xz,iJk} = \sigma_{zx,iJk} &= \eta_{iJk}\left(\frac{w_{IJk} - w_{I-1Jk}}{\Delta x} + \frac{u_{iJK} - u_{iJK-1}}{\Delta z}\right), \\
\sigma_{yz,Ijk} = \sigma_{zy,Ijk} &= \eta_{Ijk}\left(\frac{v_{IjK} - v_{IjK-1}}{\Delta z} + \frac{w_{IJk} - w_{IJ-1k}}{\Delta y}\right).
\end{aligned} \tag{3.17}$$

The discretised x-momentum equation can be constructed by balancing forces on a control volume centered at the x-velocity points:

$$\begin{aligned}
&\left((\sigma_{xx,IJK} - p_{IJK}) - (\sigma_{xx,I-1JK} - p_{I-1JK})\right)\Delta y\Delta z \\
&+ \left(\sigma_{yx,ij+1K} - \sigma_{yx,ijK}\right)\Delta x\Delta z + \left(\sigma_{zx,iJk+1} - \sigma_{zx,iJk}\right)\Delta x\Delta y = 0.
\end{aligned} \tag{3.18}$$

Dividing by $\Delta x\Delta y\Delta z$ and rearranging leads to a finite difference formula:

$$\frac{\sigma_{xx,IJK} - \sigma_{xx,I-1JK}}{\Delta x} + \frac{\sigma_{yx,ij+1K} - \sigma_{yx,ijK}}{\Delta y} + \frac{\sigma_{zx,iJk+1} - \sigma_{zx,iJk}}{\Delta z} - \frac{p_{IJK} - p_{I-1JK}}{\Delta x} = 0. \tag{3.19}$$

Substituting the expressions for stresses leads to the following form

$$\begin{aligned}
&2\frac{\eta_{I-1JK}u_{i-1JK} + \eta_{IJK}u_{i+1JK}}{\Delta x^2} + \frac{\eta_{ij+1K}u_{iJ+1K} + \eta_{ijK}u_{iJ-1K}}{\Delta y^2} + \frac{\eta_{iJk+1}u_{iJK+1} + \eta_{iJk}u_{iJK-1}}{\Delta z^2} \\
&- \left(2\frac{\eta_{I-1JK} + \eta_{IJK}}{\Delta x^2} + \frac{\eta_{ij+1K} + \eta_{ijK}}{\Delta y^2} + \frac{\eta_{iJk+1} + \eta_{iJk}}{\Delta z^2}\right)u_{iJK} \\
&+ \frac{\eta_{ij+1K}(v_{Ij+1K} - v_{I-1j+1K}) - \eta_{ijK}(v_{IjK} - v_{I-1jK})}{\Delta x\Delta y} \\
&+ \frac{\eta_{iJk+1}(w_{IJk+1} - w_{I-1Jk+1}) - \eta_{iJk}(w_{IJk} - w_{I-1Jk})}{\Delta x\Delta z} - \frac{p_{IJK} - p_{I-1JK}}{\Delta x} = 0,
\end{aligned} \tag{3.20}$$

which for the case of constant viscosity, simplifies to:

$$\frac{u_{i-1JK}+u_{i+1JK}}{\Delta x^2}+\frac{u_{iJ+1K}+u_{iJ-1K}}{\Delta y^2}+\frac{u_{iJK+1}+u_{iJK-1}}{\Delta z^2}$$
$$-\left(\frac{2}{\Delta x^2}+\frac{2}{\Delta y^2}+\frac{2}{\Delta z^2}\right)u_{iJK}-\frac{p_{IJK}-p_{I-1JK}}{\Delta x}=0. \tag{3.21}$$

Generalising this into a 17-point stencil yields

$$a^{C}_{iJK}u_{iJK}+a^{uxm}_{iJK}u_{i-1JK}+a^{uxp}_{iJK}u_{i+1JK}+a^{uyp}_{iJK}u_{iJ+1K}+a^{uym}_{iJK}u_{iJ-1K}+a^{uzp}_{iJK}u_{iJK+1}$$
$$+a^{uzm}_{iJK}u_{iJK-1}+a^{vxpyp}_{iJK}v_{Ij+1K}+a^{vxmyp}_{iJK}v_{I-1j+1K}+a^{vxpym}_{iJK}v_{IjK}+a^{vxmym}_{iJK}v_{I-1jK}$$
$$+a^{wxpzp}_{iJK}w_{IJk+1}+a^{wxmzp}_{iJK}w_{I-1Jk+1}+a^{wxpzm}_{iJK}w_{IJk}+a^{wxmzm}_{iJK}w_{I-1Jk}$$
$$+a^{pp}_{iJK}p_{IJK}+a^{pm}_{iJK}p_{I-1JK}=0. \tag{3.22}$$

The y- and z-momentum equations are discretised similarly, with the z-momentum equation containing an additional buoyancy term.

Calculation of the different stress components requires knowledge of the viscosity at several different locations in the unit cell. Normal stresses require the viscosity at the pressure point, which is typically where the viscosity is calculated because it is where temperature is defined, while shear stresses require values of the viscosity at the three centres of the cell edges (or, in 2-D, the corner of the cell). The question of how viscosity should be calculated or interpolated from the temperature/pressure points is important. Several lines of reasoning have been applied to this. Arithmetic averaging is the most straightforward. However, in rocks, the viscosity is typically exponentially dependent on temperature, so if temperature is assumed to vary linearly from node to node, the physically appropriate averaging of viscosity would be geometric. Another argument, based on continuity of stresses, is that the appropriate average is a harmonic one. This was recently tested by Deubelbeiss and Kaus (2008) for a simple test case in which an analytic solution exists. They found that harmonic and geometric interpolation give much more accurate results than arithmetic (linear) interpolation, with harmonic averaging being slightly superior to geometric.

The discretised *continuity equation* is given by

$$\frac{u_{i+1JK}-u_{iJK}}{\Delta x}+\frac{v_{Ij+1K}-v_{IjK}}{\Delta y}+\frac{w_{IJk+1}-w_{IJk}}{\Delta z}=0, \tag{3.23}$$

or, in stencil form,

$$a^{xp}_{IJK}u_{i+1JK}+a^{xm}_{IJK}u_{iJK}+a^{yp}_{IJK}v_{Ij+1K}+a^{ym}_{IJK}v_{IjK}+a^{zp}_{IJK}w_{IJk+1}+a^{zm}_{IJK}w_{IJk}=0. \tag{3.24}$$

Typical *boundary conditions* for geodynamic problems are rigid, free slip, periodic, free surface or permeable. On the staggered grid, the external boundary is normally taken to lie along a plane of velocity points perpendicular to that boundary, for example, the top and bottom boundaries pass through a plane of w points. This makes it straightforward to enforce boundary conditions where there is no flow through the boundary, as in the case of rigid or free-slip conditions.

Two methods exist for the numerical implementation of boundary conditions. One is to modify the discretised equations at points near the boundary to take into account the appropriate form of stress or velocity at the boundary. This method is most appropriate when the equations are being solved using a direct matrix solver. The second method is to include virtual 'ghost' points outside the domain, and to set the velocities at these points such as to satisfy the appropriate condition at the boundary. This ghost point method is most appropriate when an iterative method is being used to solve the equations, because then the near-boundary equations are no different from the interior equations and the ghost points can be simply set after each iteration. On a parallel computer, the boundaries of internal sub-domains can be treated in the same way, i.e. the ghost points contain copies of points in adjacent sub-domains, which must be updated after each iteration.

Consider the case of a rigid (no-slip) top boundary:

$$u_{top} = v_{top} = w_{top} = 0, \tag{3.25}$$

where w nodes occur at the boundary so it is straightforward to set $w=0$. There are no u or v nodes at the top boundary, but this constraint can be used to define the shear stress components at the boundary:

$$\sigma_{xz,top} = \eta_{iJk}\frac{-u_{top-\Delta z/2}}{(\Delta z/2)}, \quad \sigma_{yz,top} = \eta_{Ijk}\frac{-v_{top-\Delta z/2}}{(\Delta z/2)}, \tag{3.26}$$

which can then be used to derive modified weights in the discretised x- and y-momentum equations. Alternatively, the standard discretised equations can be used and after each iterative update of the velocities, ghost point values are set as follows:

$$u_{top+\Delta z/2} = -u_{top-\Delta z/2}, \quad v_{top+\Delta z/2} = -v_{top-\Delta z/2}. \tag{3.27}$$

For free-slip conditions,

$$\sigma_{xz} = \sigma_{yz} = 0, \tag{3.28}$$

which again can be used to derive modified stencil weights, or the ghost points set as:

$$u_{top+\Delta z/2} = u_{top-\Delta z/2}, \quad v_{top+\Delta z/2} = v_{top-\Delta z/2}. \tag{3.29}$$

Similar methods can be used to set the boundary velocity or stress to a fixed or spatially varying value other than zero. Periodic boundaries, as might be used on the sides, are straightforward to implement by using either the equations, or the ghost points.

Permeable boundaries are best implemented by specifying the velocity field along the boundary. Conditions such as zero velocity gradient tend to lead to problems with the solution. Specifying that the tangential velocity is zero and that the normal stress is proportional to velocity (accounting for the 'resistance' of material outside the domain) can be successful. A free surface, which deflects vertically, cannot be implemented in a straightforward manner like the other boundary conditions can. This issue is discussed in Chapter 10.

3.5.2 Solution methods

The coupled set of linear equations arising from the above discretised equation can be inserted into a matrix and solved using a direct solver (Section 6.2), as is done for example by Gerya and Yuen (2003). In this case, the equations for points near the boundaries need to be appropriately modified to account for the boundary conditions. Another complexity is that the absolute pressure is undetermined to a constant because only derivatives of pressure appear in the governing flow equations, and therefore the matrix will be singular. To avoid this, the continuity equation for one cell can be replaced by a constraint on the absolute pressure, for example, setting P to a particular value. Periodic side boundaries combined with free-slip top and bottom boundaries can also lead to a singular matrix, because in that case the horizontal velocity components are undetermined to a constant value. Again, this can be remedied by replacing one of the x-velocity equations and one of the y-velocity equations with a constraint on the velocity.

In practice, for large systems with millions of unknowns, direct solution methods are too slow and have excessive memory requirements, particularly in 3-D, so an iterative solution is preferable (Section 6.3), with a multigrid solver (Section 6.4) being optimal. However, iterative solution of these equations immediately encounters the problem that while it is straightforward to calculate corrections to velocity based on the momentum equations, there is no equation for correcting the pressure. Several schemes exist for calculating pressure corrections, and have been implemented successfully.

One iterative approach is to update each pressure simultaneously with the six surrounding velocity components, by solving a 7×7 matrix equation containing the continuity equation and the momentum equations at the six surrounding velocity points. This scheme was proposed by Vanka (1986) and implemented by Auth and Harder (1999) in a 2-D code. They used a version of the matrix from which some coefficients were dropped to make part of it diagonal. Tackley (2000a) implemented this in 3-D using the full matrix. Using this scheme, the solution converges in fewer iterations than with the point-wise schemes discussed below, but it takes several times as much CPU time per iteration, so the overall solution time is longer.

The most widely used method for iterating on velocity and pressure is SIMPLE (Semi-Implicit Method for Pressure-Linked Equation) (Patankar, 1980), the mathematical details of which are given in Section 6.5.2. Here, a physical interpretation is given. The 'purpose' of the pressure is to enforce incompressibility: if there is convergence in a control volume then the pressure at the node needs to be increased, whereas if there is divergence then the pressure needs to be decreased. From the discretised equations it is straightforward to derive a suitable pressure correction to accomplish this. Changing the pressure at one node affects velocities everywhere in the grid. In the SIMPLE method, however, it is assumed that changing the pressure at one node affects only the velocity components immediately surrounding it, and the influence of the velocity corrections on each other is neglected. Therefore, the calculated pressure correction is only an approximation. In a revised version of the SIMPLE algorithm, named SIMPLER (Patankar, 1980), an exact discretised Poisson-like equation for pressure was derived, which can lead to faster convergence. Here, however, we focus on the basic SIMPLE algorithm.

The SIMPLE algorithm also includes the possibility of iterating on the advection–diffusion equation at the same time as the momentum and continuity equations, which allow diffusion and advection to be treated implicitly. In most geodynamic codes, however, with some exceptions (Albers, 2000; Trompert and Hansen, 1996), advection and diffusion are treated as a separate step. This approach allows modern advection methods to be used (as implicit advection is very diffusive), with iterations being performed only on the momentum–continuity equations.

One iteration can be summarised as follows.

(1) Improve velocity field according to the momentum equations
  (a) x-velocity field according to the x-momentum equations
  (b) y-velocity field according to the y-momentum equations
  (c) z-velocity field according to the z-momentum equations
(2) Update pressure field to reduce flow divergence
  (a) Calculate pressure correction
  (b) Calculate velocity correction caused by pressure correction

The correction to each velocity component (step 1) depends on the stencil weight and a relaxation parameter α_m:

$$\delta u_{iJK} = -\alpha_m R_{iJK} / a^C_{iJK}, \delta v_{IjK} = -\alpha_m R_{IjK} / a^C_{IjK}, \delta w_{IJk} = -\alpha_m R_{IJk} / a^C_{IJk}, \tag{3.30}$$

where R_{iJK}, R_{IjK} and R_{IJk} are the residue (error) of the x-, y- and z-momentum equations in unit cell (i, j, k), and the velocities and a coefficients are as defined in Eqs. (3.17)–(3.24). For multigrid purposes, under-relaxation should be used (Brandt, 1982; Wesseling, 1992), i.e. $\alpha_m < 1$; a typical value is 0.7. In the absence of multigrid, over-relaxation is optimal, i.e. $\alpha_m > 1$.

When updating the velocity fields, either each component is updated over the entire grid in turn (i.e. step 1a over the entire grid, then step 1b, then step 1c) as is done in Stag3D/StagYY (Tackley 1993, 2008), or the three velocity components can be updated in the same sweep. The first approach has the advantage that the coupling between the velocity components is immediately taken into account, but in practice this seems to make little difference to the convergence rate of the iterative method. As usual, Jacobi or Gauss–Seidel iterations may be used. StagYY uses 'red-black' iterations, which give the fastest convergence rate (Press *et al.*, 2007) and have the advantage that the 'red' and 'black' points are independent, so the results are identical when the domain is decomposed onto different nodes of a parallel computer, than on a single CPU.

The pressure correction (step 2a) is calculated using a coefficient that describes how much changing the pressure at a point changes the residue of the continuity equation (i.e. divergence of velocity) at that point:

$$\delta P_{IJK} = -\alpha_{cont} \frac{R_{IJK}}{\partial R_{IJK} / \partial P_{IJK}}, \tag{3.31}$$

where the symbols have similar meanings to those in Eq. (3.24), and α_{cont} is a relaxation parameter that can be taken to be 1.0. Note that $(\partial R / \partial P)$ is not a stencil weight because the continuity equation does not include pressure, but it is related to the stencil weights of the momentum and continuity equations. In principle, changing the pressure at one point affects velocities and pressures in the entire domain, requiring a global solution. It has been found, however, that the lowest-order approximation is sufficient, and this is what is used in SIMPLE. This means that the effect of the pressure on the six neighbouring velocity points is taken into account, but its effect on more distant velocity points is not considered, neither is the effect of a change in the velocity at one point on the velocities at other points. Specifically, stencil weights of the momentum equations at the six surrounding velocity points give the amount by which velocities at those points are change when P_{IJK} is changed. Combining these with the stencil weights for the continuity equation leads to the desired approximation to the derivative:

$$\frac{\partial R_{IJK}}{\partial P_{IJK}} \approx a^{xp}_{IJK}\frac{a^{pm}_{i+1JK}}{a^{C}_{i+1JK}} + a^{xm}_{IJK}\frac{a^{pp}_{iJK}}{a^{C}_{iJK}} + a^{yp}_{IJK}\frac{a^{pm}_{Ij+1K}}{a^{C}_{Ij+1K}} + a^{ym}_{IJK}\frac{a^{pp}_{IjK}}{a^{C}_{IjK}} + a^{zp}_{IJK}\frac{a^{pm}_{IJk+1}}{a^{C}_{IJk+1}} + a^{zm}_{IJK}\frac{a^{pp}_{IJk}}{a^{C}_{IJk}}, \tag{3.32}$$

where the various stencil weights are defined in Eqs. (3.22) and (3.24).

A quick examination of $(\partial R/\partial P)$ reveals that it scales as the inverse of viscosity. If h represents the grid spacing, then $a_{continuity} \approx 1/h$, $a^{P}_{momentum} \approx 1/h$, and $a^{C}_{momentum} \approx \eta/h^2$. Thus, the pressure correction in a cell can be approximated as $-\eta\nabla\cdot\mathbf{v}$, which was used in the original Cartesian version of Stag3D (Tackley, 1993), although the latest version calculates the coefficient according to the above equation.

Another conceptual view of the iteration process is that of pseudo-compressibility: the iterations are taking pseudo-timesteps towards an eventual incompressible state. Kameyama *et al.* (2005) derived a pressure correction using this logic and also found that it is proportional to the viscosity and velocity divergence.

In step 2b, each velocity component is adjusted based on the pressure correction at the two adjacent pressure points multiplied by the appropriate stencil weight, for example for the x-velocities:

$$\delta u_{iJK} = \frac{1}{a^{C}_{iJK}}\left(\delta P_{iJK}a^{pp}_{iJK} + \delta P_{i-1JK}a^{pm}_{iJK}\right). \tag{3.33}$$

With this approach, an explicit equation for the pressure is not necessary (only pressure correction), and it is not necessary to apply explicitly any boundary conditions to pressure P. Additionally, it is never necessary to form a matrix, as all the iterations are done on a point-by-point basis.

3.5.3 Multigrid

The multigrid method, in which the residue is relaxed on a hierarchy of nested grids with different grid spacing, dramatically accelerates the convergence rate of iterative solvers because in principle it relaxes all wavelengths of the residue simultaneously, resulting in a solution time that scales in proportion to the number of unknowns (Brandt, 1982; Wesseling,

1992). This is discussed in detail in Section 6.4, with the particular example of the Poisson equation illustrated. For the Stokes equation on a staggered velocity–pressure grid, some special complexities and problems arise, which will be discussed here. Typical implementations have used the SIMPLE or SIMPLER method as a smoother inside multigrid cycles. The first issue is that the spatial relationship between coarse-grid variables and fine-grid variables is not straightforward as illustrated in Fig. 3.6. Related to this, the boundaries of the fine-grid control volumes do not coincide with the boundaries of the coarse-grid control volumes, except for the continuity equation. Nevertheless, linear interpolation between coarse and fine grids usually suffices.

The main problem with applying the multigrid method to mantle convection simulations is a lack of robustness with respect to large viscosity variations, i.e. the iterative methods converge very slowly or diverge. Broadly speaking this is because the coarse grids do not correctly 'see' the fine-grid problem, so corrections calculated at coarse levels may actually degrade the solution at finer levels rather than improving it. For mantle convection, the first application of a multigrid solver on a staggered velocity–pressure grid was made by Tackley (1993), and that study reached viscosity contrasts of only 10^3. Over the years, several researchers have proposed improvements to the multigrid algorithm to address this problem.

In the general multigrid literature, the accepted approach to deriving coarse-grid operators, particularly in the case of strongly varying coefficients, is to use matrix-dependent prolongation and restriction operators combined with the Galerkin coarse-grid approximation, as discussed in Section 6.4.2. Matrix-dependent operators and the Galerkin coarse-grid were implemented in a 2-D finite-element mantle convection code by Yang and Baumgardner (2000) with apparently astonishing results, easily handling viscosity contrasts of 10^{10} between adjacent points. Unfortunately, similar robustness was not obtained when the method was implemented in the related 3-D spherical-shell finite-element code TERRA. The method has not yet been successfully applied to a staggered-grid mantle convection code because of the complexity.

So far, mantle convection implementations of staggered-grid multigrid have instead rediscretised the equations on the coarse grid using a viscosity field that is averaged from the fine grid. Several authors have proposed improvements to the arithmetic averaging

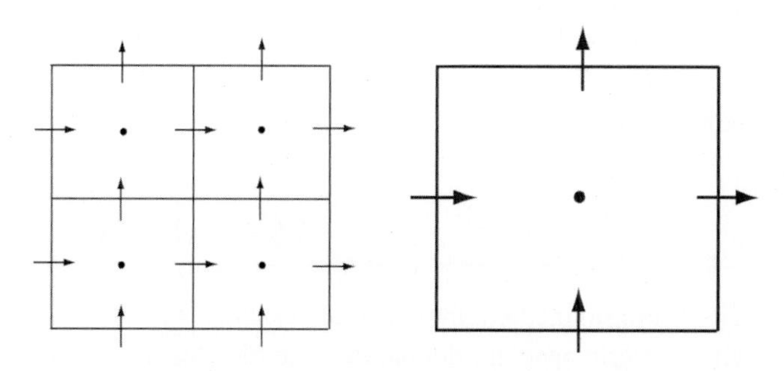

Fig. 3.6. **Fine-grid and coarse-grid cells for a multigrid scheme on a two-dimensional staggered grid. None of the coarse-grid variables is in the same location as a fine-grid variable.**

used in the scheme of Tackley (1993). Firstly, Trompert and Hansen (1996) introduced a new averaging scheme for the coarse-grid viscosities, in which anisotropic viscosities are used to calculate the coarse grid shear stresses. They also found that taking additional iterations on the pressure term helped overall convergence. Auth and Harder (1999) introduced pressure-coupled relaxations, found that convergence can be greatly improved by using F-cycles instead of V-cycles, and also that arithmetic averaging of viscosities to the coarse grid gives greater robustness than harmonic averaging and a similar performance to the scheme by Trompert and Hansen (1996). Albers (2000) introduced mesh refinement, and also found that robustness is greatly improved by using multigrid cycles that conduct more iterations on the coarse grids such as F-cycles, W-cycles and V-cycles with more coarse iterations. Kameyama *et al.* (2005) introduced a new way of conceptualising the iteration process, namely 'pseudo-compressibility', and again found that taking additional coarse-grid iterations greatly improves robustness to large viscosity variations. With these improvements, viscosity contrasts in the range 10^5–10^6 can be routinely handled, but this is still much lower than Earth-like. A recently implemented scheme greatly improves the allowable viscosity contrast, and is described below.

Tackley (2008) introduced a new pressure interpolation scheme that greatly improves multigrid convergence in the presence of large viscosity contrasts, compared to linear interpolation. This uses the philosophy behind matrix-dependent operators (Section 6.4.2) but without implementing the full matrix-dependent transfers and the Galerkin coarse-grid approximation.

Experiments indicate that the main cause of non-convergence with large viscosity variations is due to pressure corrections prolongated from the coarse to fine grids. The pressure correction is approximately proportional to the local viscosity, as discussed earlier. If a fine-grid cell has a much lower viscosity than the coarse-grid cell that contains it, then the prolongated pressure correction can be much too large, making the fine-grid solution worse. This is probably why Trompert and Hansen (1996) found that taking additional pressure iterations is helpful: the additional iterations are needed to repair the damage done by the 'correction' from the coarse grid. Adjusting fine-grid pressure corrections by the ratio of fine grid to coarse grid viscosity was tried, but did not give a robust improvement. It is much more effective to adjust the prolongated pressure according to the term $(\partial R_{continuity}/\partial P)$ introduced earlier, which contains a type of weighted average of the local viscosity values rather than the viscosity at an individual point. Specifically:

$$\delta P_{fine} = C\delta P_{coarse} \Big/ \left(\frac{\partial R_{continuity}}{\partial P} \right)_{fine}, \tag{3.34}$$

where C is a constant. Noting that in 3-D one coarse-grid control volume maps to eight fine-grid control volume, C is computed using the criterion that the average pressure must be conserved, i.e.

$$\frac{1}{8} \sum \delta P_{fine} = \delta P_{coarse} \tag{3.35}$$

leading to

$$C = \frac{8}{\left(\sum 1 \Big/ \left(\frac{\partial R_{continuity}}{\partial P}\right)_{fine}\right)}. \tag{3.36}$$

This reduces to a simple injection in the case of constant viscosity (i.e. eight fine-grid pressures are set equal to the coarse-grid pressure). This scheme is something like a matrix-dependent prolongation operator for pressure. In matrix-dependent operator theory, the restriction operator should be the transpose of the prolongation operator. Curiously, this was not found to be helpful in this application. Similar operators have been tried for the velocity components, but again, this didn't significantly improve the convergence. Convergence tests comparing the performance of this scheme to the standard linear interpolation are given in Tackley (2008), and show the dramatic improvement in robustness facilitated by this interpolation, which approximately doubles the orders of magnitude of viscosity contrast that can be handled.

3.6 Modelling convection and model extensions

3.6.1 Overall solution strategy

In earlier sections of this chapter, the major components of a convection code were introduced: time stepping the advection–diffusion equation and solving the coupled velocity–pressure solution for a given buoyancy distribution. These can be combined in different ways to model the full thermo-mechanical problem. The main choice is whether to combine time stepping and the velocity–pressure solution, as in the original SIMPLE algorithm, or whether to treat them as separate steps, first obtaining a velocity–pressure solution, then using this to advect and diffuse the relevant fields, as is done in the code StagYY (Tackley, 2008). A major advantage of solving them together is that an implicit time-stepping method can be used, allowing time steps much larger then permitted by the Courant stability condition. A disadvantage of this approach is that the resulting advection scheme is very diffusive, and not suitable for high Rayleigh number convection. More advanced advection techniques (such as TVD) require advection to be treated as a separate step. An alternative way of taking large time steps is to use a semi-Lagrangian or characteristics-based advection method (Sections 7.7 and 7.8), or a completely Lagrangian method (Section 7.10.2) (see Gerya and Yuen, 2003), but in their basic form these are not conservative like the finite volume advection methods discussed in this chapter.

Modern modelling studies typically include complexities such as compressibility, phase transitions, compositional variations, melting and spherical or cylindrical geometry. The implementation of compressibility and curvilinear geometry in a finite volume code are discussed in the next two sections, while phase transitions, compositional variations and melting can be treated with several different methods, which are discussed in Chapter 10.

3.6.2 Extension to compressible equations

A discussion of the changes and addition terms in the equations when compressibility is included can be found in Section 10.3. These can be straightforwardly implemented by using a finite volume approach, e.g. as in Tackley (1996a,b). A couple of points pertinent to the approaches described in this chapter are discussed below.

The main influence of compressibility on the momentum equation is that the normal stresses contain an additional divergence term, e.g.

$$\sigma_{xx} = 2\eta\left(\frac{\partial u}{\partial x} - \frac{1}{3}\nabla\cdot\mathbf{v}\right). \tag{3.37}$$

When an iterative solver is used, if $\nabla\cdot\mathbf{v}$ is calculated from the current estimated velocities, then instabilities can occur in the iterative solution procedure because $\nabla\cdot\mathbf{v}$ can be incorrectly very high or low during early iterations. Stability can be obtained by recognising that the continuity equation $\nabla\cdot(\bar{\rho}\mathbf{v}) = 0$ (where $\bar{\rho}$ is a reference density) leads to:

$$\bar{\rho}\nabla\cdot\mathbf{v} = -\mathbf{v}\cdot\nabla\bar{\rho}, \tag{3.38}$$

and use $-\mathbf{v}\cdot\nabla\bar{\rho}$ in the expression instead of $\nabla\cdot\mathbf{v}$ in the momentum equations.

In the compressible energy equation, the main change is the addition of viscous dissipation and adiabatic heating/cooling. These are volume sources and not fluxes, so can be calculated for each control volume and treated explicitly. Care must be taken with the viscous dissipation, because the different strain rate and stress components must be correctly calculated at different locations in the unit cell and then combined at the temperature point.

3.6.3 Extension to spherical geometry

The main question when modelling spherical geometry is how mesh a full sphere or spherical surface while retaining approximately uniform grid spacing. Appendix B discusses and illustrates the various types of spherical grid that have been used for geodynamic modelling; here those that have been used with a finite volume discretisation are mentioned.

As the finite volume method is capable of handling unstructured grids, all of these grids could, in principle, be used in conjunction with the finite volume method. In practice, however, it is most straightforward to use grids in which the grid lines are orthogonal, such as lines of longitude and latitude in spherical polar coordinates. In this case, additional terms appear in the physical equations (Appendix B), but these are easily included. Finite volume schemes using a simple (longitude, latitude) mesh have been used (Zebib *et al.*, 1980; Iwase and Honda, 1997) but these are not optimal owing to convergence of grid lines at the poles. The 'Yin-Yang' grid (Kageyama and Sato, 2004) avoids this singularity while retaining an orthogonal grid, by meshing two (longitude, latitude) patches that are centred at the equator; three finite volume mantle convection codes using this grid have been implemented (Yoshida and Kageyama, 2004; Kamayama *et al.*, 2008; Tackley, 2008). The 'cubed sphere' grid (Ronchi *et al.*, 1996) has the problem that in the basic version the grid lines are not orthogonal, which results in many additional terms in the discretised equations when expressed in terms of the local, non-orthogonal grid coordinates. Nevertheless, finite volume schemes have been successfully implemented for constant viscosity (Hernlund and Tackley, 2003) and for variable viscosity (Choblet, 2005) convection. Stemmer *et al.* (2006) found a modified cubed sphere grid in which the grid lines are almost orthogonal such that the cross terms can be neglected. At the time of writing, only one finite volume code geodynamical has used a completely unstructured mesh: that of Huettig and Stemmer (2008a), who used a spiral to generate the grid points and Voronoi diagrams to generate finite volume cells, with variable-viscosity Stokes flow discretised using the approach of Huettig and Stemmer (2008b).

4 Finite element method

4.1 Introduction

The finite element (FE) method is a computational technique for obtaining approximate solutions to the partial differential equations that arise in scientific and engineering applications and is used widely in geodynamic modelling (see Christensen, 1984, 1992; Baumgardner, 1985; Naimark and Malevsky, 1988; King *et al.*, 1990; Naimark and Ismail-Zadeh, 1995; Moresi and Solomatov, 1995; Moresi *et al.*, 2003; Ismail-Zadeh *et al.*, 1998, 2001a,b , 2004a,b, 2006, 2007). Introduced in the middle of the twentieth century (Hrennikoff, 1941; McHenry, 1943; Courant, 1943) the FE method has emerged as one of the most powerful numerical methods so far devised. Rather than approximating the partial differential equation directly as with finite difference methods (see Chapter 2), the FE method utilises a *variational problem* that involves an integral of the differential equation over the model domain. This domain is divided into a number of sub-domains called *finite elements*, and the solution of the partial differential equation is approximated by a simple polynomial function on each element. These polynomials have to be pieced together so that the approximate solution has an appropriate degree of smoothness over the entire domain. Once this has been done, the variational integral is evaluated as a sum of contributions from each finite element. The result is a set of algebraic equations for the approximate solution having a finite size rather than the original infinite-dimensional partial differential equation. Therefore, like FD methods, the FE process discretises the partial differential equation but, unlike FD methods, the approximate solution is known throughout the domain as a piecewise polynomial function and not just at a set of points.

Among the basic advantages of the method, which have led to its widespread acceptance and use, are the ease in modelling complex irregular regions, the use of non-uniform meshes to reflect solution gradations, the treatment of boundary conditions involving fluxes, and the construction of high-order approximations. Estimates of discretisation errors may be obtained for reasonable costs. These may be used to verify the accuracy of the computation, and also to control an adaptive process whereby meshes are automatically refined and coarsened and/or the degrees of polynomial approximations are varied so as to compute solutions to desired accuracies in an optimal fashion (see Babuska and Rheinboldt, 1978; Babuska *et al.*, 1983, 1986; Bern *et al.*, 1999 for more details).

The application of the FE method for solving a geodynamical problem requires a certain number of basic ingredients that we shall discuss in this chapter.

4.2 Lagrangian versus Eulerian description of motion

An important consideration with FE methods (as well as other numerical methods) is the choice of an appropriate kinematical description of motion. The algorithms of continuum mechanics make use of two principal and distinct types of the description: the *Lagrangian* and *Eulerian* formulations.

The *Lagrangian* description, in which each individual node of the computational mesh follows the associated material particle motion, is mainly used in structural geology and solid geomechanics. Figure 4.1 illustrates graphically the Lagrangian formulation of the motion. The motion of the material points relates the material coordinates, Ξ, to the spatial coordinates, $\mathbf{x}$. It is defined by an application of function ϕ such that $\phi : (t, \Xi) \mapsto (t, \mathbf{x})$, which allows to link Ξ and $\mathbf{x}$ during time as $\mathbf{x} = \mathbf{x}(t, \Xi)$. For every fixed instant t, the mapping ϕ defines a configuration in the spatial domain. The fact that the material points coincide with the same grid points during the motion, and each finite element of a Lagrangian grid contains the same material particles, represents a significant advantage from a computational point of view. The Lagrangian description allows easy tracking of free surfaces and material interfaces. Meanwhile, its weakness is its inability to follow large distortions of the computational domain without recourse to frequent remeshing operations. In the absence of remeshing, when large deformations occur, Lagrangian algorithms will undergo a loss of accuracy and may even be unable to conclude a calculation due to excessive distortions of the computational grid linked to the material.

The difficulties caused by excessive distortion of the finite element grid are overcome in the *Eulerian* formulation. The basic idea in the Eulerian description of motion, which is popular in fluid mechanics and in geodynamics, consists of examining, as time evolves, the physical quantities associated with the fluid particles passing through a fixed region of space (Fig. 4.2). In the Eulerian description, the finite element grid is fixed and the continuum moves and deforms with respect to the computational grid. The conservation equations are formulated in terms of the spatial coordinates $\mathbf{x}$ and the time t. Therefore, the Eulerian description of motion involves only variables and functions having an instantaneous significance in a fixed region of space. The material velocity $\mathbf{v}$ at a given mesh node corresponds to the velocity of the material point coincident at the considered time t with the considered node. The velocity is consequently expressed with respect to the fixed element mesh without any reference to the initial configuration of the continuum and the

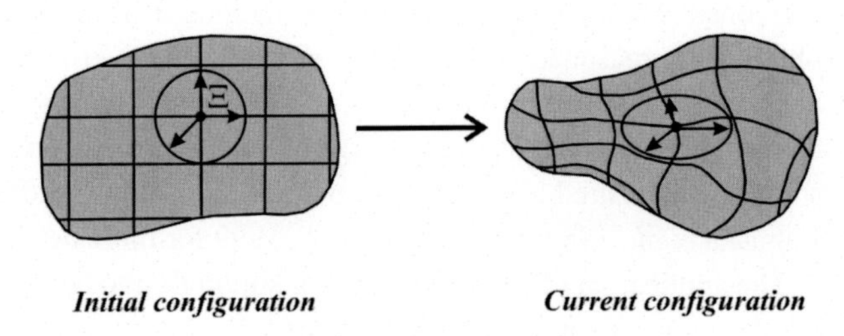

Fig. 4.1. Lagrangian description of motion.

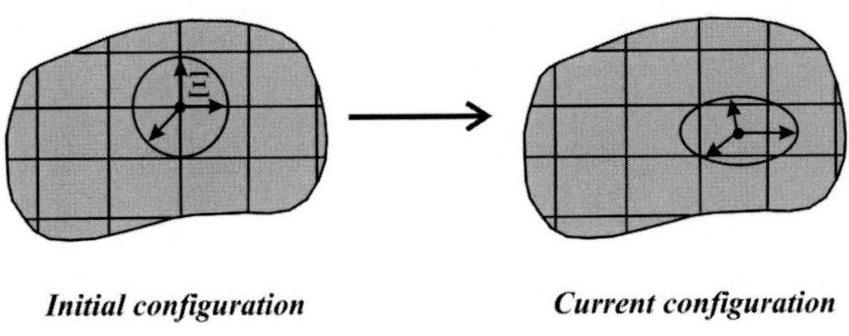

Fig. 4.2. Eulerian description of motion.

material coordinates Ξ : $\mathbf{v} = \mathbf{v}(t, \mathbf{x})$. The Eulerian formulation facilitates the treatment of large distortions in the motion, because remeshing of the computational grid is not required, and is indispensable for the simulation of turbulent flows. Its handicap is the difficulty to follow free surfaces and material interfaces.

Compared with the classical Lagrangian and Eulerian formulations, the *Arbitrary Lagrangian Eulerian* (ALE) description presents a combination of useful features of the two principal formulations of motion, while minimising as far as possible their drawbacks. The ALE method is particularly useful in flow problems involving large distortions in the presence of mobile and deforming boundaries, and is particularly useful for treating a free surface boundary. Typical examples can be the problems describing an interaction of subducting lithosphere and surrounding mantle. The key idea in the ALE formulation is the introduction of a computational grid, which can move with a velocity independent of the velocity of the material particles. With this additional freedom with respect to the Eulerian and Lagrangian descriptions, the ALE description succeeds to a certain extend in minimising the problems encountered in the classical kinematical descriptions, while combining at best their respective advantages.

4.3 Mathematical preliminaries

The process of spatial discretisation by the FE method rests on the discrete representation of a *weak integral form* of the partial differential equation to be solved. Consider a spatial domain $\Omega \subset R^n$ with piecewise smooth boundary Γ, where $n = 1, 2$ or 3 denotes the number of space dimensions. We use the notation $f : \bar{\Omega} \to R$ to state that for each spatial point $\mathbf{x} = (x_1, x_2, \ldots, x_n) \in \bar{\Omega}$, $f(\mathbf{x}) \in R$, $\bar{\Omega} = \Omega \cup \Gamma$. A function f is said to be of class $C^m(\Omega)$ if all its derivatives up to order m exist and are continuous functions.

In the FE analysis we work with integral equations, and hence we are interested in functions belonging to larger spaces than C^m. As we see, instead of requiring the mth derivative to be a continuous function, we require that its square is integrable. In fact, FE functions should belong to so called *Sobolev spaces*. We denote by $L_2(\Omega)$ the space of functions that are square integrable over the domain Ω. This space is equipped with the standard inner product $(u, v) = \int_\Omega uv d\Omega$ and norm $\|v\| = (v, v)^{1/2}$. A detailed description of Sobolev spaces can be found in the book by Adams (1975). Also we introduce another

Sobolev space $H^k(\Omega)$ of square integrable functions and their derivatives:

$$H^k(\Omega) = \left\{ u \in L_2(\Omega) \;\middle|\; \frac{\partial^{|\alpha|} u}{\partial x_1^{\alpha_1} \partial x_2^{\alpha_2} \cdots \partial x_n^{\alpha_n}} \in L_2(\Omega) \right\}, \tag{4.1}$$

for all $|\alpha| \leq k$, where $\alpha = (\alpha_1, \alpha_2, \ldots, \alpha_n)$, $|\alpha| = \alpha_1 + \alpha_2 + \cdots + \alpha_n$, and α_i is a natural number. The space $H^k(\Omega)$ is equipped with the norm

$$\|u\|_k = \left(\sum_{j=0}^{k} \sum_{|\alpha|=j} \left\| \frac{\partial^{|\alpha|} u}{\partial x_1^{\alpha_1} \partial x_2^{\alpha_2} \cdots \partial x_n^{\alpha_n}} \right\|^2 \right)^{1/2}. \tag{4.2}$$

We can note that $H^0(\Omega) = L_2(\Omega)$ and $H^1(\Omega) = \left\{ v \in L_2(\Omega) \;\middle|\; \frac{\partial v}{\partial x_i} \in L_2(\Omega),\ i = 1, \ldots, n \right\}$. This space is equipped with the inner product

$$(u, v)_1 = \int_\Omega \left(uv + \sum_{i=1}^{n} \frac{\partial u}{\partial x_i} \frac{\partial v}{\partial x_i} \right) d\Omega, \tag{4.3}$$

and its induced norm $\|u\|_1 = ((u,u)_1)^{1/2}$. We shall use the space $H_0^1(\Omega) = \{v \in H^1(\Omega) | v = 0 \text{ on } \Gamma\}$, which is a subspace of $H^1(\Omega)$ with functions vanishing on the boundary of domain Ω.

In the FE analysis, not only scalar functions (such as pressure or temperature), but also vector functions (such as velocity or velocity potential) may be considered. For vector functions with two or three components, the procedure is essentially the same as for scalar functions (for more detail we refer to the book by Donea and Huerta, 2003).

To define the *weak*, or *variational*, form of the boundary-value problems, we introduce here two classes of functions: the *test* (or *weight*) functions and the *trial* (or *admissible*) solutions. The first class of functions consists of all functions belonging to $H_0^1(\Omega)$. The second class of functions is similar to the test functions, except that the admissible functions are required to satisfy the Dirichlet conditions on the model boundary: $\{u \in H^1(\Omega) \mid u = u_* \text{ on } \Gamma\}$. For homogeneous boundary conditions ($u = 0$), the trial and test spaces coincide.

4.4 Weighted residual methods: variational problem

The methods of weighted residuals are general techniques for developing approximate solutions of operator equations. In all of these the unknown solution is approximated by a set of local basis functions containing adjustable constants or functions. These constants or functions are chosen by various criteria to give the best approximation for the selected family. A general discussion of weighted residual methods is found in Ames (1965, 1972) and Finlayson (1972).

We introduce in this section the basic principles and tools of the FE method using the following boundary value problem

$$\Im(u) \equiv -\frac{d}{dx}\left(p(x)\frac{du}{dx}\right) + q(x)u = f(x), \quad 0 < x < 1, \quad u(0) = u(1) = 0. \qquad (4.4)$$

We assume that $p(x)$ is a positive and continuously differentiable function for $x \in [0, 1]$, $q(x)$ is non-negative and continuous on $[0, 1]$, and $f(x)$ is continuous on the same interval $[0, 1]$.

By focusing on this simple problem, we hope to introduce the fundamental concepts without the geometric complexities encountered in two and three dimensions. Problems like (4.4) arise in many geodynamic problems, e.g. deformation of an elastic lithospheric plate and heat conduction in the lithosphere (Turcotte and Schubert, 2002). Even problems of this simplicity cannot in general be solved analytically. With FD methods, derivatives in (4.4) are approximated by finite differences with respect to a mesh introduced on $[0, 1]$ (see Chapter 2). With the FE method, the *method of weighted residuals* is used to construct an integral formulation of (4.4) called a *variational problem*.

Let us now consider problem (4.4). Multiplying the equation in (4.4) by a test function v and integrating over (0,1), we obtain

$$(v, \Im(u) - f) = 0, \quad \text{for all } v \in L_2([0, 1]), \qquad (4.5)$$

where u is a trial solution to (4.5). The solution of (4.4) is also a solution to (4.5) for all functions $v \in L_2([0, 1])$. Equation (4.5) is referred to as a *variational form* of problem (4.4). The variational formulation of the problem is a relatively easy way to construct the discrete equations, provides some additional insight into the problem and gives an independent check on the formulation of the problem. For approximate solutions, a larger class of trial functions can be employed in many cases if the researcher operates on the variational formulation rather than on the differential formulation of the problem (e.g. low-order test functions can be employed, because the order of derivatives is lower in the variations problem).

Using the method of weighted residuals, we now construct approximate solutions by replacing u and v with the functions U and V and solving (4.5) relative to these choices. Specially, we consider approximations of the form

$$u(x) \approx U(x) = \sum_{j=1}^{N} c_j \phi_j(x), \; v(x) \approx V(x) = \sum_{j=1}^{N} d_j \psi_j(x). \qquad (4.6)$$

The functions $\phi_j(x)$ and $\psi_j(x), j = 1, 2, \ldots, N$, are chosen, and the main goal is to determine the coefficients c_j, so that U is a good approximation of u. The approximations U and V are also called a *trial* and *test* functions respectively. Note that U and V are defined in finite-dimensional subspaces S^U (trial space) and S^V (test space) of $H_0^1([0, 1])$, respectively. Replacing v and u in (4.5) by their approximations V and U (Eq. (4.6)), we have

$$(V, \Im(U) - f) = 0, \text{ for all } V \in S^V. \qquad (4.7)$$

The *residual* $r(x) \equiv \Im(U) - f(x)$ clarifies the name of the method as 'weighted residuals'. The fact that the inner product vanishes in (4.7) implies that the residual is *orthogonal* to all functions V in the test space S^V. Substituting (4.6) into (4.7) and interchanging the sum and integral yields

$$\sum_{j=1}^{N} d_j \left(\psi_j, r\right) = 0. \tag{4.8}$$

Having selected the *basis* $\psi_j, j = 1, 2, \ldots, N$, the requirement that (4.7) be satisfied for all $V \in S^V$ implies that (4.8) be satisfied for all possible choices of $d_j, j = 1, 2, \ldots, N$. This implies that

$$(\psi_j, r) = 0,\ j = 1, 2, \ldots, N. \tag{4.9}$$

Let us now integrate the second derivative terms in (4.5) by parts. This leads to the following equation

$$\int_0^1 v \left[-\frac{d}{dx}\left(p\frac{du}{dx}\right) + qu - f \right] dx = \int_0^1 \left[-\frac{dv}{dx}p\frac{du}{dx} + vqu - vf \right] dx - vp\left.\frac{du}{dx}\right|_0^1 = 0. \tag{4.10}$$

The treatment of the boundary integral term (last term in (4.10)) needs some attention. Here we consider that v satisfies the same trivial boundary conditions (4.4) as u. In this case, the boundary term vanishes and Eq. (4.10) becomes

$$A(v, u) = (v, f),\ \text{for all } v \in H_0^1, \tag{4.11}$$

where

$$A(v, u) = \int_0^1 \left[\frac{dv}{dx}p\frac{du}{dx} + vqu \right] dx. \tag{4.12}$$

The bilinear form $A(v, u)$ is called the *strain energy*, and it frequently relates to the stored, or internal, energy in the physical system. Note that the integration by parts has eliminated the second derivative from the formulation. Thus, solutions of (4.11) might have less continuity than those satisfying either (4.4) or (4.5). For this reason, they are called *weak solutions* in contrast to the *strong solutions* of (4.4) or (4.5). Weak solutions may lack the continuity to be strong solutions, but strong solutions are always weak solutions.

Now we replace u and v by their approximations U and V according to (4.6). Both U and V are regarded as belonging to the same finite-dimensional subspace S_0^N of $H_0^1([0, 1])$, and $\phi_j, j = 1, 2, \ldots, N$ form a basis for S_0^N. Thus, U is determined as the solution of

$$A(V, U) = (V, f),\ \text{for all } V \in S_0^N. \tag{4.13}$$

The substitution of (4.6) with ψ_j replaced by ϕ_j in (4.13) reveals the more explicit form

$$A(\phi_j, U) = (\phi_j, f),\ j = 1, 2, \ldots, N. \tag{4.14}$$

Finally, to make (4.14) totally explicit, we eliminate U by using (4.6) and interchange a sum and integral to obtain

$$\sum_{k=1}^{N} c_k A(\phi_j, \phi_k) = (\phi_j, f), j = 1, 2, \ldots, N. \tag{4.15}$$

Therefore, the coefficients c_k of the approximate solution (4.6) are determined as the solution of a set of the linear algebraic equations (4.15). Different choices of the basis ϕ_j make integrals involved in the strain energy (4.12) easy or difficult to evaluate. They also affect the accuracy of the approximate solution. The term $A(\phi_j, \phi_k)$ is referred to as the *stiffness matrix*.

The various weighted residual methods differ in the criteria that they employ to calculate the coefficients c_i in (4.6) such that the residual $r(x)$ is small. However, in all methods c_i are determined so as to make a weighted average of $r(x)$ vanish. If the test space S^V is selected to be the same as the trial space S^U, and the same basis for each space is used (e.g. $\psi_j(x) = \phi_j(x), k = 1, 2, \ldots, N$), this choice leads to the *Galerkin method*, sometimes called *Bubnov–Galerkin method* (Bubnov, 1913; Galerkin, 1915)

$$(\phi_j, \Im(U) - f) = 0, \quad j = 1, 2, \ldots, N. \tag{4.16}$$

In the *least squares method* (Gauss–Legendre; see Hall, 1970), the square of the norm of the residual $r(x)$ is minimised with respect to the parameters c_k:

$$\frac{\partial}{\partial c_k} \|\Im(U) - f\|^2 = 0, \quad k = 1, 2, \ldots, N. \tag{4.17}$$

In the *collocation method* (Frazer *et al.*, 1937), the residual $r(x)$ is set to zero at n distinct points in the solution domain to obtain n simultaneous equations for the parameters c_k. The location of the n points can be somewhat arbitrary, and a uniform pattern may be appropriate, but usually researchers should use some judgment to select 'appropriate' locations.

An important step in using weighted residual methods is the solution of the simultaneous equations required to determine the parameters c_k. In the Galerkin method, the stiffness matrix is symmetric and positive defined, if $\Im$ is a symmetric and positive define operator. In the least squares method a symmetric stiffness matrix is always generated irrespective of the properties of the operator $\Im$. However, in the collocation method, a non-symmetric stiffness matrix may be generated. Therefore, in practical analysis, the Galerkin and least squares methods are usually preferable.

4.5 Simple FE problem

Finite element (FE) methods are in fact weighted residual methods that use bases of polynomials having a compact support. Thus, the functions ϕ_j and $\psi_j, j = 1, 2, \ldots, N$, are non-zero on only a small portion of model domain. Since continuity may be difficult to impose, bases will typically use the minimum continuity necessary to ensure the existence of integrals and solution accuracy. The use of piecewise polynomial functions simplify the evaluation of integrals involved in the strain energy (4.12). Choosing bases with a compact support leads to a *sparse* (and *well-conditioned* in many cases) linear algebraic system (4.15) for the solution. (Note that a system of equations is considered to be *well-conditioned* if a small change in the coefficient matrix or a small change in the right-hand side results in a small change in the solution vector.)

Let us introduce the simplest continuous piecewise polynomial approximation of u and v (see Eqs. (4.6)). This would be a piecewise linear polynomials with respect to a mesh $0 = x_0 < x_1 < \cdots < x_N = 1$ introduced on [0, 1]. Each subinterval (x_{j-1}, x_j), $j = 1, 2, \ldots, N$, is called a *finite element*. The basis is created from the 'hat function'

$$\phi_j(x) = \begin{cases} \dfrac{x - x_{j-1}}{x_j - x_{j-1}}, & \text{if } x_{j-1} \le x < x_j, \\ \dfrac{x_{j+1} - x}{x_{j+1} - x_j}, & \text{if } x_j \le x < x_{j+1}, \\ 0, & \text{otherwise.} \end{cases} \tag{4.18}$$

As shown in Fig. 4.3, ϕ_j is non-zero only on the two elements containing the *node* x_j. It rises and descends linearly on these two elements and has a maximal unit value at $x = x_j$. Indeed, it vanishes at all nodes except x_j, and hence

$$\phi_j(x_i) = \delta_{ij} = \begin{cases} 1, & \text{if } x_i = x_j, \\ 0, & \text{otherwise,} \end{cases} \tag{4.19}$$

where δ_{ij} is the Kronecker delta. Using this basis in (4.18) with (4.6), we consider approximations of the form

$$U(x) = \sum_{j=1}^{N-1} c_j \phi_j(x). \tag{4.20}$$

Since each $\phi_j(x)$ is a continuous piecewise linear function of x, the summation U is also continuous and piecewise linear function. Evaluating U at a node x_k of the mesh using (4.19) yields $U(x_k) = \sum_{j=1}^{N-1} c_j \phi_j(x_k) = c_k$. Thus, the coefficients c_k, $k = 1, 2, \ldots, N-1$, are the values of U at the interior nodes of the mesh (see Fig. 4.4). By selecting the lower and upper summation indices as 1 and $N - 1$ we have ensured that the Eq. (4.20) satisfies

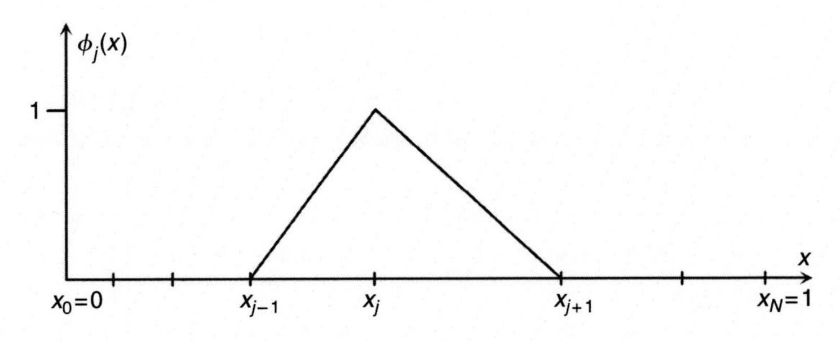

Fig. 4.3. **One-dimensional finite element mesh and piecewise linear hat function $\phi_j(x)$.**

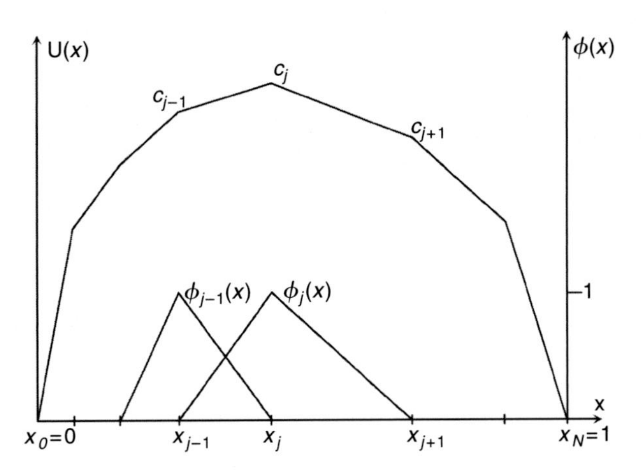

Fig. 4.4. **Piecewise linear finite element solution *U(x)*.**

the prescribed boundary conditions $U(0) = U(1) = 0$. As an alternative, we could have added basis elements $\phi_0(x)$ and $\phi_N(x)$ to the approximation and written the approximate trial function as

$$U(x) = \sum_{j=0}^{N} c_j \phi_j(x). \tag{4.21}$$

Since (4.19) is true, $U(x_0) = c_0$ and $U(x_N) = c_N$, thus the boundary conditions are satisfied by requiring $c_0 = c_N = 0$. The representations (4.20) or (4.21) are thus identical; however, (4.21) would be useful with non-trivial boundary conditions. The restriction of the finite element solution (4.21) to the element $[x_{j-1}, x_j]$ is the linear function

$$U(x) = c_{j-1}\phi_{j-1}(x) + c_j\phi_j(x), \qquad x \in [x_{j-1}, x_j], \tag{4.22}$$

since ϕ_{j-1} and ϕ_j are the only non-zero basis functions on $[x_{j-1}, x_j]$ (Fig. 4.4). Now using the Galerkin method, Eqs. (4.15) must be solved. The equations can be evaluated in a straightforward manner by substituting ϕ_k and ϕ_j using (4.18) and (4.19) and by evaluating the inner product and strain energy according to equation (4.12) (for more detail see Johnson, 1987).

4.6 The Petrov–Galerkin method for advection-dominated problems

The Galerkin FE method is not ideally suited to solve advection-dominated problems. Consider the 1-D stationary heat advection–diffusion problem:

$$u\frac{dT}{dx} - \kappa\frac{\partial^2 T}{\partial x^2} = f(x), \quad x \in (0, H),$$
$$T = 0, \quad \text{at } x = 0 \text{ and } x = H, \tag{4.23}$$

where the velocity u and the coefficient of thermal diffusion κ are constant, and f is the heat source. The weak form associated with this mathematical problem is given (after integration by parts of the diffusion term) by

$$\int_0^H \left(wu\frac{dT}{dx} + \frac{dw}{dx}\kappa\frac{dT}{dx} \right) dx = \int_0^H wf\,dx. \tag{4.24}$$

The weak form can be discretised by using a uniform mesh of linear elements of size h, which are defined by two nodes (see Section 4.5). With the linear trial and weight functions over the elements and after relevant transformations, the Eq. (4.24) can be written in the following discrete form at an interior node j:

$$u\frac{T_{j+1} - T_{j-1}}{2h} - \kappa\frac{T_{j+1} - 2T_j + T_{j-1}}{h^2} = \frac{f_{j+1} + 4f_j + f_{j-1}}{6}. \tag{4.25}$$

Notice that the left hand-side of (4.25) produced with linear elements coincides with the equation of second-order central differences. In this respect, the Galerkin method based on linear elements and the finite difference method based on central differences appear to be closely related.

To characterise the relative importance of advective and diffusive effects in a given flow problem, we introduce the mesh Péclet number ($Pe = 0.5uh/\kappa$), which expresses the ratio of advective to diffusive transport. This allows us to rewrite the discrete equation (4.25) in the form:

$$\frac{u}{2h}\left(\frac{Pe - 1}{Pe}T_{j+1} + \frac{2}{Pe}T_j - \frac{Pe + 1}{Pe}T_{j-1} \right) = \frac{f_{j+1} + 4f_j + f_{j-1}}{6}. \tag{4.26}$$

It is shown that the Galerkin solution is corrupted by non-physical oscillations when the Péclet number is larger than one. The Galerkin method loses its best approximation property when the non-symmetric advection operator dominates the diffusion operator in the transport heat equation, and consequently spurious node-to-node oscillations appear (Donea *et al.*, 2000).

To avoid the spurious oscillations, at least two modifications of the Galerkin scheme (4.26) can be considered. We clarify this by considering the exact solution to the problem (4.23) with a constant heat source $f = 1$, and $H = 1$:

$$T(x) = \frac{1}{u}\left(x - \frac{1 - \exp(\mu x)}{1 - \exp\mu} \right), \tag{4.27}$$

where $\mu = u/\kappa$. To obtain an exact scheme, we identify the value of three coefficients, say α_1, α_2 and α_3, such that

$$\alpha_1 T_{j-1} + \alpha_2 T_j + \alpha_3 T_{j+1} = 1 \tag{4.28}$$

for all nodal coordinates T_j, mesh dimensions h and Péclet numbers Pe. From the exact solution (4.27) we have

$$\begin{cases} T_{j-1} = \frac{1}{u}\left(x_j - h - \frac{1-\exp(\mu x_j)\exp(-2Pe)}{1-\exp\mu}\right) \\ T_j = \frac{1}{u}\left(x_j - \frac{1-\exp(\mu x_j)}{1-\exp\mu}\right) \\ T_{j+1} = \frac{1}{u}\left(x_j + h - \frac{1-\exp(\mu x_j)\exp(2Pe)}{1-\exp\mu}\right) \end{cases}. \tag{4.29}$$

Introducing these expressions in (4.28) and solving for α_1, α_2 and α_3, we obtain the following relation:

$$\frac{u}{2h}\left((1-\coth Pe)T_{j+1} + 2\coth Pe T_j - (1+\coth Pe)T_{j-1}\right) = 1. \tag{4.30}$$

We rewrite (4.30) in two alternative forms. First, we have a form similar to the original Galerkin scheme (4.25):

$$u\frac{T_{j+1} - T_{j-1}}{2h} - (\kappa + \tilde{\kappa})\frac{T_{j+1} - 2T_j + T_{j-1}}{h^2} = 1, \tag{4.31}$$

where $\tilde{\kappa} = \beta u h/2 = \beta Pe\kappa$ ($\beta = \coth Pe - 1/Pe$) is an added numerical diffusion. The second numerical scheme is:

$$\frac{1-\beta}{2}u\frac{T_{j+1} - T_j}{h} + \frac{1+\beta}{2}u\frac{T_j - T_{j-1}}{h} - \kappa\frac{T_{j+1} - 2T_j + T_{j-1}}{h^2} = 1, \tag{4.32}$$

where the discretisation of the advective term appears as a weighted average of the fluxes (advection) of the solution to the left and to the right of node j. Such schemes are called *upwind* schemes. Therefore, in the first scheme (4.31), an artificial diffusion was added in order to counterbalance the negative numerical diffusion introduced by the Galerkin approximation based on linear elements. In the second scheme (4.32), an upwind approximation of the advective term is used, because the centred scheme employed is not ideal in advection-dominated problems. Precisely, the early remedies were based on these two philosophies. In fact, both methodologies are equivalent, i.e. an upwind approximation introduces numerical diffusion and vice versa.

In an FE framework, several techniques can be utilised to achieve the upwind effect. The basic idea is to replace the standard Galerkin formulation with a so-called *Petrov–Galerkin* weighted residual formulation in which the weight function may be selected from a different class of functions than the approximate solution. The first upwind finite element formulations were based on modified weight functions such that the element upstream of a node is weighted more heavily than the element downstream of a node (Christie *et al.*, 1976; Heinrich *et al.*, 1977; Hughes, 1978; Heinrich and Zienkiewicz, 1979; Griffiths and Mitchell, 1979). Another approach discussed above in this section is to introduce artificial diffusion to counteract the negative dissipation introduce by the Galerkin formulation (with linear elements). To explain this approach let us consider the following equation to replace the equation in (4.23):

$$u\frac{dT}{dx} - (\kappa + \tilde{\kappa})\frac{\partial^2 T}{\partial x^2} = 0, \tag{4.33}$$

where the heat source is taken as zero (for simplification), $\tilde{\kappa} = 0.5\beta uh$, and β is a free parameter, which governs the amplitude of the added numerical diffusion. Hughes and Brooks (1979) suggested replacing the usual weak formulations (Eq. (4.24) with $f = 0$) by the following:

$$\int_0^H \left(wu\frac{dT}{dx} + \frac{dw}{dx}(\kappa + \tilde{\kappa})\frac{dT}{dx} \right) dx = 0, \tag{4.34}$$

where the magnitude of the added diffusion depends on $\beta (0 \le \beta \le 1)$ with the optimal value $\beta = \coth Pe - 1/Pe$ and the value $\beta = 1$ corresponding to full upwind differencing. Sometimes the added numerical diffusion is referred to as *balancing diffusion*. Equation (4.34) can be rewritten in the form

$$\int_0^H \left(\left[w + 0.5\beta h\frac{dw}{dx} \right] u\frac{dT}{dx} + \frac{dw}{dx}\kappa\frac{dT}{dx} \right) dx = 0, \tag{4.35}$$

which shows that the balancing diffusion method uses a modified weight function, given by $\tilde{w} = w + 0.5\beta h\frac{dw}{dx}$ for the advective term only. Since these weight functions give more weight to the element upstream of a node, the modified functions are upwind-type weight functions. The relevant scheme/method is called the *streamline-upwind* (SU) scheme/method.

Hughes and Brooks (1982) subsequently proposed to apply the modified weight function to all terms in Eq. (4.23) in order to obtain a consistent formulation. Moreover, they noted that for linear elements the perturbation to the standard test function could be neglected in the diffusion term. The concept of adding diffusion along the streamlines in a consistent manner has been successfully exploited in the *Streamline-Upwind Petrov–Galerkin* (SUPG) method. In order to stabilise the advective term in a consistent manner (consistent stabilisation), ensuring that the solution of the differential equation is also a solution of the weak form, Hughes and Brooks (1982) proposed to add an extra term over the element interiors to the Galerkin weak form. This term is a function of the residual of the differential equation to ensure consistency. Let explain this using one-dimensional steady heat advection–diffusion problem (4.23). The residual is defined as

$$R(T) = u\frac{dT}{dx} - \kappa\frac{\partial^2 T}{\partial x^2} - f. \tag{4.36}$$

The general form of the stabilisation techniques is

$$\int_0^H \left(wu\frac{dT}{dx} + \frac{dw}{dx}\kappa\frac{dT}{dx} \right) dx + \int_0^H Q(w)\tau R(T) dx = \int_0^H wf\, dx, \tag{4.37}$$

where $Q(w)$ is a certain operator applied to the test function, and τ is the stabilisation parameter. In the case of SUPG method, the operator Q is defined as $Q = u\frac{dw}{dx}$, which corresponds to the perturbation of the test function introduced in the SU method.

4.7 Penalty-function formulation of Stokes flow

One of the approaches to treating Stokes flow numerically is to use a penalty-function formulation, which leads to a simple and effective finite element implementation of incompressibility. To find the solution to the Stokes flow (i.e. velocity $\mathbf{u} = (u_1, u_2)$), we consider the following boundary value problem, which is composed of the equations of the momentum and mass conservations:

$$\frac{\partial \sigma_{ij}}{\partial x_j} + \rho g \mathbf{e} = 0, \qquad \frac{\partial u_1}{\partial x_1} + \frac{\partial u_2}{\partial x_2} = 0, \tag{4.38}$$

and relevant boundary conditions. The stress tensor σ_{ij} is represented as

$$\sigma_{ij} = -P\delta_{ij} + \eta\left(\frac{\partial u_i}{\partial x_j} + \frac{\partial u_j}{\partial x_i}\right), \tag{4.39}$$

where ρ is density, g is the acceleration due to gravity, $\mathbf{e}$ is a unit vector in the x_2-direction, $\mathbf{x} = (x_1, x_2)$ is the Cartesian coordinates, P is pressure, δ_{ij} is the Kronecker delta, and η is viscosity.

In the penalty-function formulation of the Stokes problem, the equation representing the stress tensor is replaced by

$$\sigma_{ij}^{(\lambda)} = -P^{(\lambda)}\delta_{ij} + \eta\left(\frac{\partial u_i^{(\lambda)}}{\partial x_j} + \frac{\partial u_j^{(\lambda)}}{\partial x_i}\right), \tag{4.40}$$

where

$$P^{(\lambda)} = -\lambda\left(\frac{\partial u_1^{(\lambda)}}{\partial x_1} + \frac{\partial u_2^{(\lambda)}}{\partial x_2}\right), \tag{4.41}$$

and $\lambda > 0$ is a parameter. The incompressibility condition is dropped in the penalty-function formulation.

The convergence of the penalty-function solution to the Stokes flow solution was proven by Temam (1977). This formulation enforces incompressibility, and at the same time it eliminates the unknown pressure field. This is useful, because the amount of computational work decreases (no pressure equation is solved). This approach was employed by King *et al.* (1990) to simulate thermal convection in the mantle.

4.8 FE discretisation

One of the important aspects of FE modelling is the discretisation of the model domain. In Section 4.5, we considered a simple one-dimensional discretisation of the domain into finite elements. Depending on the choice of the model formulation (Lagrangian or Eulerian), the finite element shape may consist of triangles and/or squares in two-dimensional space and tetrahedrons and/or rectangular parallelepipeds in the three-dimensional space. Figure 4.5

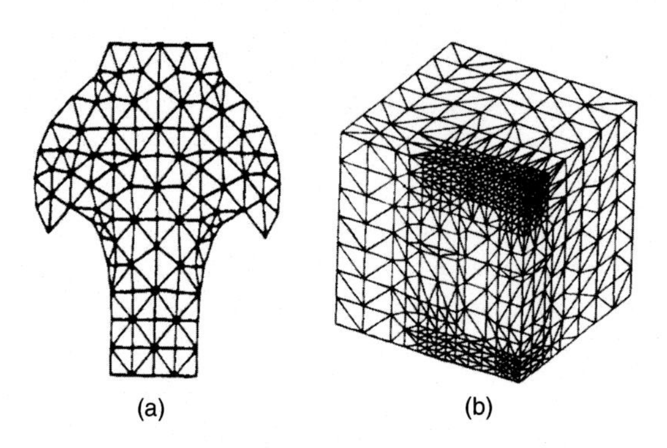

Fig. 4.5. Two-dimensional (a) and three-dimensional (b) finite elements.

shows a few examples of 2-D and 3-D finite element discretisations. The Eulerian FE method employs non-deformable elements (and a fixed mesh). On the contrary, the Lagrangian FE method works with the deformable elements. A typical FE software framework contains a pre-processing module to define the model geometry, initial and boundary conditions, and other input data. The module creates a computer model domain Ω (e.g. using a computer aided design, CAD, system); discretises Ω into a finite element mesh; creates a geometric and mesh database describing the mesh entities (vertices, edges, faces and elements) and their relationships to each other and to the model geometry; and finally defines problem-dependent data such as the coefficient functions in the differential equations, loading, initial data, and boundary conditions.

Discretising two-dimensional domains via triangular or quadrilateral FE meshes can either be a simple or difficult task depending on the geometric or solution complexities. Discretising three-dimensional domains is more complicated. Uniform meshes are appropriate for many geodynamic problems, which have model domains defined by simple geometric shapes (e.g. rectangular domains), but non-uniform meshes might provide better performance when solutions vary rapidly (e.g. in thermal boundary layers). Finite element techniques (and software) have always been associated with unstructured and non-uniform meshes. Early software left it to the users to generate FE meshes manually. This required the entry of the coordinates of all element vertices. Node and element indexing, typically, was also done manually. This is a tedious and error prone process that has now largely been automated, at least in two dimensions. Adaptive solution-based mesh refinement procedures concentrate higher element densities in regions of rapid solution variation and attempt to automate the task of modifying (refining/coarsening) an existing mesh. Domain discretisation is not a subject for this chapter, and readers are referred to Kikuchi (1986); Flaherty *et al.* (1989); Babuska *et al.* (1995); Bathe (1996); Verfürth (1996); Carey (1997); and Bern *et al.* (1999) for details on FE discretisation.

4.9 High-order interpolation functions: cubic splines

The FE method is not limited to piecewise linear approximations, and its extension to higher-degree polynomials is straightforward. To increase the accuracy of the FE solution one can

Table 4.1. Coefficients of cubic splines.

	$\alpha(y)$				$\beta(y)$			
n	c_0	c_1	c_2	c_3	c_0	c_1	c_2	c_3
1	0	0	1/2	−11/36	0	1	0	−1/3
2	7/36	1/12	−5/12	7/36	2/3	0	−1	1/2
3	1/18	−1/6	1/6	−1/18	1/6	−1/2	1/2	−1/6

	$\delta(y)$				$\gamma(y)$			
n	c_0	c_1	c_2	c_3	c_0	c_1	c_2	c_3
1	1	0	0	−1/6	0	0	0	1/4
2	5/6	−1/2	−1/2	1/3	1/4	3/4	3/4	−3/4
3	1/6	−1/2	1/2	−1/6	1	0	−3/2	3/4
4	0	0	0	0	1/4	−3/4	3/4	−1/4

	$\delta^*(y)$				$\beta^*(y)$					$\alpha^*(y)$		
n	c_0	c_1	c_2	c_3	c_0	C_0	c_1	c_2	c_3	c_1	c_2	c_3
1	0	0	0	1/6	0	0	0	1/6	0	0	0	1/18
2	1/6	1/2	1/2	−1/3	1/6	1/2	1/2	−1/2	1/18	1/6	1/6	−7/36
3	5/6	1/2	−1/2	1/6	2/3	0	−1	1/3	7/36	−1/12	−5/12	11/36

either increase the number of linear elements used in the FE analysis or use higher-order interpolation functions. For example, a quadratic or cubic polynomial can be employed as a basis function. Compared to the linear shape functions, high-order polynomials are typically implemented by increasing the number of nodes in each element, but in order to increase the order of continuity between elements they can also be defined at several neighbouring elements (i.e. with a larger support). The quadratic or cubic shape functions possess properties similar to those of the linear shape functions, namely: a shape function has a value of unity at its corresponding node, and a value of zero at the other adjacent nodes. The quadratic and cubic interpolation functions offer good results in FE formulations of geodynamic problems (e.g. Christensen, 1992; Naimark and Ismail-Zadeh, 1995; Naimark *et al.*, 1998; Ismail-Zadeh *et al.*, 1998, 2001a). However, if additional accuracy is needed, fourth or even higher-order polynomials can be employed as basis functions.

In this section we consider cubic shape functions (specifically cubic splines) as an example of polynomials used in FE modelling to represent the spatial variation of a given variable. Cubic splines can be constructed by the following manner. Consider a segment $0 \le y \le L$ divided into N small sub-segments by points $y_n = (n-1)h, h = L/(N-1), n = 1, 2, \ldots, N$. Let us now introduce seven functions: $\alpha(y), \beta(y), \delta(y), \delta^*(y), \beta^*(y)$, and $\alpha^*(y)$ defined for $0 \le y \le 3h$ and the function $\gamma(y)$ defined for $0 \le y \le 4h$, with each being a cubic $c_0 + c_1(y - y_n)/h + c_2((y - y_n)/h)^2 + c_3((y - y_n)/h)^3$ in a small segment $y_n \le y \le y_{n+1}$, $n = 1, 2, 3$ and 4. The values of c_i are listed in Table 4.1, and the functions are plotted in Fig. 4.6.

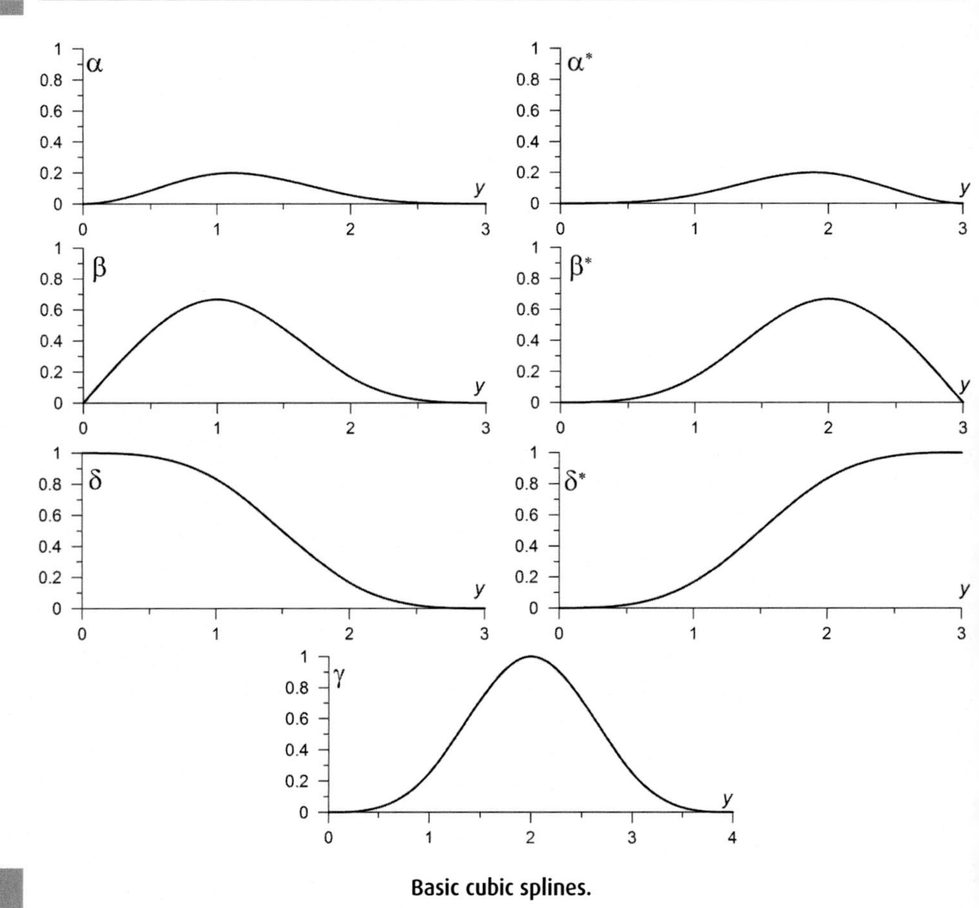

Fig. 4.6. **Basic cubic splines.**

These seven standard functions so defined have the following properties. The functions and both their first and second derivatives are continuous over their support domain, so that these functions are splines. The functions also satisfy the following conditions (signs $'$ and $''$ denote the first and the second derivative sign, respectively).

At $y = 0$:

$$\alpha(y) = \beta(y) = \delta^*(y) = \beta^*(y) = \alpha^*(y) = 0, \delta(y) = 1,$$
$$\alpha'(y) = \delta'(y) = \delta^{*\prime}(y) = \beta^{*\prime}(y) = \alpha^{*\prime}(y) = 0, \beta'(y) = 1/h,$$
$$\beta''(y) = \delta''(y) = \delta^{*\prime\prime}(y) = \beta^{*\prime\prime}(y) = \alpha^{*\prime\prime}(y) = 0, \alpha''(y) = 1/h^2,$$

at $y = 3h$:

$$\alpha(y) = \beta(y) = \delta(y) = \gamma(y) = \beta^*(y) = \alpha^*(y) = 0,\ \delta^*(y) = 1,$$
$$\alpha'(y) = \beta'(y) = \delta'(y) = \gamma'(y) = \delta^{*\prime}(y) = \alpha^{*\prime}(y) = 0,\ \beta^{*\prime}(y) = -1/h,$$
$$\alpha''(y) = \beta''(y) = \delta''(y) = \gamma''(y) = \delta^{*\prime\prime}(y) = \beta^{*\prime\prime}(y) = 0,\ \alpha^{*\prime\prime}(y) = 1/h^2,$$

at $y = 4h$:

$$\gamma(y) = \gamma'(y) = \gamma''(y) = 0,$$

at $y = 2h$:

$$\gamma(y) = 1.$$

Basis splines on the interval $0 \le y \le L$ are functions $s_1(y), s_2(y), \ldots, s_N(y)$ chosen from the above standard splines:

$s_1(y)$ and $s_2(y)$ (boundary splines) are selected from $\alpha(y), \beta(y)$ and $\delta(y)$ to satisfy boundary conditions at $y = 0$, e.g. $s_1(y) = \delta(y)$ and $s_2(y) = \beta(y)$ to approximate a function $f(y)$ such that $f(0) = a$ and $f''(0) = 0$;

$s_{N-1}(y)$ and $s_N(y)$ (boundary splines) are selected from $\alpha^*(y), \beta^*(y)$ and $\delta^*(y)$ to satisfy boundary conditions at $y = L$ in the same manner as at $y = 0$;

$$s_i(y) = \gamma(y - (i-2)h) \text{ for } (i-2)h \le y \le (i+2)h,\ i = 2, 3, \ldots, N-2.$$

4.10 Two- and three-dimensional FE problems

The main objective of this section is to introduce the FE formulations for problems of slow viscous flow, which are used intensively in numerical modelling of geodynamic processes (e.g. thermal and thermo-chemical mantle convection, lithosphere dynamics, flow in the lower crust etc.). We consider here the Eulerian formulation of the motion and hence present an Eulerian FE approach. Readers are referred to Bathe (1996) and Zienkiewicz and Taylor (2000) for general implementations of the Lagrangian FE approach.

4.10.1 Two-dimensional problem of gravitational advection

Mathematical statement. We present a numerical approach for solving the two-dimensional Stokes flow problems where physical properties (density and viscosity) change discontinuously across advected boundaries. The approach combines the Galerkin method with a method of integration over advected layers, where a finite-dimensional space of spline weights is used together with a Cartesian coordinate representation of the terms with a discontinuous viscosity. This approach allows us to approximate a natural shape of a free surface, instead of *a posteriori* calculation of its topography from the normal stress at the upper free-slip boundary.

We consider the rectangular model region Ω (Fig. 4.7): $0 \le x \le H_x, -H_z \le z \le 0$, where H_x and H_z are the model width and depth, respectively. A Newtonian fluid with variable density ρ and viscosity η fills this region. Curves L_e, $e = 1, 2, \ldots, E$ divide the model region Ω into several sub-regions Ω_e, $e = 1, 2, \ldots, E + 1$. We assume that each curve L_e is

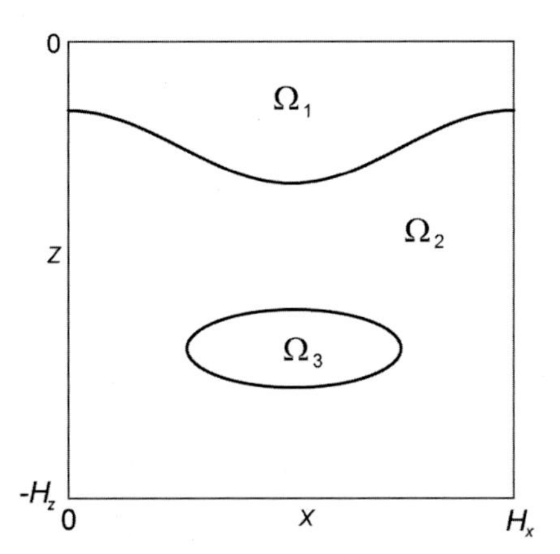

Fig. 4.7. **Geometry of the 2-D model for the case of two interfaces.**

closed or starts and terminates at the boundary of Ω, and has no self-intersections. Figure 4.7 shows two curves, L_1 and L_2, and three sub-regions Ω_1, Ω_2 and Ω_3. In what follows, we consider one curve L for simplicity, though the number of curves can be arbitrary. We also use a dimensionless form of equations governing the model, so that after the appropriate change of variables, the model region Ω occupies the square $0 \leq x \leq 1,\ 0 \leq z \leq 1$.

Introduce the following notation:

$$\begin{aligned}
D_x &= \partial/\partial x, D_z = \partial/\partial z, D_{xx} = D_x D_x, \\
D_{zz} &= D_z D_z, D_{xz} = D_x D_z, D_t = \partial/\partial t, \\
\Im(\eta)\psi &= 4D_{xz}(\eta D_{xz}\psi) + (D_{zz} - D_{xx})[\eta(D_{zz} - D_{xx})\psi], \\
\Im(\eta;\psi,\varphi) &= \eta\,[4D_{xz}\psi D_{xz}\varphi + (D_{zz}\psi - D_{xx}\psi)(D_{zz}\varphi - D_{xx}\varphi)]\,, \text{ and} \\
D(A,\psi) &= D_x\psi D_z A - D_z\psi D_x A,
\end{aligned}$$

where $\psi(t,x,z)$, $\varphi(t,x,z)$ and $A(t,x,z)$ are functions having continuous derivatives entering in the notation.

We seek the stream function $\psi(t,x,z)$, density $\rho(t,x,z)$, viscosity $\eta(t,x,z)$, and the family of curves $L : x = x(t,q), z = z(t,q)$ (q is a parameter of points on a curve, $0 \leq q \leq Q$) satisfying the differential equations (g is the acceleration due to gravity)

$$\begin{aligned}
\Im(\eta)\psi &= -gD_x\rho, \\
D_t\rho &= D(\rho,\psi),\ D_t\mu = D(\eta,\psi), \\
dx/dt &= D_z\psi,\ dz/dt = -D_x\psi,
\end{aligned} \tag{4.42}$$

the impenetrability and free-slip boundary conditions

$$\begin{aligned}
\psi &= D_{xx}\psi = 0 \text{ at } x = 0 \text{ and } x = 1, \\
\psi &= D_{zz}\psi = 0 \text{ at } z = 0 \text{ and } z = 1,
\end{aligned} \tag{4.43}$$

and initial conditions at $t = t_0$

$$\rho = \rho^0(x,z), \quad \eta = \eta^0(x,z),$$
$$x(q) = x^0(q), \quad z(q) = z^0(q). \tag{4.44}$$

The first equation is the two-dimensional Stokes equation represented in terms of the stream function ψ. Velocity $\mathbf{v}(t,x,z) = (u(t,x,z), w(t,x,z))$ can be obtained from the stream function as $u = \partial\psi/\partial z$ and $w = -\partial\psi/\partial x$. The second and third equations describe the advection of density and viscosity with the flow, and the remaining equations determine the trajectories of points $x(t,q)$ and $z(t,q)$ located at $t_0 = 0$ on the curve $L_0 = L(t_0)$.

The Galerkin method with tracking interfaces. We define a *weak solution* of the problem. Let us multiply the first equation in (4.42) by a function $\varphi(t,x,z)$ satisfying the same boundary conditions (4.43) as $\psi(t,x,z)$, integrate by parts the left- and right-hand sides of the product twice and once, respectively, and observe that the integral over the model boundary vanishes. Multiply the second and third equations in (4.42) by functions ϑ and ζ, respectively, and integrate the results. A weak solution of the problem stated above is the set of functions $\psi(t,x,z)$, $\rho(t,x,z)$, $\eta(t,x,z)$, $x(t,q)$ and $z(t,q)$ satisfying the above boundary and initial conditions and the following equations:

$$\iint\limits_{\Omega} \Im(\eta;\psi,\varphi)dxdz = g\iint\limits_{\Omega} \rho D_x\varphi dxdz,$$
$$\iint\limits_{\Omega} (D_t\rho)\vartheta dxdz = \iint\limits_{\Omega} D(\rho,\psi)\vartheta dxdz,$$
$$\iint\limits_{\Omega} (D_t\eta)\zeta dxdz = \iint\limits_{\Omega} D(\eta,\psi)\zeta dxdz,$$
$$dx/dt = D_z\psi, \ dz/dt = -D_x\psi, \tag{4.45}$$

where φ, ϑ, and ζ are test functions.

Numerical solutions are obtained in the form of weighted sums of basic bicubic splines. However, bicubic splines, being excellent for the case of smooth unknown functions, become inadequate when these functions are discontinuous. To preserve the accuracy of spline representations for cases of discontinuous unknowns, Naimark *et al.* (1998) suggested the following approach.

Let us represent the unknown functions $\rho(t,x,z)$ and $\eta(t,x,z)$ as sums of two functions, one smooth and the other constant over Ω_1 and Ω_2:

$$\rho(t,x,z) = \rho_0(t,x,z) + \rho_1(t,x,z), \qquad \eta(t,x,z) = \eta_0(t,x,z) + \eta_1(t,x,z), \tag{4.46}$$

where $\rho_1(t,x,z)$ and $\eta_1(t,x,z)$ have the first and second continuous derivatives, whereas $\rho_0(t,x,z)$ and $\eta_0(t,x,z)$ take on constant values in Ω_1 and Ω_2:

$$\rho_0 = \begin{cases} \rho_0^{01}, \text{ if } (x,z) \in \Omega_1, \\ \rho_0^{02}, \text{ if } (x,z) \in \Omega_2, \end{cases} \qquad \eta_0 = \begin{cases} \eta_0^{01}, \text{ if } (x,z) \in \Omega_1, \\ \eta_0^{02}, \text{ if } (x,z) \in \Omega_2, \end{cases} \tag{4.47}$$

where ρ_0^{01}, ρ_0^{02}, η_0^{01} and η_0^{02} are functions of time, but do not depend on x and z. Let us substitute the representation (4.46) for the density and viscosity into the first relation in (4.45) and obtain the result

$$\iint_{\Omega} \Im(\eta_1; \psi, \varphi)dxdz + \eta_0^{01} \iint_{\Omega_1} \Im(1; \psi, \varphi)dxdz + \eta_0^{02} \iint_{\Omega_2} \Im(1; \psi, \varphi)dxdz$$

$$= g\left(\iint_{\Omega} \rho_1 D_x\varphi dxdz + \rho_0^{01} \iint_{\Omega_1} D_x\varphi dxdz + \rho_0^{02} \iint_{\Omega_2} D_x\varphi dxdz \right); \tag{4.48}$$

$$\iint_{\Omega} (D_t\rho_1)\vartheta\, dxdz = \iint_{\Omega} D(\rho_1, \psi)\vartheta\, dxdz,$$

$$\iint_{\Omega} (D_t\eta_1)\zeta\, dxdz = \iint_{\Omega} D(\eta_1, \psi)\zeta\, dxdz, \tag{4.49}$$

because $D_t\rho_0 = D_t\eta_0 = D(\rho_0, \psi) = D(\eta_0, \psi) = 0$ in the interior. These equations, together with

$$dx/dt = D_z\psi, \qquad dz/dt = -D_x\psi, \tag{4.50}$$

and with boundary and initial conditions described above define a weak solution for the case of discontinuous density and viscosity.

Approximations of the unknown functions ψ, ρ_1 and η_1 are represented as linear combinations of basic bicubic splines with unknown coefficients (here and below we assume summation over repeated subscripts taking on the following values, $i, k, m = 1, \ldots, I$; $j, l, n = 1, \ldots, J$):

$$\psi = \psi_{ij}(t)\mathrm{s}_i(x)\mathrm{s}_j(z),\ \rho_1 = \rho_{ij}(t)\hat{\mathrm{s}}_i(x)\hat{\mathrm{s}}_j(z),\ \eta_1 = \eta_{ij}(t)\hat{\mathrm{s}}_i(x)\hat{\mathrm{s}}_j(z),$$

where $\mathrm{s}_i(x)$, $\mathrm{s}_j(z)$, $\hat{\mathrm{s}}_i(x)$ and $\hat{\mathrm{s}}_j(z)$ are the basic cubic splines satisfying the required boundary conditions. The curve L is approximated by a polygon whose vertices have coordinates $x_\beta(t), z_\beta(t), \beta = 1, \ldots, B$. These vertices are located on L_0 at $t = t_0$. Let us substitute the above representations into Eqs. (4.48) and (4.49) and integrate forms involving products of basic splines and their derivatives. This results in a set of linear algebraic equations for the unknowns ψ_{ij}, and in a set of ordinary differential equations for ρ_{ij}, η_{ij}, $x(t, x^0, z^0)$ and $z(t, x^0, z^0)$:

$$\psi_{ij}(t)C_{ijkl} = \rho_{ij}(t)F_{ijkl} + \Psi_{kl}(t),$$

$$\frac{\partial \rho_{ij}}{\partial t} G_{ijkl} = \rho_{ij}(t)E_{ijkl}, \qquad \frac{\partial \eta_{ij}}{\partial t} G_{ijkl} = \eta_{ij}(t)E_{ijkl},$$

$$\frac{dx}{dt} = \psi_{ij}(t)s_i(x)\frac{ds_j(z)}{dz}, \qquad \frac{dz}{dt} = -\psi_{ij}(t)\frac{ds_i(x)}{dx}s_j(z). \tag{4.51}$$

Coefficients C_{ijkl} are sums of three terms: $C_{ijkl} = C_{ijkl}^1 + C_{ijkl}^{01} + C_{ijkl}^{01}$, where the first term is obtained from η_1 by substituting its spline representation into the first integral in (4.48), rearranging sums, and integrating products of splines and their derivatives. The result takes the form

$$C^1_{ijkl} = \eta_{mn}\left(4A^{110}_{ikm}B^{110}_{jln} + A^{000}_{ikm}B^{220}_{jln} - A^{200}_{ikm}B^{020}_{jln} - A^{020}_{ikm}B^{200}_{jln} + A^{220}_{ikm}B^{000}_{jln}\right), \tag{4.52}$$

where

$$A^{pqr}_{ikm} = \int_0^1 s_i(x)^{(p)} s_k(x)^{(q)} \hat{s}_m(x)^{(r)} dx, \quad B^{pqr}_{jln} = \int_0^1 s_j(z)^{(p)} s_l(z)^{(q)} \hat{s}_n(z)^{(r)} dz. \tag{4.53}$$

Here $(\ldots)^{(p)}$ denotes the derivative of order p of a function $(\ldots)$ and the zero-order derivative is the function itself. The terms C^{01}_{ijkl} and C^{02}_{ijkl} are obtained by integrating products of splines and their derivatives over regions Ω_1 and Ω_2, which results in the forms

$$C^{01}_{ijkl} = \eta^{01}_0 \iint_{\Omega_1} \Im(1; s_i(x)s_j(z), s_k(x)s_l(z))dxdz,$$

$$C^{02}_{ijkl} = \eta^{02}_0 \iint_{\Omega_2} \Im(1; s_i(x)s_j(z), s_k(x)s_l(z))dxdz. \tag{4.54}$$

We see that elements C^1_{ijkl} depend on the continuous term η_1, but are independent of the curve L. On the other hand, elements C^{01}_{ijkl} and C^{02}_{ijkl} depend on the curve L and on the constants η^{01}_0 and η^{02}_0, but are independent of the continuous term η_1. Coefficients F_{ijkl} in the right-hand side of the first equation in (4.51) are obtained by integration:

$$F_{ijkl} = \int_0^1 (\hat{s}_i(x))^{(1)} s_l(x)dx \int_0^1 \hat{s}_j(z)s_l(z)dz. \tag{4.55}$$

The term Ψ_{kl} is obtained from the last two integrals in the right-hand side of (4.48), where φ is set to $s_k(x)s_l(z)$. The sum of these integrals takes the form

$$\Psi_{kl} = g(\rho^{02}_0 - \rho^{01}_0) \int_L s_k(\xi)s_l(\xi)d\xi, \tag{4.56}$$

as explained in detail by Naimark and Ismail-Zadeh (1995).

Coefficients G_{ijkl} and E_{ijkl} entering the second and third equations in (4.51) are also calculated by integrating the basic splines and their derivatives:

$$G_{ijkl} = \int_0^1 \hat{s}_i(x)\hat{s}_k(x)dx \int_0^1 \hat{s}_j(z)\hat{s}_l(z)dz, \quad E_{ijkl} = \psi_{mn}\left(\hat{A}^{001}_{ikm}\hat{B}^{100}_{jln} - \hat{A}^{100}_{ikm}\hat{B}^{001}_{jln}\right), \tag{4.57}$$

where $\hat{A}^{pqr}_{ikm}$ and $\hat{B}^{pqr}_{jln}$ are obtained from A^{pqr}_{ikm} and B^{pqr}_{jln} in (4.53) with $s_i(x), s_k(x), \hat{s}_m(x), s_j(z), s_l(z)$ and $\hat{s}_n(z)$ replaced by $\hat{s}_i(x), \hat{s}_k(x), s_m(x), \hat{s}_j(z), \hat{s}_l(z)$ and $s_n(z)$, respectively.

The unknowns to be found from (4.51) are the following: $\rho_{ij}(t_s), \eta_{ij}(t_s), \psi_{ij}(t_s), x_\beta(t_s)$, and $z_\beta(t_s), s = 1, 2, \ldots, S$. The second, third, fourth, and fifth relationships in (4.51) constitute the set of ordinary differential equations (ODEs) for unknowns $\rho_{ij}, \eta_{ij}, x_\beta$ and z_β. We solve this set of equations by the fourth-order *Runge–Kutta method*. The right-hand sides of these equations include unknowns ψ_{ij} found from the first set of equations in (4.51). Initial values $\rho_{ij}(t_0)$ and $\eta_{ij}(t_0)$ are derived from the conditions $\rho_1(0,x,z) = \rho_{ij}(0)\hat{s}_i(x)\hat{s}_j(z)$ and $\eta_1(0,x,z) = \eta_{ij}(0)\hat{s}_i(x)\hat{s}_j(z)$ by using spline interpolation.

Let us describe the calculation of the right-hand sides. We assume that the unknowns have been calculated at $t = t_s$ and use Eqs. (4.52)–(4.54) to find the stiffness matrix C_{ijkl} and Eqs. (4.55) and (4.56) to compute the right-hand sides of the first set in (4.51). We solve this set for ψ_{ij} using the *Cholesky method* and use the values so found, together with (4.57), to calculate the right-hand sides of the above ODE.

Coefficients (4.53), (4.55) and (4.57) can be computed once and used in all calculations. Certain difficulties arise in (4.54). The integrals in (4.54) depend on the curve L changing with time. Calculations of forms (4.54) can be reduced to direct integration of polynomials over regions bounded by the curve L and model boundaries; these polynomials are products of splines and their derivatives (see Naimark *et al.*, 1998, for details).

An exact solution of Eqs. (4.48)–(4.50) is unknown, even for the simplest cases and boundary conditions. The numerical approach described in this section was verified by comparing numerical, theoretical (Chandrasekhar, 1961) and experimental (Ramberg, 1968) results from the linear theory of the Rayleigh–Taylor instability (see Naimark *et al.*, 1998). Also the accuracy of the numerical results was compared by Naimark *et al.* (1998) to that of the results obtained by numerical approaches by Christensen (1992) and Naimark and Ismail-Zadeh (1995).

Model example. To illustrate an implementation of the numerical approach, consider a simple evolutionary model of a dense fluid sinking due to gravity into less-dense fluid, which can approximate an evolution of a lithospheric slab. The rectangular domain ($0 \le x \le 2000$ km, $0 \le z \le 700$ km) is filled by a viscous fluid, and a 100 km thick horizontal layer approximating the lithosphere is introduced in the model domain. A small stepwise perturbation is prescribed at the bottom of the layer (see Fig. 4.8). Note that the perturbation is not symmetric with respect to the line $x = 1000$ km. The density of the layer is higher than the density of the ambient fluid (3300 kg m^{-3}) by 3%. The viscosity of the fluid is constant (10^{21} Pa s) in the model domain in experiment 1, whereas the viscosity of the layer is higher than that of the ambient fluid by two orders of magnitude (10^{23} Pa s) in experiment 2.

Figures 4.8 and 4.9 illustrate the evolution of the dense upper layer in experiments 1 and 2, respectively. Because of the Rayleigh–Taylor instability the small perturbation of the dense layer overlying the less-dense fluid gives rise to the descent of the layer at the place of the perturbation. Another two downwellings form later at the lateral boundaries of the model. Once the dense fluid reaches the bottom of the domain it spreads over the lower boundary of the model pushing the less-dense fluid upward. In experiment 1, uprising diapirs evolve at the lower boundary as a result of being pushed by the dense fluid. In experiment 2, the shapes of downwellings distinguishes from that in experiment 1, and the process of descending of the dense layer is slower compared to that in experiment 2.

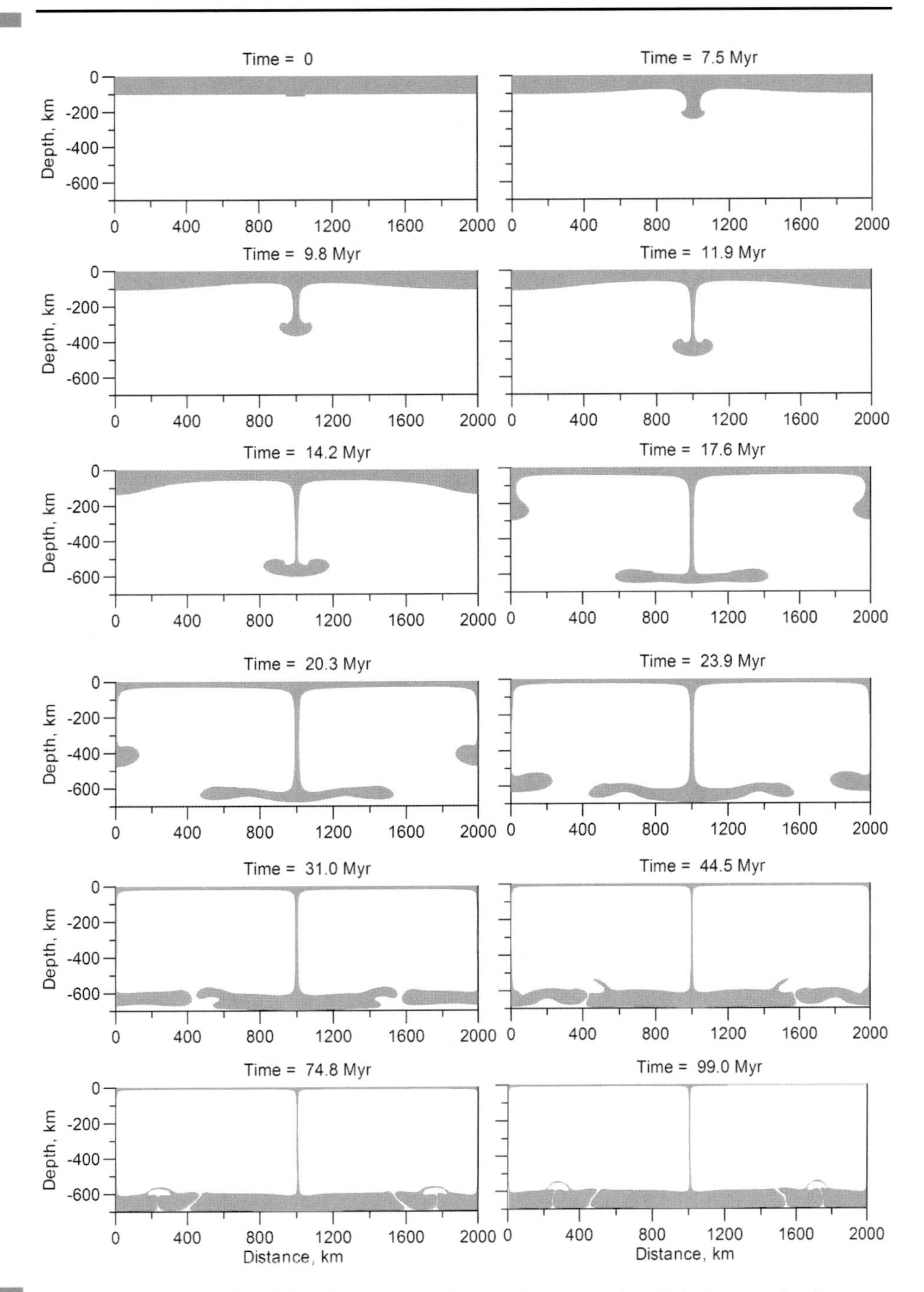

Fig. 4.8. **A model of descending lithosphere in experiment 1 (constant viscosity) at successive times.**

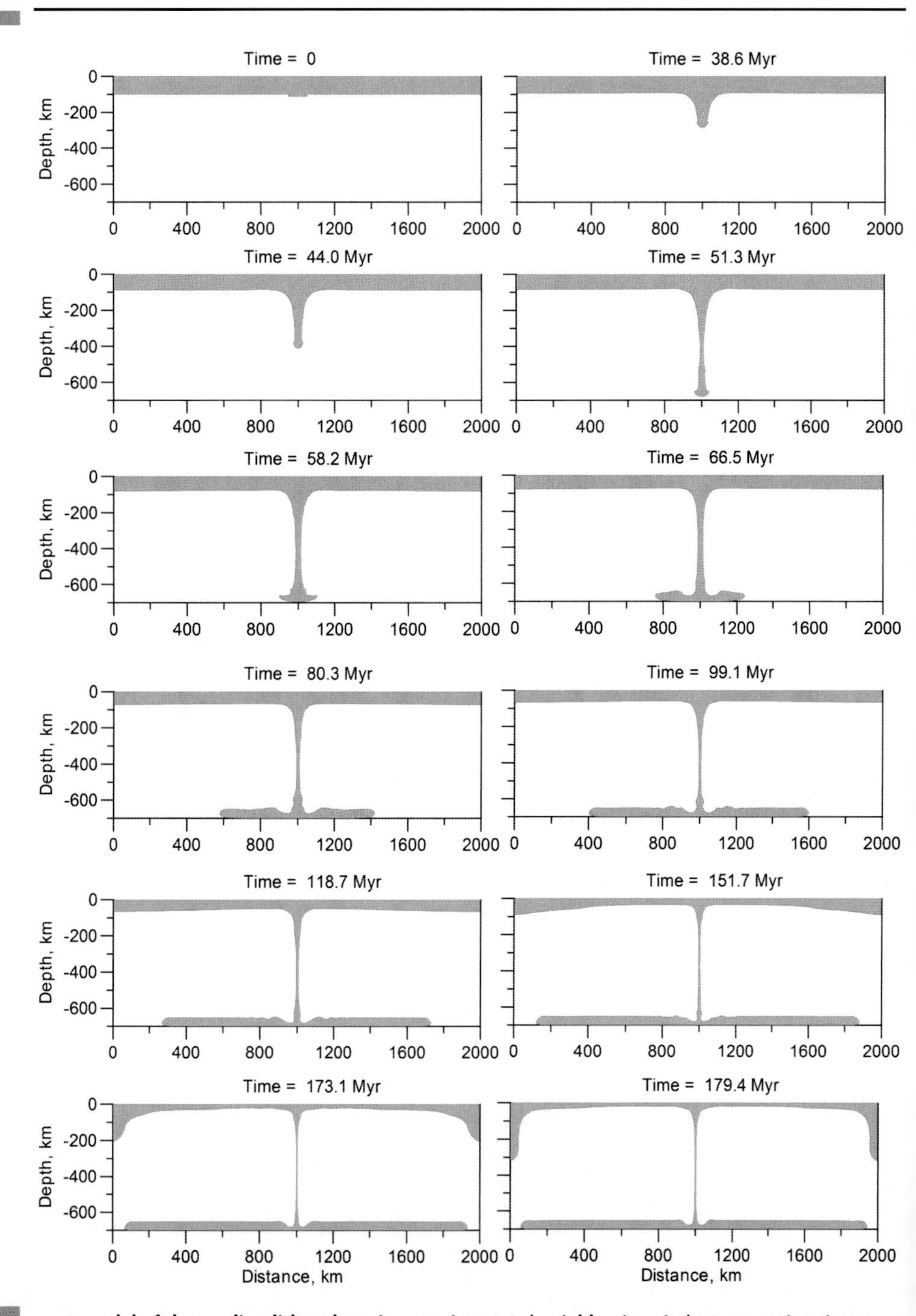

Fig. 4.9. **A model of descending lithosphere in experiment 2 (variable viscosity) at successive times.**

4.10.2 Three-dimensional problem of gravitational advection

Mathematical statement. We consider the problem of the slow flow of an incompressible viscous fluid of variable density and viscosity in the rectangular region $\Omega = (0, x_1 = l_1) \times (0, x_2 = l_2) \times (0, x_3 = l_3)$, where x_1, x_2 and x_3 are the Cartesian coordinates of a spatial point $\mathbf{x}$, and the x_3-axis is pointing upward. The following governing equations describe the flow (Ismail-Zadeh *et al*., 1998; 2001a): momentum conservation

$$\nabla P = \mathrm{div}(\eta \mathbf{E}) + F, \tag{4.58}$$

continuity for incompressible fluid

$$\mathrm{div}\mathbf{u} = \partial u_1/\partial x_1 + \partial u_2/\partial x_2 + \partial u_3/\partial x_3 = 0, \tag{4.59}$$

and advection of density and viscosity with the flow

$$\partial \rho/\partial t + \mathbf{u} \cdot \nabla \rho = 0, \qquad \partial \eta/\partial t + \mathbf{u} \cdot \nabla \eta = 0. \tag{4.60}$$

Equations (4.58)–(4.60) contain the following variables and parameters: time t; velocity $\mathbf{u} = (u_1(t,\mathbf{x}), u_2(t,\mathbf{x}), u_3(t,\mathbf{x}))$; pressure $P = P(t,\mathbf{x})$; density $\rho = \rho(t,\mathbf{x})$; viscosity $\eta = \eta(t,\mathbf{x})$; and the body force per unit volume $F = (0, 0, -g\rho)$, where g is the acceleration due to gravity. Here, ∇, div, and $\mathbf{E}$ denote the gradient operator, divergence operator, and strain rate tensor $\mathbf{E} = \{e_{ij}(\mathbf{u})\} = \{\partial u_i/\partial x_j + \partial u_j/\partial x_i\}$, respectively, and

$$\mathrm{div}(\eta \mathbf{E}) = \left(\sum_{m=1}^{3} \frac{\partial(\eta e_{m1})}{\partial x_m}, \sum_{m=1}^{3} \frac{\partial(\eta e_{m2})}{\partial x_m}, \sum_{m=1}^{3} \frac{\partial(\eta e_{m3})}{\partial x_m} \right). \tag{4.61}$$

Equations (4.58)–(4.60) make up a closed set of equations that determine the unknown $\mathbf{u}$, P, ρ and η as functions of independent variables t and $\mathbf{x}$.

The number of unknowns is reduced by introducing the two-component representation of the velocity potential $\boldsymbol{\Psi} = (\psi_1, \psi_2, \psi_3 = 0)$, from which the velocity is obtained as

$$\mathbf{u} = \mathrm{curl}\, \boldsymbol{\Psi}; \quad u_1 = -\frac{\partial \psi_2}{\partial x_3}, \; u_2 = -\frac{\partial \psi_1}{\partial x_3}, \; u_3 = \frac{\partial \psi_2}{\partial x_1} - \frac{\partial \psi_1}{\partial x_2}. \tag{4.62}$$

The two-component representation of the vector velocity potential (4.62) is computationally advantageous as compared to the representation of the velocity field by scalar poloidal and toroidal potentials (Section 1.3.8). We refer readers to Ismail-Zadeh *et al*. (2001a) for details on the two-component representation of the velocity potential.

Applying the curl operator to (4.58) and using the identities $\mathrm{curl}(\nabla P) = 0$, we derive the following equations from (4.58) and (4.59):

$$\begin{aligned} &D_{2i}(\eta e_{i3}) - D_{3i}(\eta e_{i2}) = g D_2 \rho, \\ &D_{3i}(\eta e_{i1}) - D_{1i}(\eta e_{i3}) = -g D_1 \rho, \\ &D_{1i}(\eta e_{i2}) - D_{2i}(\eta e_{i1}) = 0, \quad i = 1, 2, 3. \end{aligned} \tag{4.63}$$

Hereinafter we assume a summation over repeated subscripts. The strain rate components e_{ij} are defined in terms of the vector velocity potential as

$$\begin{aligned} e_{11} &= -2D_{13}\psi_2, e_{22} = 2D_{23}\psi_1, e_{33} = 2(D_{31}\psi_2 - D_{32}\psi_1), \\ e_{12} &= D_{13}\psi_1 - D_{23}\psi_2,\ e_{13} = D_{11}\psi_2 - D_{33}\psi_2 - D_{12}\psi_1,\ e_{23} \\ &= D_{33}\psi_1 - D_{22}\psi_1 + D_{21}\psi_2. \end{aligned} \tag{4.64}$$

We set the initial time at zero $t_0 = 0$ and assume the density and viscosity to be known at the initial time. On the boundary Γ of Ω, which consists of the faces $x_i = 0$ and $x_i = l_i$ ($i = 1, 2, 3$), we consider the condition of impenetrability with perfect slip:

$$\partial \mathbf{u}_\tau / \partial \mathbf{n} = 0, \quad \mathbf{u} \cdot \mathbf{n} = 0. \tag{4.65}$$

Here, $\mathbf{n}$ is the outward unit normal vector at a point on the boundary Γ, and $\mathbf{u}_\tau$ is the projection of the velocity vector onto the tangent plane at the same point on Γ.

In terms of the vector velocity potential the boundary conditions (4.65) take the following forms:

$$\begin{aligned} &\psi_2 = D_1\psi_1 = D_{11}\psi_2 = 0 && \text{at } \Gamma_1(x_1 = 0) \text{ and } \Gamma_1(x_1 = l_1), \\ &\psi_1 = D_2\psi_2 = D_{22}\psi_1 = 0 && \text{at } \Gamma_2(x_2 = 0) \text{ and } \Gamma_2(x_2 = l_2), \\ &\psi_1 = \psi_2 = D_{33}\psi_1 = 0 && \text{at } \Gamma_3(x_3 = 0) \text{ and } \Gamma_3(x_3 = l_3). \end{aligned} \tag{4.66}$$

Thus, the problem of gravitational advection is to determine functions $\psi_1 = \psi_1(t, \mathbf{x})$, $\psi_2 = \psi_2(t, \mathbf{x})$, $\rho = \rho(t, \mathbf{x})$ and $\eta = \eta(t, \mathbf{x})$ satisfying (4.60) and (4.63) in Ω at $t \geq t_0$, the prescribed boundary (4.66) and the initial conditions.

The Galerkin method. To solve numerically (4.63), we use an Eulerian FEM (Galerkin method) and replace the equations with an equivalent variational equation. Consider any arbitrary admissible test vector function $\mathbf{\Phi} = (\varphi_1, \varphi_2, \varphi_3 = 0)$ satisfying the same conditions as for the vector function $\mathbf{\Psi}$ and multiply the first two equations of Eq. (4.63) by φ_1 and φ_2, respectively. Taking the result and integrating by parts over Ω, and using the boundary conditions for the desired and test vector functions, we obtain the variational equation

$$\begin{aligned} &\aleph(\eta; \mathbf{\Psi}, \mathbf{\Phi}) = \Re(\eta, \rho; \mathbf{\Phi}), \\ &\aleph(\eta; \mathbf{\Psi}, \mathbf{\Phi}) = \iiint_\Omega \eta[2e_{11}\tilde{e}_{11} + 2e_{22}\tilde{e}_{22} + 2e_{33}\tilde{e}_{33} + e_{12}\tilde{e}_{12} + e_{13}\tilde{e}_{13} + e_{23}\tilde{e}_{23}]d\mathbf{x}, \\ &\Re(\eta, \rho; \mathbf{\Phi}) = \iiint_\Omega g\rho\left(\frac{\partial \varphi_1}{\partial x_2} - \frac{\partial \varphi_2}{\partial x_1}\right) d\mathbf{x}, \end{aligned} \tag{4.67}$$

and the expressions for $\tilde{e}_{ij}$ in terms of $\mathbf{\Phi}$ are identical to the expressions for e_{ij} in terms of the function $\mathbf{\Psi}$.

We represent the components of the vector velocity potential as a sum of tricubic splines ω^s_{ijk}

$$\psi_s(t,\mathbf{x}) \approx \psi^s_{ijk}(t)\omega^s_{ijk}(\mathbf{x}), \quad s = 1,2 \tag{4.68}$$

with the unknown functions $\psi^s_{ijk}(t)$. Hereinafter, we take $i,l,p = 1,2,\ldots,N_1; j,m,q = 1,2,\ldots,N_2$; and $k,n,r = 1,2,\ldots,N_3$. Density and viscosity are approximated by linear combinations of appropriate trilinear basis functions:

$$\rho(t,\mathbf{x}) \approx \rho_{ijk}(t)\tilde{s}^1_i(x_1)\tilde{s}^2_j(x_2)\tilde{s}^3_k(x_3), \quad \eta(t,\mathbf{x}) \approx \eta_{ijk}(t)\tilde{s}^1_i(x_1)\tilde{s}^2_j(x_2)\tilde{s}^3_k(x_3), \tag{4.69}$$

where $\tilde{s}^1_i(x_1)$, $\tilde{s}^2_j(x_2)$ and $\tilde{s}^3_k(x_3)$ are linear basis functions. The trilinear basis functions provide good approximations for step functions (such as density or viscosity that change abruptly from one layer to another).

Substituting approximations (4.68)–(4.69) into the variational equation (4.67) we arrive at a system of linear algebraic equations (SLAE) for the unknown $\psi^s_{ijk}(t)$, which defines a positive definite band stiffness matrix:

$$\psi^s_{ijk}C^{lmn}_{sijk}(\eta_{ijk}) = g\rho_{ijk}F^{lmn}_{ijk}. \tag{4.70}$$

The coefficients C^{lmn}_{sijk} and F^{lmn}_{ijk} in (4.70) are the integrals of various products of cubic splines and their derivatives. Namely,

$$C^{lmn}_{sijk} = \sum \eta_{pqr} w_{a_1a_2b_1b_2c_1c_2} A^{a_1a_2}_{silp} B^{b_1b_2}_{sjmq} C^{c_1c_2}_{sknr}, \tag{4.71}$$

where the sum is taken over all non-negative integers a_1, a_2, b_1, b_2, c_1, and c_2 such that each of them does not exceed 2 and $a_1 + a_2 + b_1 + b_2 + c_1 + c_2 = 4$. The values of $w_{a_1a_2b_1b_2c_1c_2}$ are readily obtained by collecting similar terms in the sums. Coefficients $A^{a_1a_2}_{silp}$, $B^{b_1b_2}_{sjmq}$ and $C^{c_1c_2}_{sknr}$ are integrals of the form

$$A^{a_1a_2}_{silp} = \int_0^{l_1} \left(D_{a_1}\gamma^s_i(x_1)\right)\left(D_{a_2}\gamma^s_l(x_1)\right)\tilde{s}^1_p(x_1)dx_1,$$

$$B^{b_1b_2}_{sjmq} = \int_0^{l_2} \left(D_{b_1}\zeta^s_j(x_2)\right)\left(D_{b_2}\zeta^s_m(x_2)\right)\tilde{s}^2_q(x_2)dx_2,$$

$$C^{c_1c_2}_{sknr} = \int_0^{l_3} \left(D_{c_1}\vartheta^s_k(x_3)\right)\left(D_{c_2}\vartheta^s_n(x_3)\right)\tilde{s}^3_r(x_3)dx_3, \tag{4.72}$$

where $\{\mu\}$, $\{\zeta\}$ and $\{\vartheta\}$ are cubic splines and $\{\tilde{s}\}$ are linear basis functions. Coefficients F_{ijk}^{lmn} take the following forms:

$$
\begin{aligned}
F_{ijk}^{lmn} &= P_{il}^{01} Q_{jm}^{00} R_{kn}^{00} - P_{il}^{00} Q_{jm}^{01} R_{kn}^{00}, \\
P_{il}^{ab} &= \int_0^{L_1} \left(D_a \tilde{s}_i^1(x_1) \right) \left(D_b \mu_l^1(x_1) \right) dx_1, \\
Q_{jm}^{ab} &= \int_0^{L_2} \left(D_a \tilde{s}_j^2(x_2) \right) \left(D_b \zeta_m^1(x_2) \right) dx_2, \\
R_{kn}^{ab} &= \int_0^{L_3} \left(D_a \tilde{s}_k^3(x_3) \right) \left(D_b \vartheta_n^1(x_3) \right) dx_3.
\end{aligned} \tag{4.73}
$$

The SLAE is solved by the *conjugate gradient method* designed specially for multi-processor computers (Golub and Van Loan, 1989). Approximations of the density and viscosity for a prescribed velocity can be computed by the *method of characteristics*, i.e. by advecting the initial density and viscosity along the characteristics of (4.48).

The accuracy of the numerical method was tested by Ismail-Zadeh *et al.* (1998, 2001a) using the analytical solution to the coupled Stokes and density advection equations (Truskov, 2002), and verifying the conservation of mass at each time step, and the accuracy of the vector velocity potential $\boldsymbol{\Psi}$.

Model example. We show the implementation of the method described here on a model of viscous flow in the crust, namely, a model of the salt diapirism. The model domain is a rectangular region ($l_1 = l_2 = 30$ km, $l_3 = 10$ km) divided into $38 \times 38 \times 38$ rectangular

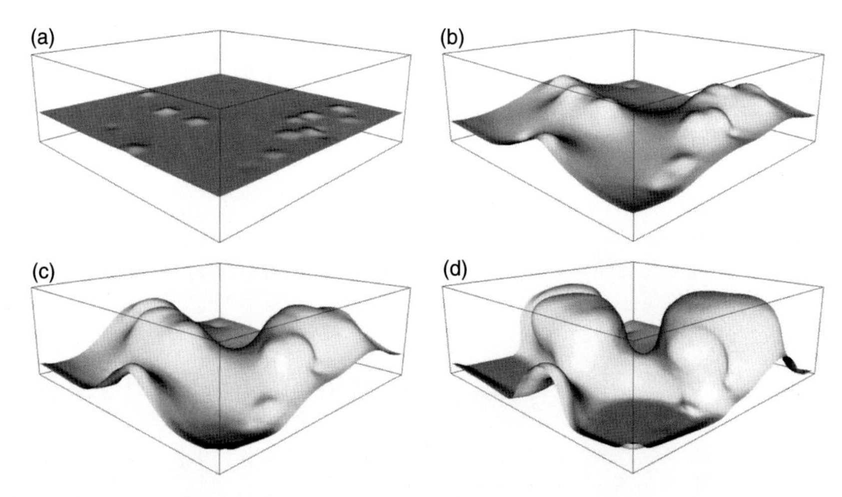

Fig. 4.10. A model of the evolution of salt diapirs toward increasing maturity. Interfaces between salt and its overburden are presented at successive times: initial position (a), after 17.7 Myr (b), 19.2 Myr (c), and 21.3 Myr (d).

elements in order to approximate the vector velocity potential and viscosity. Density is represented on a grid three times finer, $112 \times 112 \times 112$. The model viscosities and densities are assumed to be 10^{20} Pa s and 2.65×10^3 kg m^{-3} for the overburden layer and 10^{18} Pa s and 2.24×10^3 kg m^{-3} for the salt layer, respectively.

A rise of salt diapirs was modelled through an overburden deposited prior to the interface perturbation. A salt layer of 3 km thickness at the bottom of the model box is overlain by a sedimentary overburden of 7 km thickness at time $t = 0$. The interface between the salt and its overburden was disturbed randomly with an amplitude ~100 m. Figure 4.10 (a–d) shows the positions of the interface between salt and overburden in the model at successive times over a period of about 21 My. The evolution clearly shows two major phases: an initial phase resulting in the development of salt pillows lasting about 18 Myr (a, b) and a mature phase resulting in salt dome evolution lasting about 3 Myr (c, d).

4.11 FE solution refinements

As we could observe from the previous sections, finite element solutions are approximate solutions to the exact solutions of mathematical problems. Here we discuss the methods by which the FE solution results could be made more accurate, reducing the errors once a FE solution has been obtained. As the process depends on previous results, it is called *adaptive*. Such adaptive methods were introduced to FE calculations by Babuska and Rheinboldt (1979).

Various procedures exist for the refinement of FE solutions. Broadly these fall into two categories.

(1) The h-refinement in which the same class of elements continue to be used but they are changed in size. In some locations, elements are made larger and in others made smaller, in order to provide maximum economy in reaching the desired solution.
(2) The p-refinement in which we continue to use the same element size and simply increase, generally hierarchically, the order of the polynomial used in their definition (see Section 4.9).

Each of the two categories can be subdivided into typical methods. Namely, for h-refinement we should mention (i) the method of element subdivision (enrichment), which is based on the division of existing elements into smaller ones keeping the original element boundaries intact; (ii) the method of a complete mesh regeneration or remeshing; and (iii) r-refinement, which keeps the total number of nodes constant and adjusts their position to obtain an optimal approximation.

With p-refinement, the situation is different. There are two subclasses of the refinement: (i) an increase of the polynomial order uniformly throughout the whole solution domain: and (ii) an increase of the polynomial order locally using hierarchical refinement. For neither of these approaches has a direct procedure been developed that allows the prediction of the best refinement to be used to obtain a given error. The procedures generally require more resolutions and tend to be more costly. However, the convergence for a given number of variables is more rapid with the p-refinement and it has much to recommend it. There also exists the hp-refinement in which both methods of adaptivity are combined.

So far the mantle dynamics community has nearly always used meshes that are spatially uniform and do not change with time. Grid refinement, particularly adaptive grid refinement, offers several advantages to this, as demonstrated in 2-D and 3-D finite element codes by Davies (2008) and Davies *et al.* (2007), and is a promising technology to pursue in the future. Readers are also referred to Honda (1996) and Burstedde *et al.* (2008), as examples of the application of adaptive grid refinement procedures to geodynamic problems.

4.12 Concluding remarks

In this chapter we have introduced the basic elements of the FE method and presented a few cases of FE approximations of mathematical problems used to numerically model geodynamic processes. Finally, we summarise the basic steps, which are involved in FE analysis of a numerical model.

Step 1. Pre-processing phase

- Select the Lagrangian or Eulerian formulation (sometimes ALE formulation) for FE modelling.
- Create and discretise the solution domain into finite elements; that is, partition the problem domain into nodes and elements. When the Lagrangian FE method is employed, the discretisation of the model domain is sometimes not a simple task, and modern software can assist users to create a complex FE domain.
- Select basis functions, develop equations for elements, and construct the stiffness matrix considering boundary and initial conditions and the vector of external forces (function $f(x)$ at the nodes).

Step 2. Solution phase

- Solve a set of linear (or non-linear) algebraic equations simultaneously to obtain the solution (such as, e.g. velocity) to the discrete equations.

Step 3. Post-processing phase

- Store and display (visualise) solution information.
- Obtain other important information, that is, calculate additional quantities. At this point, you may be interested in values of principal stresses, heat fluxes, *a posteriori* error estimates, etc.

It is important to note that the basic steps involved in any FE analysis, regardless of how you generate the FE model, will be the same as those listed above.

5 Spectral methods

5.1 Introduction

Spectral methods have been widely used in geophysical modelling of different fluids including the atmosphere, ocean, outer core and mantle. Variables are expanded as a sum of orthogonal global basis functions, typically trigonometric or polynomial, in contrast to finite difference, finite element and finite volume methods in which the basis functions are local. The convergence of the method is faster than spatial methods, meaning that high mathematical accuracy can be obtained with relatively few basis functions when representing smoothly varying fields, although sharp gradients or discontinuities can cause problems. Spherical geometry is easily treated by using spherical harmonics as basis functions, giving approximately uniform resolution over the sphere.

Basis functions are typically different in the horizontal (azimuthal) and vertical (radial) directions because of the differing boundary conditions (side boundaries are often periodic). In the vertical (radial) direction, Chebyshev polynomials or finite differences are typically used. Once expanded in harmonics, spatial derivatives are given by exact analytic expressions. For linear equations, the equations for different harmonics decouple in spectral space, and each mode can be solved independently. The method is thus ideally suited for equations in which the coefficients in front of dependent variables (e.g. viscosity, thermal diffusivity, wave velocity) are spatially constant. Non-linear products such as advection terms are typically calculated in spatial (grid) space, and fast transformation of variables between grid space and spatial space is possible if the Fast Fourier Transform (FFT) algorithm is used. Even so, the execution time does not scale optimally with problem size as it does with multigrid methods (Section 6.4). Nevertheless, spectral methods are competitive for problems with up to millions of grid points, if physical properties vary only in the radial direction.

The popularity of spectral methods in the solid Earth geodynamics (e.g. mantle convection) community has declined in recent years, probably because of their limited ability to handle lateral variations in viscosity, but they are still dominant in the geodynamo modelling community owing to the natural treatment of the magnetic boundary conditions.

5.2 Basis functions and transforms

5.2.1 Overview

For periodic domains, it is common to use trigonometric functions as basis functions (i.e. a Fourier series), whereas in non-periodic domains (such as the radial direction) Chebyshev or Legendre polynomials are preferred. For spherical domains, spherical harmonics are the natural choice, and these are a combination of trigonometric and associated Legendre functions. In many geodynamic codes, the spectral expansion is performed only in the horizontal (azimuthal) directions and a grid-based discretisation such as finite differences is used in the vertical (radial) direction.

5.2.2 Trigonometric

For periodic domains, Fourier series are a natural choice. Non-periodic domains with zero or zero-gradient boundary conditions can also be treated by using sine or cosine expansions respectively. For the periodic case, representation in complex numbers can be used:

$$f(x) = \sum_{n=0}^{N} F_n \exp(i2\pi nx/L), \tag{5.1}$$

where $f(x)$ is a complex function (although in most geophysical applications it will be real), L is the periodic domain length (hence the fundamental wavelength of the system), N is the maximum frequency number used, and F_n are complex coefficients.

Derivatives are easily calculated in spectral space, for example:

$$\frac{df(x)}{dx} = \sum_{n=0}^{N} \frac{i2\pi nF_n}{L} \exp(i2\pi nx/L). \tag{5.2}$$

In the discretised version, f is known at N equally spaced grid points $x_k = kL/N$, $k = 0, 1, 2, \ldots, N-1$ (the maximum k is $N-1$ rather than N because of the periodicity $x_N = x_0$). The spectral coefficients can be calculated by the discrete Fourier transform (DFT):

$$F_n = \sum_{k=0}^{N-1} f_k \exp(2\pi ikn/N), \quad n = 0, 1, 2, \ldots, N-1, \tag{5.3}$$

and the grid points can be uniquely recovered by the inverse DFT:

$$f_k = \frac{1}{N} \sum_{n=0}^{N-1} F_n \exp(-2\pi ikn/N). \tag{5.4}$$

The maximum spatial frequency that can be represented has one oscillation every two grid points, i.e. $n = N/2$, and is known as the Nyquist frequency. Frequencies $n = N/2 + 1, \ldots, N-1$ are often viewed as 'negative' frequencies, equivalent to $n = -(N/2-1), \ldots, -1$, respectively. If f is real then there is half as much information content as for complex f, the result of which is that the coefficients for negative frequencies are the complex conjugate of the coefficients for positive frequencies, so only the positive frequencies $n = 0, \ldots, N/2$ need to be calculated and stored; furthermore F_0 and $F_{N/2}$ are real numbers.

A Fourier expansion has the advantage that a fast transform exists between spatial and grid space, the Fast Fourier Transform (FFT), for which the number of operations scales as $N \log N$, where N is the number of points (for details see Press *et al.* (2007) and for a modern implementation see Frigo and Johnson (2005)). This works best for N a power of two, although other small number factors are also possible. Special versions of this for real functions, or for sine or cosine series, are available.

5.2.3 Chebyshev polynomials

Chebyshev polynomials of the first kind are commonly used for representing non-periodic domains, such as exists in the vertical (radial) direction of the mantle. This is because of their excellent convergence properties and because the resolution becomes higher near the boundaries, which is good for resolving thermal or mechanical boundary layers near boundaries. Although there are several ways to write them, Chebyshev polynomials are most conveniently written by using trigonometic functions as:

$$T_n(x) = \cos\left(n \arccos x\right), \tag{5.5}$$

where x is in the range $(-1, 1)$ and $n \geq 0$. A function $f(x)$ can then be expanded as:

$$f(x) = \sum_{n=0}^{N} F_n T_n(x). \tag{5.6}$$

Because of their relationship to cosines, the appropriate grid points to use are evenly spaced in arccos space, i.e.

$$x_j = \cos \frac{\pi j}{N}, \quad j = 0, 1, 2, \ldots, N. \tag{5.7}$$

Although Chebyshev polynomials are quite different from cosines, from one perspective they can be viewed as cosines 'in disguise', because when viewed from the perspective of the unevenly spaced grid points, they appear to be cosines. Accordingly, a cosine FFT can be used to transform fields between spectral and grid space. To illustrate their grid refinement property: if an expansion up to $N = 64$ is used to represent the mantle depth of 2890 km, then the grid spacing ranges from 141.5 km in the centre of the mantle to 3.5 km next to the boundaries.

Despite their apparent similarity to cosines, derivatives are more difficult to evaluate in spectral space, and a recursion relationship must be used. If we write the derivative of the above function as (noting the $N - 1$ maximum order):

$$\frac{\partial f}{\partial x} = \sum_{n=0}^{N-1} F'_n T_n(x), \tag{5.8}$$

then from the relationship

$$2T_n(x) = \frac{1}{n+1}\frac{d}{dx}T_{n+1}(x) - \frac{1}{n-1}\frac{d}{dx}T_{n-1}(x) \tag{5.9}$$

the coefficients for the derivative can be found with this recursion relationship:

$$\begin{aligned} F'_N &= 0,\ F'_{N-1} = 2NF_N, \\ F'_{n-1} &= F'_{n+1} + 2nF_n,\ n = 2, 3, \ldots, N-1, \\ F'_0 &= \frac{1}{2}F'_2 + F_1. \end{aligned} \tag{5.10}$$

As it will be seen later, derivatives of Chebyshev polynomials are often required at particular points. A convenient way of calculating these from the functions already calculated is by using the relationships:

$$\begin{aligned} T'_{n+1}(z) &= 2zT'_n(z) + 2T_n(z) - T'_{n-1}(z),\ T'_0(z) = 0,\ T'_1(z) = 1, \\ T''_{n+1}(z) &= 2zT''_n(z) + 4T'_n(z) - T''_{n-1}(z),\ T''_1(z) = 0,\ T''_2(z) = 0. \end{aligned} \tag{5.11}$$

5.2.4 Spherical harmonics

Spherical harmonics are solutions of Laplace's equation on a sphere and are ideal for expansion of data on a spherical surface. They satisfy

$$\nabla_h^2 Y_\ell^m(\theta,\phi) = -\ell(\ell+1)Y_\ell^m(\theta,\phi) \tag{5.12}$$

on the unit sphere (i.e. with radius = 1), where ∇_h^2 is the azimuthal (θ, ϕ) component of the Laplacian on a spherical surface. Spherical harmonics Y_ℓ^m consist of a Fourier expansion in the ϕ direction and associated Legendre functions P_ℓ^m in the θ direction:

$$Y_\ell^m(\theta,\phi) = \sqrt{\frac{2\ell+1}{4\pi}\frac{(\ell-m)!}{(\ell+m)!}}P_\ell^m(\cos\theta)\exp(im\phi), \tag{5.13}$$

where ℓ is the degree, m is the order (from $-\ell$ to $+\ell$), ϕ is longitude and θ is colatitude. The degree ℓ can be thought of as the total number of cycles (oscillations) over the sphere, while m is the number of cycles in longitude. The number of cycles in latitude is $(\ell - m)$. The normalisation factor in Eq. (5.13) is such that the integral over the sphere of the spherical harmonic multiplied by its complex conjugate is equal to 1. Other normalisation conventions exist so care must be taken when using coefficients obtained from another source.

A field on a spherical surface can be expanded as:

$$f(\theta,\phi) = \sum_{\ell=0}^{L}\sum_{m=-\ell}^{\ell} F_\ell^m Y_\ell^m(\theta,\phi), \tag{5.14}$$

where L is the maximum spherical harmonic degree and F_ℓ^m are coefficients. Both f and F are complex. If f is real, then $F_\ell^{-m} = (-1)^m (F_\ell^m)^*$, where the $*$ denotes complex conjugate, so only the coefficients for $m \geq 0$ need to be calculated and stored. The transform between grid and spectral space is given by

$$F_\ell^m = \int_0^{2\pi} \int_0^{\pi} f(\theta, \phi) Y_\ell^{m*}(\theta, \phi) \sin\theta d\theta d\phi. \tag{5.15}$$

Grid points. The appropriate grid points for representing the discretised version of f are $2L$ evenly spaced points in the ϕ direction, whereas in the θ direction the Gaussian quadrature points for the Legendre integrals should be used. These quadrature points are the zeros of the Legendre polynomial of degree L. As an example, a spherical harmonic expansion up to $L = 128$ would map to 256 ϕ-points by 128 θ-points, with the θ-points at the zeros of Legendre polynomial P_{128}. Degree L is the Nyquist frequency on this grid so the sine components do not exist; this degree is often neglected and the expansion truncated at one degree less (e.g. Glatzmaier, 1988). Because the ϕ transform can be performed using the FFT algorithm L is typically chosen to be a power of two.

Transform. When transforming fields between grid space and spectral space or back, use can be made of the FFT in the ϕ direction. The transform process from grid space to spectral space is:

- use the FFT algorithm to transform from (θ, ϕ) to (θ, m);
- Legendre transform from (θ, m) to (ℓ, m).

The reverse sequence is used to transform from spectral space to grid space. Legendre transforms from intermediate (θ, m) space to (ℓ, m) are evaluated by Gaussian quadrature:

$$F_\ell^m = \sum_{i=0}^{L-1} w_{i\ell}^m \tilde{F}_i^m, \tag{5.16}$$

where $w_{i\ell}^m$ are the relevant coefficients, which are products of Gaussian quadrature weights and associated Legendre polynomials. In the opposite direction a similar form is used:

$$\tilde{F}_i^m = \sum_{i=0}^{L-1} y_{i\ell}^m F_\ell^m, \tag{5.17}$$

where $y_{i\ell}^m$ are values of $Y_\ell^m(\theta_i, 0)$. These transforms are typically performed using a matrix multiplication. While this does not scale well with the number of grid points, the number of operations can be minimised by noting that

- if the fields are real, only the coefficients for $m \geq 0$ need to be calculated, as mentioned above;
- the computation time for the spectral to grid transform can be reduced by almost 50% by noting that half of the Y_ℓ^m are symmetric about the equator ($\theta = \pi/2$) while the other half are antisymmetric.

If the symmetric and anti-symmetric parts are calculated separately, then the total field for $0 \leq \theta \leq \pi/2$ is constructed by adding them, whereas the total field for $\pi/2 \leq \theta \leq \pi$ is constructed by subtracting them. Although some fast Legendre transform methods have been proposed, they have not yet proven useful in geodynamo or mantle applications (Lesur and Gubbins, 1999), although there have been some recent advances in this area (Spotz and Swarztrauber, 2001; Healy *et al.*, 2004).

Derivatives. Derivatives can be evaluated very accurately in spectral space, which is one of the advantages of the method. While the ϕ-derivative is straightforward:

$$\frac{\partial}{\partial \phi} Y_\ell^m = -imY_\ell^m, \tag{5.18}$$

the θ-derivative couples harmonics with different degrees. Various useful identities are available including:

$$\begin{aligned} \sin\theta \frac{\partial}{\partial \theta} Y_\ell^m &= \ell C_{\ell+1}^m Y_{\ell+1}^m - (\ell+1) C_\ell^m Y_{\ell-1}^m, \\ \frac{\partial}{\partial \theta}\left(Y_\ell^m \sin\theta\right) &= (\ell+1) C_{\ell+1}^m Y_{\ell+1}^m - \ell C_\ell^m Y_{\ell-1}^m, \\ C_\ell^m &= \left(\frac{(\ell+m)(\ell-m)}{(2\ell+1)(2\ell-1)}\right)^{1/2}. \end{aligned} \tag{5.19}$$

For second derivatives:

$$\sin\theta \frac{\partial}{\partial \theta}\left(\sin\theta \frac{\partial Y_\ell^m}{\partial \theta}\right) = Y_\ell^m \left(m^2 - \ell(\ell+1)\sin^2\theta\right), \tag{5.20}$$

which is another way of writing Eq. (5.12).

The above information refers to scalar spherical harmonics. For treating vector or tensor fields directly, it can be more practical to use *generalised spherical harmonics*, as used by, for example, Ricard and Vigny (1989) and Forte and Peltier (1991); for details the reader is referred to Phinney and Burridge (1973) and Jones (1985).

5.3 Solution methods

5.3.1 Poisson's equation

We start with an example, for which a solution can be obtained trivially. In two dimensions:

$$\frac{\partial^2 \psi}{\partial x^2} + \frac{\partial^2 \psi}{\partial z^2} = r(x,z), \tag{5.21}$$

where ψ is an unknown field and r is a known source distribution. The first step is to choose a spectral expansion for r and ψ. If the boundary conditions happen to be periodic, then a

Fourier expansion works well:

$$r(x,z) = \sum_{\ell}\sum_{m} R_{\ell m} \exp\left[i(k_\ell x + k_m z)\right],$$
$$\psi(x,z) = \sum_{\ell}\sum_{m} \Psi_{\ell m} \exp\left[i(k_\ell x + k_m z)\right], \tag{5.22}$$

where $R_{\ell m}$ and $\Psi_{\ell m}$ are the spectral coefficients; $k_\ell = 2\pi\ell/L_x$, $k_m = 2\pi m/L_z$, and L_x and L_z are the domain width and height, respectively. By substituting these into the original equation, a set of equations for the spectral coefficients is obtained. Each mode is perfectly decoupled, resulting in one equation for each (ℓ, m):

$$-\left(k_\ell^2 + k_m^2\right)\Psi_{\ell m} = R_{lm}, \tag{5.23}$$

hence

$$\Psi_{\ell m} = -\frac{R_{\ell m}}{k_\ell^2 + k_m^2}. \tag{5.24}$$

The facts that the basis functions fit the boundary conditions and that the derivatives of the basis functions do not involve any other frequencies, allowed this perfect decoupling. Boundary conditions of zero, or of zero gradient, can also be fit using sine or cosine expansions, respectively. The solution algorithm is thus rather simple.

(1) Start with known $r(x,y)$.
(2) Perform Fast Fourier Transform to obtain $R_{\ell m}$.
(3) Calculate $\Psi_{\ell m}$ using equation (5.24).
(4) Inverse Fast Fourier Transform to obtain the solution ψ.

5.3.2 Galerkin, Tau and pseudo-spectral methods

For less straightforward problems, for example ones in which the basis functions do not fit the boundary conditions in all directions, a more complicated method must be used to obtain a system of discretised equations to solve for the spectral coefficients. The choice to be made is where the equations are forced to be satisfied, somewhat analogous to the choice of weight function in the finite element method. In the spectral method, there are three common choices of weight function. These may also be thought of in terms of the residue, i.e. the error in satisfying the equations.

In the *Galerkin method*, new basis functions that satisfy the boundary conditions are constructed from the initial basis functions. As with the Galerkin method applied to finite elements, the discretised equations are integrated over the domain using the new basis functions as weight functions, and the residue is required to be zero, leading to a set of linear discretised equations to be solved. This approach requires that the residue be orthogonal to the new basis functions.

In the *Tau method*, the original basis functions are used and the boundary conditions appear as separate equations. The discretised equations are integrated over the domain by using the basis functions as weight functions. This is equivalent to requiring the residue to be orthogonal to the basis functions.

In the *collocation* or *pseudo-spectral* method, the equations are required to be satisfied at a number of grid points, and the boundary conditions are required to be satisfied at the boundary points.

In recent geodynamic spectral codes the pseudo-spectral method is almost always used, so the rest of the chapter focuses on this approach.

5.4 Modelling mantle convection

5.4.1 Constant viscosity, three-dimensional Cartesian geometry

In order to illustrate by using straightforward algebra how the pseudo-spectral method can be applied to model mantle convection, solution of the simplest set of equations is detailed in this section: those of infinite Prandtl number, Boussinesq convection with constant physical properties in Cartesian geometry. Subsequent sections review how the approach is modified to treat various physical complexities that would be needed in a modern research code.

The governing dimensionless equations describing conservation of mass, momentum and energy are presented in the form:

$$\begin{aligned} \nabla \cdot \mathbf{v} &= 0, \\ -\nabla P + \nabla^2 \mathbf{v} &= -Ra\, \Theta \mathbf{e}, \\ \frac{\partial\, \Theta}{\partial t} &= \nabla^2 \Theta - \mathbf{v} \cdot \nabla \Theta + Q, \end{aligned} \tag{5.25}$$

where $\mathbf{v}$ is velocity, P is pressure, Ra is the Rayleigh number, Θ is temperature, t is time, Q is internal heating rate and $\mathbf{e}$ is a unit vector in the vertical direction. Here the domain is taken to have a depth of 1.0 with impermeable top and bottom boundaries, and be periodic in the x- and y-directions with lengths L_x and L_y respectively.

Because there is no time-derivative in the momentum equation, the problem may be split into two steps: (i) solution of the momentum and continuity equations for velocity and pressure for a given temperature field, then (ii) time-stepping the temperature field for a given velocity field. Alternatively, the two steps may be coupled, but here the decoupled approach is taken for simplicity.

A way of simplifying the above equations is to express the velocity field in terms of poloidal and toroidal potentials,

$$\mathbf{v} = \nabla \times \nabla \times (W\mathbf{e}) + \nabla \times (Z\mathbf{e}), \tag{5.26}$$

where W is the poloidal potential and Z is the toroidal potential. As discussed in Section 1.3.8, the continuity equation is thus eliminated and in the case of homogeneous boundary conditions and laterally constant viscosity the toroidal term is zero, allowing the momentum equation to be reduced to the simple form:

$$\nabla^4 W = Ra\,\Theta. \tag{5.27}$$

The pressure has been eliminated, so the number of variables has been reduced from four (pressure and three velocity components) to one. Θ and W can now be expanded horizontally in Fourier series using coefficients that vary in the z-direction:

$$\Theta(x,y,z) = \sum_{m=0}^{M}\sum_{n=0}^{N} \tilde{\Theta}_{mn}(z)\exp\left[i\left(k_m x + k_n y\right)\right],$$

$$W(x,y,z) = \sum_{m=0}^{M}\sum_{n=0}^{N} \tilde{W}_{mn}(z)\exp\left[i\left(k_m x + k_n y\right)\right], \tag{5.28}$$

where M and N are the maximum frequencies in the x- and y-directions, respectively, $k_m = 2\pi m/L_x$ and $k_n = 2\pi n/L_y$.

Because Eq. (5.27) is linear, each (m,n) decouples and can be treated independently, leading to $(M+1)(N+1)$ one-dimensional ordinary differential equations (ODEs) in the z-direction. Herein lies one of the major advantages of the spectral method: reducing a three-dimensional problem to a large number of one-dimensional problems that are relatively easy and quick to solve. The momentum equation for one (m,n) can be written as:

$$\left(\frac{\partial^2}{\partial r^2} - \left(k_m^2 + k_n^2\right)\right)\left(\frac{\partial^2}{\partial r^2} - \left(k_m^2 + k_n^2\right)\right)\tilde{W}_{mn}(z) = Ra\,\tilde{\Theta}_{mn}(z). \tag{5.29}$$

The next step is to determine the appropriate boundary conditions for W. The velocity components can be obtained from Eq. (5.26) by noting that $\mathbf{v} = \nabla\times\nabla\times(W\mathbf{e}) = \nabla\nabla\cdot(W\mathbf{e}) - \nabla^2(W\mathbf{e}) = \nabla(\partial W/\partial z) - \mathbf{e}\nabla^2 W$, and are given by:

$$v_x(x,y,z) = \sum_{m=0}^{M}\sum_{n=0}^{N} \frac{\partial \tilde{W}_{mn}}{\partial z} ik_m \exp\left[i\left(k_m x + k_n y\right)\right],$$

$$v_y(x,y,z) = \sum_{m=0}^{M}\sum_{n=0}^{N} \frac{\partial \tilde{W}_{mn}}{\partial z} ik_n \exp\left[i\left(k_m x + k_n y\right)\right],$$

$$v_z(x,y,z) = \sum_{m=0}^{M}\sum_{n=0}^{N} \left(k_m^2 + k_n^2\right)\tilde{W}_{mn}(z)\exp\left[i\left(k_m x + k_n y\right)\right]. \tag{5.30}$$

Hence, impermeable upper and lower boundaries imply that $\tilde{W}_{mn} = 0$, and free-slip (shear stress free) means that $\partial^2\tilde{W}_{mn}/\partial z^2 = 0$. If rigid (no-slip), the appropriate condition would be $\partial\tilde{W}_{mn}/\partial z = 0$. The meaning of these boundary conditions is discussed in Section 1.4.

It is most convenient to solve Eq. (5.27) in two Poisson steps, so that the maximum derivative is second order rather than fourth order, i.e. solve

$$\nabla^2 H = Ra\, \Theta, \quad \nabla^2 W = H, \tag{5.31}$$

where H is an intermediate field.

The problem is now to solve Poisson's equation for a general field S given a source term R, with boundary conditions $S = 0$:

$$\left(\frac{\partial^2}{\partial z^2} - \left(k_m^2 + k_n^2\right)\right) \tilde{S}_{mn}(z) = \tilde{R}_{mn}(z). \tag{5.32}$$

Two main methods have been used to discretise the vertical direction: finite differences (see Gable *et al.*, 1991; Christensen and Harder, 1991; Young, 1974; Harder, 1998, Machetel *et al.*, 1986, Zhang and Christensen, 1993) and Chebyshev polynomials (see Glatzmaier, 1988; Balachandar and Yuen, 1994). Using second-order finite differences, the final discretised equation can be written as:

$$\frac{2}{\Delta z^2}\tilde{S}_{mn}^{i+1} + \frac{2}{\Delta z^2}\tilde{S}_{mn}^{i-1} - \left(k_m^2 + k_n^2 + \frac{2}{\Delta z^2}\right)\tilde{S}_{mn}^{i} = \tilde{R}_{mn}^{i},\ i = 1, 2, 3, \ldots, K-1,$$
$$\tilde{S}_{mn}^{0} = \tilde{S}_{mn}^{K} = 0, \tag{5.33}$$

where the vertical grid points are numbered $0, \ldots, K$. Inserted into a matrix, this leads to a $(K+1) \times (K+1)$ tridiagonal matrix for each (m, n), which can be solved very efficiently to obtain the S coefficients. Variable vertical grid spacing is straightforward to implement with finite differences.

Alternatively, a Chebyshev expansion gives a more accurate result for a given number of vertical grid points, but leads to a dense matrix. Expressing S in terms of Chebyshev polynomials, and introducing a function ξ to map from the Chebyshev domain of $(-1, 1)$ to the physical domain of $(0, 1)$:

$$\tilde{S}_{mn}(z) = \sum_{k=0}^{K} \breve{S}_{mn}^{k} T_k(\xi(z)),\ \xi = 2z - 1. \tag{5.34}$$

Hence at each vertical collocation point (Eq. (5.7)):

$$\sum_{k=0}^{K} \breve{S}_{mn}^{k} \left(\frac{\partial^2 T_k(\xi(z_i))}{\partial z^2} - T_k(\xi(z_i))\left(k_m^2 + k_n^2\right)\right) = \tilde{R}_{mn}(z_i),$$
$$\sum_{k=0}^{K} \breve{S}_{mn}^{k} T_k(\xi(0)) = 0, \quad \sum_{k=0}^{K} \breve{S}_{mn}^{k} T_k(\xi(1)) = 0. \tag{5.35}$$

This set of discretised equations, one for each vertical point i, can be inserted into a $(K + 1) \times (K + 1)$ matrix (each row holding the equation for one radial level), which will be dense, because the Chebychev polynomials exist at all levels. Nevertheless, because each matrix only holds the one-dimensional vertical problem for a particular (m, n), the problem is much easier to solve than the original 3-D problem. Solution of the matrix equation leads to the fully spectral coefficients $\breve{S}^k_{mn}$, which can be Chebyshev transformed to get the coefficients at each vertical grid point. The above equation involves second derivatives of Chebyshev polynomials, which may be calculated using a recurrence relationship given in Section 5.2.3. If Chebyshev collocation is used for equations with vertically varying coefficients then aliasing may occur (see below), which can be reduced by truncating the expansion. Vertical grid refinement at places other than the upper and lower boundaries, as might be needed if one or more phase changes are included, can be implemented either using a mapping, as done by S. Honda *et al.* (1993a,b) based on Bayliss and Turkel (1992), or by using multiple Chebyshev expansions matched at the phase change(s). Two Chebyshev expansions were used by Glatzmaier and Schubert (1993) and Tackley *et al.* (1993), while three were used by Tackley *et al.* (1994). This does, however, lead to a very small advective time step owing to the small grid spacing near the interface ($\sim$ few kilometres) and may lead to stability problems for strongly advection-dominated flows, i.e. at high Rayleigh number.

Now that W has been calculated from the current temperature distribution, attention turns to time-stepping the temperature equation. Expanding the third equation in (5.25) leads to:

$$\frac{\partial \Theta}{\partial t} = \frac{\partial^2 \Theta}{\partial x^2} + \frac{\partial^2 \Theta}{\partial y^2} + \frac{\partial^2 \Theta}{\partial z^2} - v_x \frac{\partial \Theta}{\partial x} - v_y \frac{\partial \Theta}{\partial y} - v_z \frac{\partial \Theta}{\partial z} + Q. \tag{5.36}$$

Treating the time derivative as a simple first-order Euler explicit time step (Section 7.2) and expanding in terms of horizontal mode (m,n) leads to:

$$\begin{aligned}
\tilde{\Theta}^{t+\Delta t}_{mn}(z) &\equiv \tilde{\Theta}^t_{mn}(z) + \Delta t \left[LIN^t_{mn}(z) + NL^t_{mn}(z) \right], \\
LIN^t_{mn}(z) &= -\left(k_m^2 + k_n^2\right) \tilde{\Theta}^t_{mn}(z) + \frac{\partial^2 \tilde{\Theta}^t_{mn}(z)}{\partial z^2} + \tilde{Q}_{mn}(z), \\
NL^t_{mn}(z) &= -\left(v_x \frac{\partial \Theta}{\partial x} + v_y \frac{\partial \Theta}{\partial y} + v_z \frac{\partial \Theta}{\partial z} \right)^t_{mn} (z).
\end{aligned} \tag{5.37}$$

The linear term LIN^t_{mn}, including diffusion and internal heating, can be straightforwardly calculated in spectral (m, n) space, with the second z-derivative calculated using the same method as for finding the vertically dependent coefficients, i.e. finite differences or Chebyshev polynomials. If Chebyshev polynomials, then the most accurate method is to transform from z-space to Chebyshev space, use the recurrence relations discussed earlier to obtain the spectral coefficients of the derivative, then transform back to z-space. Most likely Q is spatially constant so its spectral coefficients are zero except for mode (0,0).

The non-linear advection term NL^t_{mn} cannot be easily calculated in spectral space because it couples together all (m, n). It is most efficiently calculated in grid space using the *spectral transform method*, which involves (i) calculating the spectral coefficients of v_x, v_y, v_z, $\partial\Theta/\partial x$, $\partial\Theta/\partial y$ and $\partial\Theta/\partial z$, (ii) transforming these into grid space, and (iii) performing the necessary products and sum in grid space, (iv) transforming the resulting non-linear term into spectral space and using it in the above equation.

Aliasing is an important problem that occurs if two fields that contain the full spatial frequency range are multiplied together in grid space: the resulting non-linear product will suffer from aliasing of its frequency components. For example, if each variable contains Fourier modes $0, \ldots, M$ in the x-direction, then the product will contain Fourier modes $0, \ldots, 2M$, which cannot be represented on M grid points. The result is that modes $(M+1), \ldots, 2M$ will be aliased into degrees $M-1, \ldots, 0$, respectively. To avoid this, it is common practice to use 50% more grid points in each direction, in this case, $3M/2$ in the x-direction (technically the requirement is $(3M+1)/2$). Then aliasing still occurs, but only in the coefficients for modes $M, \ldots, 3M/2$, which can be 'thrown away' after the inverse transform to spectral space. Often, insufficient attention is paid to the problem of aliasing.

To summarise, the overall solution algorithm can be represented as follows.

(1) Initialise the temperature field, specifically the coefficients Θ_{mni}, i.e. for each (m, n) at each vertical level i.
(2) Given Θ_{mni}, calculate W_{mni} by solving Poisson's equation twice. This involves solving $(M+1)(N+1)$ one-dimensional ODEs in the z-direction.
(3) Given Θ_{mni} and W_{mni}, take a time step by doing the following.
    (a) Calculating the velocities and temperature derivatives in spectral space.
    (b) Transforming them to grid space and multiplying to get the $\mathbf{v} \cdot \nabla\Theta$ term.
    (c) Transforming $\mathbf{v} \cdot \nabla\Theta$ back to spectral space.
    (d) De-aliasing by throwing away the coefficients for the higher modes.
    (e) Forming the diffusion derivatives in spectral space and taking a time step using the first equation in (5.37).
(4) Repeat steps (2) and (3) as required.

There are many variations on the above approach. If additional physical complexity is included it is necessary to use a more complicated version of the equations and solve for more variables in the vertical ODEs, as discussed in the next sections. Regarding the energy equation, some authors time step it entirely in grid space, for example by using a finite difference or finite volume approach (see Gable *et al.*, 1991; Christensen and Harder, 1991; Monnereau and Quere, 2001). Semi-implicit time-stepping can be implemented for the linear terms by solving the energy and momentum equations simultaneously instead of sequentially (Glatzmaier, 1988). More accurate time integration methods (Chapter 7) are typically used; for example Balachandar and Yuen (1994) use the fourth-order Runge–Kutta method for the non-linear terms and the implicit Crank–Nicolson method for the linear diffusion term, whereas Glatzmaier (1988) uses the second-order Adams–Bashforth method for the non-linear terms and the semi-implicit Crank–Nicolson method for the linear terms. Some researchers prefer to work with dimensional equations rather than dimensionless ones.

Another possible solution method for solving the vertical ODEs is to use *propagator matrices* (Hager and O'Connell, 1978). These describe how the solution at one radial level is related to the solution at another radial level, using an analytical solution to go between levels. The domain is divided into a number of radial shells, with a propagator matrix for each shell. By combining the matrices, the global flow solution can be constructed. The use of this method has mostly been restricted to instantaneous flow problems, in particular to calculate the geoid for certain mass distributions and viscosity profiles (see Richards and Hager, 1984) not for time-dependent convection, so is not further discussed here. Another, related, approach is to use *Green's functions*, but again this has mostly been used to calculate instantaneous surface observables from internal density anomalies (see Forte and Peltier, 1994).

5.4.2 Constant viscosity, spherical geometry

For convection in spherical geometry with constant properties, the above method can be used with spherical harmonics replacing Fourier series in the horizontal direction, for example:

$$W(\theta,\phi) = \sum_{\ell=0}^{L} \sum_{m=-\ell}^{\ell} \tilde{W}_{\ell m} Y_{\ell m}(\theta,\phi). \tag{5.38}$$

The biharmonic equation that results from substituting the poloidal velocity potential into the momentum equation is also slightly modified, taking the form (Chandrasekhar, 1961):

$$\nabla^4 \left(\frac{W}{r} \right) = \frac{Ra}{r} \Theta, \tag{5.39}$$

which can also be written as:

$$D_r^2 W = Ra\Theta, \tag{5.40}$$

where

$$D_r = \frac{\partial^2}{\partial r^2} + \frac{1}{r^2 \sin\theta} \frac{\partial}{\partial \theta} \sin\theta \frac{\partial}{\partial \theta} + \frac{1}{r^2 \sin^2\theta} \frac{\partial^2}{\partial \phi^2}. \tag{5.41}$$

Such equations are used in Harder (1998) and Machetel *et al.* (1986).

Spherical harmonic expansions also suffer from aliasing when two fields are multiplied together. If each variable contains degrees $0, \ldots, L$ then the product will contain degrees $0, \ldots, 2L$, which cannot be represented on $L \times 2L$ grid points, resulting in degrees $(L+1), \ldots, 2L$ being aliased into degrees $L, \ldots, 0$ respectively. The remedy is to use $3L/2 \times 3L$ grid points (technically the requirement is $(3L+1)/2 \times (3L+1)$) then 'throw away' the high degree coefficients after the inverse transform to spectral space. This anti-aliasing method is used in the code of Glatzmaier (1988).

5.4.3 Compressibility

A discussion of the approximations and equations used for modelling compressible flow in the mantle is given in Section 10.3; here we highlight implementation issues relevant to the present numerical approach.

If the anelastic approximation is assumed, then density in the continuity equation is assumed to vary with depth and the continuity equation becomes:

$$\nabla \cdot (\bar{\rho}\mathbf{v}) = 0, \tag{5.42}$$

where $\bar{\rho}$ is a reference state density that does not change with time and velocity potentials, if used, become mass flux potentials:

$$\bar{\rho}\mathbf{v} = \nabla \times \nabla \times (W\mathbf{e}) + \nabla \times (Z\mathbf{e}). \tag{5.43}$$

The anelastic approximation includes the effect of dynamic pressure on temperature and density anomaly (in the buoyancy term), the first of which is ignored in the anelastic liquid approximation, and both of which are ignored in the truncated anelastic approximation. This requires that pressure be solved for, rather than eliminated as in Eq. (5.40). In the code of Glatzmaier (1988), velocities are still replaced by W, but pressure is also solved for by simultaneously solving the radial momentum equation and the divergence of the momentum equation by using Chebyshev collocation. The code also simultaneously solves the energy equation and gravitational potential equation, giving four equations in four unknowns (W, pressure, gravitational potential and entropy). A thorough exploration of the effects of compressibility was made by Bercovici *et al.* (1992) using this code. Balachandar and Yuen (1994) use a different approach in their Cartesian compressible code, keeping velocities as variables and deriving an explicit equation for pressure, which is solved together with other equations by using Chebyshev collocation.

Compressibility also implies the variation of other material properties, particularly thermal expansivity, thermal conductivity and viscosity. If these vary with depth only, then they do not cause any coupling of different harmonics and can straightforwardly be incorporated into the radial ODEs.

The additional terms of adiabatic heating and viscous dissipation appear in the energy equation. Both of these terms are non-linear, so need to be evaluated in grid space by using the spectral transform method. Adiabatic heating involves the product of the variables vertical velocity and temperature, whereas viscous dissipation involves the product of stress and strain rate.

5.4.4 Self-gravitation and geoid

The perturbation in gravitational potential Φ is calculated using a simple Poisson's equation,

$$\nabla^2 \Phi = -4\pi G \delta\rho, \tag{5.44}$$

where $\delta\rho$ is the density anomaly, and G is the universal gravitational constant, and is thus ideally suited for a spectral solution approach, particularly because this also leads to a natural treatment of the potential boundary conditions, which depend on spherical harmonic degree ℓ:

$$\left[\frac{\partial \Phi_\ell^m}{\partial r}\right]_{Surface} = \left[\frac{\ell+1}{r}\Phi_\ell^m\right]_{Surface}, \quad \left[\frac{\partial \Phi_\ell^m}{\partial r}\right]_{CMB} = \left[\frac{\ell}{r}\Phi_\ell^m\right]_{CMB}. \tag{5.45}$$

Thus, a spectral approach has been used in numerous instantaneous flow calculations designed to study the geoid, starting with Ricard *et al.* (1984) and Richards and Hager (1984). To do this it is necessary to solve simultaneously for the flow (including self-gravitation), gravitational potential and dynamic surface and core–mantle boundary (CMB) topography. This calculation has also been incorporated into some spectral codes designed for convection (Zhang and Christensen, 1993; Tackley *et al.*, 1994). Even with a grid-based convection code, it is often more convenient to calculate the geoid using a spectral technique, as done by Zhong *et al.* (2008). For full details the reader is referred to these publications.

5.4.5 Tectonic plates and laterally varying viscosity

Tectonic plates at the surface introduce toroidal motion even when viscosity does not vary laterally. Several authors have implemented spectral flow calculations including rigid plates. Most of these are for calculating instantaneous flow, not convection. Hager and O'Connell (1981) imposed observed plate velocities as a boundary condition and used a propogator matrix technique applied to a toroidal and poloidal decomposition to calculate flow in the mantle resulting from plate motions and internal density anomalies. Other models calculate plate motion based on the net torque (force) on a plate being zero. Ricard and Vigny (1989) consider the total flow in spherical geometry to be the superposition of the flow induced by density anomalies with free-slip boundaries, and the flow induced by plate motions. Forte and Peltier (1991, 1994) use a Green's function based approach in a model where the net torque on plates is not necessarily zero. Rigid plates have also been included in full convection calculations. In the Cartesian code MC3D of Gable *et al.* (1991) the flow equations are written in terms of toroidal and poloidal velocities, with (for each horizontal mode) four coupled equations in the vertical coordinate describing the velocities and stresses for the poloidal component and two for the toroidal component, solved iteratively using finite differences. Monnereau and Quere (2001) in spherical geometry also separate the toroidal and poloidal equations and solve using finite differences in radius.

Laterally varying viscosity, i.e. varying in the direction of the spectral expansion, introduces cross terms between derivatives of viscosity and derivatives of velocity, which couple different modes together. Thus, the problem no longer decouples for each spectral mode, requiring a much more complex solution technique. As viscosity varies very strongly in the solid Earth and terrestrial planets this makes the spectral method no longer the optimal method; nevertheless several researchers have successfully applied it to model mantle convection with moderate viscosity variations (e.g. three orders of magnitude).

A typical approach is to decompose viscosity into a mean (horizontally averaged) part and a fluctuating (laterally varying) part. Terms in the momentum equation that involve the mean part still decouple for each mode so can be kept on the left-hand side of the equation and solved in the usual way. Terms that involve the fluctuating part are moved to the right-hand side and treated in an iterative manner, calculated by using a spectral transform method after each update in the velocity field. Iterations are taken until the scheme converges. Examples using this approach are as follows.

- Christensen and Harder (1991) implemented such a scheme in Cartesian geometry using toroidal and poloidal potentials, with the toroidal potential equation obtained by taking the z-component of the curl of momentum equation and the poloidal potential equation obtained by taking the z-component of double curl of momentum equation. They used a finite difference approach to solve the resulting z-equations and for the energy equation.

- Zhang and Christensen (1993) used a different approach in spherical geometry, choosing six unknowns that are related to stresses or combinations of velocity times a transform function that is chosen to reduce lateral gradients in the product. This resulted in six first-order differential equations in radius, which were solved by using uneven finite difference points. Again, iterations were used to deal with the non-linear terms.
- Bercovici (1993, 1995) used a spectral transform method to solve for toroidal and poloidal velocity potentials in a two-dimensional sheet representing the lithosphere, with flow driven by specified sources and sinks that create flow divergence, in Cartesian and spherical geometry, respectively.
- Balachandar *et al.* (1995) used the approach discussed earlier.
- Cadek and Fleitout (2003) introduced a new variable, which is the product of velocity and viscosity, to stabilise the iteration scheme to viscosity contrasts of $10^2 - 10^3$.
- Schmalholz *et al.* (2001) use such an approach to model viscoelastic folding in two dimensions and were able to reach viscosity contrasts of factor 5×10^5.

Instead of putting the coupled terms on the right-hand side, it is also possible to write and solve a set of linear equations that includes all the couplings, either using an iterative technique or direct solver. Cadek and Matyska (1992) and Martinec *et al.* (1993) use a variational approach to iteratively minimise the dissipative energy for non-linear or linear rheology respectively. Forte and Peltier (1994) used a variational approach to derive the appropriate matrix equations and Green's functions for the fully coupled problem, expressed in terms of poloidal and toroidal potentials expanded by using spherical harmonics azimuthally and trigonometic functions in the radial direction. They used this to calculate coupling between different modes. Moucha *et al.* (2007) extended this approach to solve for flow up to degree 32. This results in a single, very large matrix problem, which requires large computational resources (memory, CPU time) to solve.

6 Numerical methods for solving linear algebraic equations

6.1 Introduction

A discretisation of the partial differential equations that are used to describe the dynamics of the Earth's interior results in a system of algebraic equations. The equations are linear or non-linear depending on the nature of the partial differential equations from which they are derived. Systems of linear algebraic equations can be expressed very conveniently in terms of matrix notation. (The elementary properties of matrices are reviewed in Appendix A.)

We consider a system of linear algebraic equations presented in the following matrix form:

$$\mathbf{A}\,\mathbf{x} = \mathbf{b}, \tag{6.1}$$

where $\mathbf{A}$ is a given $n \times n$ matrix assumed to be non-singular, $\mathbf{b}$ is a given column n vector, and $\mathbf{x}$ is the solution vector to be determined. We write the system (6.1) as

$$\begin{aligned}
a_{11}x_1 + a_{12}x_2 + a_{13}x_3 + \cdots + a_{1n}x_n &= b_1, \\
a_{21}x_1 + a_{22}x_2 + a_{23}x_3 + \cdots + a_{2n}x_n &= b_2, \\
\cdots\cdots\cdots\cdots\cdots\cdots\cdots\cdots\cdots\cdots\cdots\cdots \\
a_{n1}x_1 + a_{n2}x_2 + a_{n3}x_3 + \cdots + a_{nn}x_n &= b_n,
\end{aligned} \tag{6.2}$$

where $x_j (j = 1, 2, \ldots, n)$ are the unknown elements of the vector $\mathbf{x}$, $a_{ij} (i,j = 1, 2, \ldots, n)$ are the coefficients of the matrix $\mathbf{A}$, and $b_i (i = 1, 2 \ldots, n)$ are the elements of the vector $\mathbf{b}$.

In the case of linear algebraic equations, the discrete equations can be solved by either direct or iterative methods, while in the case of non-linear equations, the discrete equations have to be solved by an iterative method. Therefore, whether the equations are linear or not, effective methods for solving linear systems of algebraic equations are required.

6.2 Direct methods

Direct methods are systematic procedures based on algebraic elimination. There are a number of methods for the direct solution of systems of linear algebraic equations. We consider in this section several efficient direct methods: Gauss elimination, LU-factorisation and the Cholesky method.

6.2.1 Gauss elimination

The basic method for solving systems of linear algebraic equations is Gauss elimination. It is based on the systematic reduction of large systems of equations to smaller ones. Consider the system (6.2). The heart of the algorithm is the technique for eliminating $a_{ij} (i > j)$, i.e. replacing them with zero. To do this, we first subtract a_{21}/a_{11} times the first equation from the second equation to eliminate the coefficient of x_1 in the second equation. Then we subtract a_{31}/a_{11} times the first equation from the third equation, a_{41}/a_{11} times the first equation from the fourth equation, and so on, until the coefficients of x_1 in the last $n-1$ equations have all been eliminated. This gives the reduced system of equations

$$\begin{aligned} a_{11}x_1 + a_{12}x_2 + a_{13}x_3 + \cdots + a_{1n}x_n &= b_1, \\ a_{22}^{(1)}x_2 + a_{23}^{(1)}x_3 + \cdots + a_{2n}^{(1)}x_n &= b_2, \\ \cdots\cdots\cdots\cdots\cdots\cdots\cdots\cdots\cdots\cdots\cdots \\ a_{n2}^{(1)}x_2 + a_{n3}^{(1)}x_3 + \cdots + a_{nn}^{(1)}x_n &= b_n, \end{aligned} \tag{6.3}$$

where

$$a_{ij}^{(1)} = a_{ij} - a_{1j}\frac{a_{i1}}{a_{11}}, \quad b_i^{(1)} = b_i - b_1\frac{a_{i1}}{a_{11}}, \quad i,j = 2,3,\ldots,n. \tag{6.4}$$

Precisely the same procedure is now applied to the last $n-1$ equations of the system (6.3) to eliminate the coefficients of x_2 in the last $n-2$ equations, and so on, until the entire system has been reduced to the *triangular form*

$$\begin{bmatrix} a_{11} & a_{12} & \ldots & a_{1n} \\ 0 & a_{22}^{(1)} & \ldots & a_{2n}^{(1)} \\ \vdots & \vdots & \ddots & \vdots \\ 0 & 0 & \ldots & a_{nn}^{(n-1)} \end{bmatrix} \begin{bmatrix} x_1 \\ x_2 \\ \vdots \\ x_n \end{bmatrix} = \begin{bmatrix} b_1 \\ b_2^{(1)} \\ \vdots \\ b_n^{(n-1)} \end{bmatrix}. \tag{6.5}$$

The superscripts indicate the number of times the elements have, in general, been changed. This completes the *forward elimination* (or *triangular reduction*) phase of the Gauss elimination algorithm. We should note here that we have assumed that a_{11} and $a_{ii}^{(i-1)}$ are all non-zero elements (otherwise we could not divide by these elements). In general, the elements can be zero and later we will consider the case of zero or small divisor elements.

The Gauss elimination method is based on the fact (usually established in an introductory linear algebra course) that replacing any equation of the original system (6.2) by a linear combination of itself and another equation does not change the solution of (6.2). Thus the triangular system (6.5) has the same solution as the original system (6.2). The purpose of the forward elimination is to reduce the original system to one, which is easy to solve; this is a common theme in much of scientific computing. The last part of the Gauss elimination method consists of the solution of (6.5) by backward substitution, in which the equations are solved in reverse order:

$$
\begin{aligned}
x_n &= \frac{b_n^{(n-1)}}{a_{nn}^{(n-1)}}, \\
x_{n-1} &= \frac{b_{n-1}^{(n-2)} - a_{n-1,n}^{(n-2)} x_n}{a_{n-1,n-1}^{(n-2)}}, \\
&\cdots\cdots\cdots\cdots\cdots\cdots \\
x_1 &= \frac{b_1 - a_{12}x_2 - \cdots - a_{1n}x_n}{a_{11}}.
\end{aligned}
\tag{6.6}
$$

6.2.2 LU-factorisation

Gauss elimination is related to a factorisation of the matrix **A**:

$$\mathbf{A} = \mathbf{L}\ \mathbf{U}. \tag{6.7}$$

Here **U** is the upper triangular matrix of (6.5) obtained in the forward reduction, and **L** is a unit lower triangular matrix (all main diagonal elements are 1) in which the subdiagonal element l_{ij} is the multiplier used for eliminating the jth variable from the ith equation. For example, if the original system of linear algebraic equations is presented as

$$\begin{bmatrix} 3 & 5 & -2 \\ 4 & -9 & 7 \\ 1 & 6 & -8 \end{bmatrix} \begin{bmatrix} x_1 \\ x_2 \\ x_3 \end{bmatrix} = \begin{bmatrix} 7 \\ 7 \\ -11 \end{bmatrix}, \tag{6.8}$$

then the system obtained by the Gauss elimination is

$$\begin{bmatrix} 3 & 5 & -2 \\ 0 & -47 & 29 \\ 0 & 0 & 1 \end{bmatrix} \begin{bmatrix} x_1 \\ x_2 \\ x_3 \end{bmatrix} = \begin{bmatrix} 7 \\ -7 \\ 3 \end{bmatrix}. \tag{6.9}$$

The multipliers used to obtain (6.9) from (6.8) are 4/3, 1/3 and −13/47. Therefore, the matrix **L** in this case can be presented in the form:

$$\begin{bmatrix} 1 & 0 & 0 \\ 4/3 & 1 & 0 \\ 1/3 & -13/47 & 1 \end{bmatrix}, \tag{6.10}$$

and the matrix **A** is the product of the matrix of (6.10) and the matrix of (6.9).

In the general case, the elimination step that produces (6.3) from (6.2) is equivalent to multiplying (6.2) by the matrix

$$\mathbf{L}_1 = \begin{bmatrix} 1 & 0 & \dots & 0 \\ -l_{21} & 1 & \dots & 0 \\ \vdots & \vdots & \ddots & \vdots \\ -l_{n1} & 0 & \dots & 1 \end{bmatrix}. \tag{6.11}$$

Continuing in this way, the reduced system (6.5) may be written as

$$\mathbf{L}\mathbf{A}\mathbf{x} = \mathbf{L}\mathbf{b}, \ \tilde{\mathbf{L}} = \mathbf{L}_{n-1}\mathbf{L}_{n-2}\cdots\mathbf{L}_2\mathbf{L}_1, \quad \text{where}$$

$$\mathbf{L}_i = \begin{bmatrix} 1 & 0 & \dots & \dots & \dots & 0 \\ \vdots & \vdots & \ddots & \ddots & \ddots & \vdots \\ 0 & 0 & \dots & 1 & \dots & 0 \\ 0 & 0 & \dots & -l_{i+1,i} & \dots & 0 \\ \vdots & \vdots & \ddots & \vdots & \ddots & \vdots \\ 0 & 0 & \dots & -l_{n,i} & \dots & 1 \end{bmatrix}. \tag{6.12}$$

Each of the matrices $\mathbf{L}_i$ has determinant equal to 1 and so is non-singular. Therefore the product $\mathbf{L}$ is non-singular.

The factorisation (6.7) is referred to as the *LU-decomposition* (or *LU-factorisation*) of the matrix $\mathbf{A}$. The Gauss elimination algorithm for solving $\mathbf{A}\mathbf{x} = \mathbf{b}$ is equivalent to the following simpler steps of calculations:

$$\begin{aligned} &(1) \text{ factorisation of A: } \mathbf{A} = \mathbf{L}\mathbf{U}; \\ &(2) \text{ solving } \mathbf{L}\mathbf{y} = \mathbf{b}; \text{ and} \\ &(3) \text{ solving } \mathbf{U}\mathbf{x} = \mathbf{y}. \end{aligned} \tag{6.13}$$

This representation of Gauss elimination is convenient for some computational variants of the elimination process.

The previous discussion we assumed that the matrix $\mathbf{A}$ of the system consists of no zero elements (or it has few zero elements). In practice, when a system of partial differential equations is discretised using the finite element, finite difference or finite volume method, the elements of the matrix of the system are primarily zero. The simplest non-trivial example of this is the *tridiagonal matrix*: there are no more than three non-zero elements in each row of the matrix regardless of the size of n. Tridiagonal matrices are special cases of *banded* matrices in which the non-zero elements are all contained in diagonals about the main diagonal. The reader is referred to Golub and Ortega (1992), for details on how to develop an efficient Gauss elimination algorithm for banded matrices.

In our discussion of the Gauss elimination process we assumed also that a_{11} and all subsequent divisors were non-zero. However, we do not need to make such an assumption provided that we revise the algorithm so as to interchange equations if necessary. For example, in the case of $a_{11} = 0$, some other elements in the first column of the matrix **A** must be non-zero (otherwise, the matrix is a singular). If $a_{k1} \neq 0$, then we interchange the first equation in the system with the kth equation and proceed with the elimination process. Similarly, an interchange can be done if any computed diagonal element that is to become a divisor in the next stage should vanish.

6.2.3 Cholesky method

In the case of a symmetric positive definite matrix there is an important variant of Gauss elimination, the Cholesky method, which is based on a factorisation (or decomposition) of the form

$$\mathbf{A} = \mathbf{L}\mathbf{L}^T. \tag{6.14}$$

Here **L** is a lower-triangular matrix but does not necessarily have numbers '1' on the main diagonal as in the LU-factorisation. The factorisation (6.14) is unique, provided that **L** is required to have positive diagonal elements. The product in (6.14) is

$$\begin{bmatrix} l_{11} & \dots & 0 & \dots & 0 \\ \vdots & \ddots & \vdots & \ddots & \vdots \\ l_{i1} & \dots & l_{ii} & \dots & 0 \\ \vdots & \ddots & \vdots & \ddots & \vdots \\ l_{n1} & \dots & l_{ni} & \dots & l_{nn} \end{bmatrix} \begin{bmatrix} l_{11} & \dots & l_{i1} & \dots & l_{n1} \\ \vdots & \ddots & \vdots & \ddots & \vdots \\ 0 & \dots & l_{ii} & \dots & l_{ni} \\ \vdots & \ddots & \vdots & \ddots & \vdots \\ 0 & \dots & 0 & \dots & l_{nn} \end{bmatrix}. \tag{6.15}$$

By equating elements of the first column of (6.15) with corresponding elements of A, we see that

$$l_{11} = (a_{11})^{1/2}, \quad l_{i1} = \frac{a_{i1}}{l_{11}}, \quad i = 2, \dots, n. \tag{6.16}$$

In general,

$$a_{ii} = \sum_{k=1}^{i} l_{ik}^2, \quad a_{ij} = \sum_{k=1}^{j} l_{ik} l_{jk}, \quad j < i, \tag{6.17}$$

which forms the basis for determining the columns of **L** in sequence. Once L is computed, the solution of the linear system can proceed just as in the LU decomposition (6.13): solve $\mathbf{Ly} = \mathbf{b}$ and then solve $\mathbf{L}^T\mathbf{x} = \mathbf{y}$.

The Cholesky factorisation enjoys three advantages over the LU-factorisation. The first advantage is that there are approximately half as many arithmetic operations. The second one is that, because of symmetry, only the lower triangular part of **A** needs to be stored. And finally, the method extends readily to banded matrices and preserves the bandwidth.

6.3 Iterative methods

A system of linear algebraic equations can be solved by the Gauss elimination, LU-decomposition or the Cholesky method. Unfortunately, the triangular factors of sparse matrices are not sparse, so the cost of these methods is quite high, and the number of operations required ($O(N^3)$ for these basic methods, although some more sophisticated algorithms specifically for sparse matrices can improve on this) scales faster than the number of unknowns. Furthermore, the discretisation error is usually much larger than the accuracy of the computer arithmetic so there is no reason to solve the system that accurately. Solution to somewhat more accuracy than that of the discretisation scheme suffices. Moreover, if the system is non-linear then direct methods are not applicable, except to solve a version of the system that is linearised about some point, such as the present iterative approximation.

This makes iterative methods more attractive. Iterative methods are more efficient and demand far less storage than direct methods, especially in three dimensions. (In two dimensions, direct methods using sparse matrix techniques can still be useful.) Iterative methods obtain the solution asymptotically by an iterative procedure in which a trial solution is assumed, the trial solution is substituted into the system of equations to determine the mismatch (or error), and an improved solution is obtained from the mismatch data; the iterative procedure is repeated until a converged result is obtained. If each iteration is cheap and the number of iterations is small, an iterative solver may cost less than a direct solver. When computing N unknowns, a method may be said to have optimal efficiency if the computing work is $O(N)$. This may be achieved by *multigrid* methods. Generally, iterative methods have the computing work $O(N^\alpha)$, $\alpha > 1$. In practice the turn-around time (that is, the elapsed wall-clock time between the start and termination of computations) is important, and it can be decreased by using a faster parallel computer.

We consider in this section several basic iterative methods for large sparse systems of equations. The reader is referred to Faddeev and Faddeeva (1963), Hageman and Young (1981), Golub and Ortega (1992), Axelsson (1996) and Saad (1996) for further details of iterative methods.

6.3.1 Jacobi method

We consider the linear system (6.1) and assume that the diagonal elements of the matrix $\mathbf{A}$ are non-zero ($a_{ii} \neq 0, i = 1, \ldots, n$). One of the simplest iterative procedures is the *Jacobi method*. Assume that an initial approximation $\mathbf{x}^{(0)}$ to the solution is chosen. Then the next iterate is given by

$$x_i^{(1)} = \frac{1}{a_{ii}} \left(b_{ii} - \sum_{j \neq i} a_{ij} x_j^{(0)} \right), \quad i = 1, \ldots n. \tag{6.18}$$

It will be useful to write this in matrix-vector notation, and for this purpose, we let $\mathbf{D} = \operatorname{diag}(a_{11}, \ldots, a_{nn})$ and $\mathbf{B} = \mathbf{D} - \mathbf{A}$. Then it is easy to verify that (6.18) may be written as

$\mathbf{x}^{(1)} = \mathbf{D}^{-1}(\mathbf{b} + \mathbf{B}\mathbf{x}^{(0)})$, and the entire sequence of Jacobi iterates is defined by

$$\mathbf{x}^{(k+1)} = \mathbf{D}^{-1}(\mathbf{b} + \mathbf{B}\mathbf{x}^{(k)}), \quad k = 0, 1, 2, \ldots \tag{6.19}$$

6.3.2 Gauss–Seidel method

A closely related iteration is derived from the following observation. After $x_i^{(1)}$ is computed in Eq. (6.18) it is available to use in the computation of $x_2^{(1)}$, and it is natural to use this updated value rather than the original estimate $x_2^{(0)}$. If we use updated values as soon as they are available, then (6.18) becomes

$$x_i^{(1)} = \frac{1}{a_{ii}}\left(b_{ii} - \sum_{j<i} a_{ij}x_j^{(1)} - \sum_{j>i} a_{ij}x_j^{(0)}\right), \qquad i = 1, \ldots n, \tag{6.20}$$

which is the first step in the *Gauss–Seidel iteration*. To write this iteration in matrix-vector form, let $-\mathbf{L}$ and $-\mathbf{U}$ denote the strictly lower and upper triangular parts of $\mathbf{A}$; that is, both $\mathbf{L}$ and $\mathbf{U}$ have zero main diagonals and

$$\mathbf{A} = \mathbf{D} - \mathbf{L} - \mathbf{U}. \tag{6.21}$$

If we multiply (6.20) through by a_{ii}, then it is easy to verify that the n equations in (6.20) can be written as

$$\mathbf{D}\mathbf{x}^{(1)} - \mathbf{L}\mathbf{x}^{(1)} = \mathbf{b} + \mathbf{U}\mathbf{x}^{(0)}. \tag{6.22}$$

Since $\mathbf{D} - \mathbf{L}$ is a lower-triangular matrix with non-zero diagonal elements, it is non-singular. Hence the entire sequence of Gauss–Seidel iterates is defined by

$$\mathbf{x}^{(k+1)} = (\mathbf{D} - \mathbf{L})^{-1}[\mathbf{U}\mathbf{x}^{(k)} + \mathbf{b}], k = 0, 1, 2, \ldots \tag{6.23}$$

The convergence of iterative methods is established for special conditions. For example, it can be shown that if the matrix $\mathbf{A}$ is strictly diagonally dominant (that is $|a_{ii}| > \sum_{j \neq i} |a_{ij}|$, $i = 1, \ldots, n$), then the Jacobi and Gauss–Seidel iterations converge to the unique solution of $\mathbf{A}\mathbf{x} = \mathbf{b}$ for any starting vector $\mathbf{x}^{(0)}$. Or, if the matrix $\mathbf{A}$ is symmetric and positive-definite, then the Gauss–Seidel iterates converge to the unique solution of $\mathbf{A}\mathbf{x} = \mathbf{b}$ for any starting vector $\mathbf{x}^{(0)}$. Meanwhile, even when the Jacobi and Gauss–Seidel methods are convergent, the rate of convergence may be so slow as to preclude their usefulness.

In certain cases it is possible to accelerate considerably the rate of convergence of the Gauss–Seidel method. Given the current approximation $\mathbf{x}^{(k)}$, we compute initially the Gauss–Seidel iterate

$$\tilde{x}_i^{(k+1)} = \frac{1}{a_{ii}}\left(b_i - \sum_{j<i} a_{ij}x_j^{(k+1)} - \sum_{j>i} a_{ij}x_j^{(k)}\right) \tag{6.24}$$

as an intermediate value, and then take the final value of the new approximation to the ith component to be

$$x_i^{(k+1)} = x_i^{(k)} + \omega \left(\tilde{x}_i^{(k+1)} - x_i^{(k)} \right). \tag{6.25}$$

Here ω is a parameter that has been introduced to accelerate the rate of convergence. Substituting (6.24) into (6.25)

$$x_i^{(k+1)} = (1-\omega)x_i^{(k)} + \frac{\omega}{a_{ii}} \left(b_i - \sum_{j<i} a_{ij}x_j^{(k+1)} - \sum_{j>i} a_{ij}x_j^{(k)} \right) \tag{6.26}$$

and rearranging the equation, we obtain

$$a_{ii}x_i^{(k+1)} + \omega \sum_{j<i} a_{ij}x_j^{(k+1)} = (1-\omega)a_{ii}x_i^{(k)} - \omega \sum_{j>i} a_{ij}x_j^{(k)} + \omega b_i. \tag{6.27}$$

This relationship between the new iterates $x_i^{(k+1)}$ and the old $x_i^{(k)}$ holds for $i = 1, \ldots, n$, and using (6.21) we can write it in matrix-vector form as

$$\mathbf{Dx}^{(k+1)} - \omega \mathbf{Lx}^{(k+1)} = (1-\omega)\mathbf{Dx}^{(k)} + \omega \mathbf{Ux}^{(k)} + \omega \mathbf{b}. \tag{6.28}$$

Since the matrix $\mathbf{D} - \omega \mathbf{L}$ is lower triangular, and, by assumption, has non-zero diagonal elements, it is non-singular, so we can write

$$\mathbf{x}^{(k+1)} = (\mathbf{D} - \omega \mathbf{L})^{-1} \left[(1-\omega)\mathbf{D} + \omega \mathbf{U} \right] \mathbf{x}^{(k)} + \omega (\mathbf{D} - \omega \mathbf{L})^{-1} \mathbf{b}. \tag{6.29}$$

This equation defines the *successive over-relaxation* (SOR) method in the case that $\omega > 1$. Note that, if $\omega = 1$, Eq. (6.29) reduces to the Gauss–Seidel iteration. If the matrix $\mathbf{A}$ is symmetric and positive-definite, than the SOR iterates (6.29) converge to the solution of $\mathbf{Ax} = \mathbf{B}$ for any $\omega \in (0, 2)$ and any starting vector $\mathbf{x}^0$. The parameter ω has to be chosen in such a way to optimise the rate of convergence of the iteration (6.29). In certain cases, such as when used in the framework of the multigrid method (Section 6.4) it is desirable to use *under-relaxation*, i.e. $\omega < 1$.

Convergence can also be improved by the use of *red–black* (also known as *odd–even*) ordering. In this, one iteration consists of two sweeps through the grid with each one updating every other point, i.e. the first sweep updates the 'red' or 'odd' points, while the second sweep updates the 'black' or 'even' points. The arrangement of points is like the black and white squares on a chessboard, i.e. every other point, alternating every row. For operators in which there is no coupling between diagonally offset points, such as a five-point two-dimensional diffusion operator, during each sweep there is no dependency on values already updated during the sweep, which makes it ideal for domain decomposition on parallel computers where each CPU is working on a different part of the global grid.

6.3.3 Conjugate gradient method

A large number of iterative methods for solving linear systems of equations can be derived as minimisation methods. If **A** is symmetric and positive-definite, then the quadratic function

$$\mathbf{q}(\mathbf{x}) = \frac{1}{2}\mathbf{x}^T\mathbf{A}\mathbf{x} - \mathbf{x}^T\mathbf{b} \tag{6.30}$$

has a unique minimiser, which is the solution of $\mathbf{Ax} = \mathbf{b}$. Therefore, methods that attempt to minimise (6.30) are also methods to solve $\mathbf{Ax} = \mathbf{b}$. Many minimisation methods for (6.30) can be written in the form

$$\mathbf{x}^{(k+1)} = \mathbf{x}^{(k)} - \alpha_k\mathbf{p}^{(k)}, \quad k = 0, 1, \ldots \tag{6.31}$$

Given the direction vector $\mathbf{p}^{(k)}$, one way to choose α_k is to minimise $\mathbf{q}$ along the line $\mathbf{x}^{(k)} - \alpha_k\mathbf{p}^{(k)}$; that is,

$$\mathbf{q}(\mathbf{x}^{(k)} - \alpha_k\mathbf{p}^{(k)}) = \min_{\alpha} \mathbf{q}(\mathbf{x}^{(k)} - \alpha\mathbf{p}^{(k)}). \tag{6.32}$$

For fixed $\mathbf{x}^{(k)}$ and $\mathbf{p}^{(k)}$, the expression $\mathbf{q}\left(\mathbf{x}^{(k)} - \alpha\mathbf{p}^{(k)}\right)$ is a quadratic function of α and may be minimised explicitly to give

$$\alpha_k = -(\mathbf{p}^{(k)}, \mathbf{r}^{(k)})/(\mathbf{p}^{(k)}, \mathbf{A}\mathbf{p}^{(k)}), \quad \mathbf{r}^{(k)} = \mathbf{b} - \mathbf{A}\mathbf{x}^{(k)}. \tag{6.33}$$

In this relation, and henceforth, we use the notation $(\mathbf{u}, \mathbf{v})$ to denote the inner product $\mathbf{u}^T\mathbf{v}$. Although there are many other ways to choose the parameter α_k we will use only (6.31) and concentrate on different choices of the direction vectors $\mathbf{p}^{(k)}$. One simple choice is $\mathbf{p}^{(k)} = \mathbf{r}^{(k)}$, which gives the *method of steepest descent*:

$$\mathbf{x}^{(k+1)} = \mathbf{x}^{(k)} - \alpha_k(\mathbf{b} - \mathbf{A}\mathbf{x}^{(k)}), \quad k = 0, 1, \ldots \tag{6.34}$$

This is also known as the *Richardson method* and is closely related to the *Jacobi method*. As with the Jacobi method, the convergence of (6.34) is usually slow. Another simple strategy is to take $\mathbf{p}^{(k)}$ as one of the unit vectors $\mathbf{e}_i$, which has a 1 in position i and is zero elsewhere. Then, if $\mathbf{p}^{(0)} = \mathbf{e}_1, \mathbf{p}^{(1)} = \mathbf{e}_2, \ldots, \mathbf{p}^{(n-1)} = \mathbf{e}_n$, and the parameters α_k are chosen by (6.33), n steps of (6.31) are equivalent to one Gauss–Seidel iteration on the system $\mathbf{Ax} = \mathbf{B}$.

A very interesting choice of direction vector arises from requiring that they satisfy

$$(\mathbf{p}^{(i)}, \mathbf{A}\mathbf{p}^{(j)}) = 0, \quad i \neq j. \tag{6.35}$$

Such vectors are called conjugate (with respect to **A**). It can be shown that, if $\mathbf{p}^{(0)}, \ldots, \mathbf{p}^{(n-1)}$ are conjugate and the parameters α_k are chosen by (6.33), then the iterates $\mathbf{x}^{(k)}$ of (6.31) converge to the exact solution in at most n steps. This property is not useful in practice because of rounding errors; moreover, for large problems n is far too many iterations. However, for many problems of computational geodynamics, a method based on conjugate directions may converge, up to a convergence criterion, in far fewer than n steps.

To use a conjugate direction method it is necessary to obtain the vectors $\mathbf{p}^k$ that satisfy Eq. (6.35). The preconditioned conjugate gradient algorithm generates these vectors as part of the overall method. The algorithm can be described as follows.

Choose $\mathbf{x}^{(0)}$. Set $\mathbf{r}^{(0)} = \mathbf{b} - \mathbf{A}\mathbf{x}^{(0)}$. Solve $\mathbf{M}\hat{\mathbf{r}}^{(0)} = \mathbf{r}^{(0)}$. Set $\mathbf{p}^{(0)} = \hat{\mathbf{r}}^{(0)}$.
For $k = 0, 1, \ldots$

$$\alpha_k = -(\hat{\mathbf{r}}^{(k)}, \mathbf{r}^{(k)})/(\mathbf{p}^{(k)}, \mathbf{A}\mathbf{p}^{(k)}),$$
$$\mathbf{x}^{(k+1)} = \mathbf{x}^{(k)} - \alpha_k \mathbf{p}^{(k)},$$
$$\mathbf{r}^{(k+1)} = \mathbf{r}^{(k)} + \alpha_k \mathbf{A}\mathbf{p}^{(k)}.$$

Test for convergence.

$$\text{Solve } \mathbf{M}\mathbf{r}^{(k+1)} = \mathbf{r}^{(k+1)}$$
$$\beta_k = (\hat{\mathbf{r}}^{(k+1)}, \mathbf{r}^{(k+1)})/(\hat{\mathbf{r}}^{(k)}, \mathbf{r}^{(k)})$$
$$\mathbf{p}^{(k+1)} = \hat{\mathbf{r}}^{(k+1)} + \beta_k \mathbf{p}^k. \tag{6.36}$$

If we assume now that $\mathbf{M}$ is the unit matrix, then $\hat{\mathbf{r}}^{(k)} = \mathbf{r}^{(k)}$ and the above algorithm defines the *conjugate gradient method*. The role of the matrix $\mathbf{M}$ is to 'precondition' the matrix $\mathbf{A}$ and reduce its condition number so as to obtain faster convergence. The reader is referred to Golub and Ortega (1992) for other choices of the matrix $\mathbf{M}$.

6.3.4 Method of distributive iterations

Let consider the linear algebraic $n \times n$ system (6.1). A stationary iterative method is defined as follows:

$$\mathbf{x}^{(k+1)} = \mathbf{B}\mathbf{x}^{(k)} + \mathbf{c}, \tag{6.37}$$

where $\mathbf{c} = (\mathbf{I} - \mathbf{B})\mathbf{A}^{-1}\mathbf{b}$, and neither $\mathbf{B}$ nor $\mathbf{c}$ depend on iterations k. Note that the Jacobi (Section 6.3.1) and Gauss–Seidel (Section 6.3.2) methods belong to the stationary methods. Then Eq. (6.37) can be rewritten as

$$\mathbf{M}\mathbf{x}^{(k+1)} = \mathbf{N}\mathbf{x}^{(k)} + \mathbf{b}, \tag{6.38}$$

with $\mathbf{M} - \mathbf{N} = \mathbf{A}$, $\mathbf{M} = \mathbf{A}(\mathbf{I} - \mathbf{B})^{-1}$, $\mathbf{N} = \mathbf{M}\mathbf{B}$, and $\mathbf{B} = \mathbf{I} - \mathbf{M}^{-1}\mathbf{A} = \mathbf{M}^{-1}\mathbf{N}$. So, we see that every stationary iterative method corresponds to a splitting of matrix $\mathbf{A}$. Usually the $\mathbf{M}$-*matrix* (see Appendix A) is chosen such a way that Eqs. (6.38) can be solved with little work.

Now let us represent Eq. (6.1) using the conditioning matrix $\mathbf{B}$:

$$\mathbf{A}\mathbf{B}\tilde{\mathbf{x}} = \mathbf{b}, \quad \mathbf{x} = \mathbf{B}\tilde{\mathbf{x}}. \tag{6.39}$$

$\mathbf{AB}$ can be an $\mathbf{M}$-*matrix*, while $\mathbf{A}$ is not. The iterative methods discussed in the previous sections can be applied to Eq. (6.39). For the iterative solution of (6.39) we split the matrix

$\mathbf{AB} = \mathbf{M} - \mathbf{T}$ (hence $\mathbf{A} = \mathbf{MB}^{-1} - \mathbf{TB}^{-1}$). This leads to the following stationary iterative method for the system (6.1): $\mathbf{MB}^{-1}\mathbf{x}^{(k+1)} = \mathbf{TB}^{-1}\mathbf{x}^{(k)} + \mathbf{b}$ or

$$\mathbf{x}^{(k+1)} = \mathbf{x}^{(k)} + \mathbf{BM}^{-1}(\mathbf{b} - \mathbf{Ax}^{(k)}). \tag{6.40}$$

The method (6.40) is called *distributive iteration*. This equation shows that the correction $\mathbf{M}^{-1}(\mathbf{b} - \mathbf{Ax}^{(k)})$ corresponding to non-distributive ($\mathbf{B} = \mathbf{I}$) iteration is distributed over the elements of $\mathbf{x}^{(k+1)}$ (Parter, 1979).

6.4 Multigrid methods

While Jacobi and Gauss–Seidel iterations are appealing because of their simplicity, low memory requirements and the fact that the amount of computational work for one iteration scales $O(N)$, where N is the total number of grid points, they have the major drawback of slow convergence, i.e. the number of iterations required to reach convergence is large and increases with grid size. Specifically, the number of Jacobi or Gauss–Seidel iterations required to reach convergence scales as $O(n_{\max}^2)$, where $n_{\max}$ is the number of grid points along the direction with the most grid points, while for SOR with an optimal relaxation parameter ω, the number of iterations scales as $O(n_{\max})$.

This poor convergence behaviour can be illustrated for Poisson's equation solved using finite differences, i.e.

$$\nabla^2 u = f, \tag{6.41}$$

where u is the desired solution and f is the source term. This can be approximated by finite differences in two dimensions as follows:

$$\frac{u_{i,j+1} + u_{i,j-1} + u_{i+1,j} + u_{i-1,j} - 4u_{i,j}}{h^2} = f_{ij}, \tag{6.42}$$

where i and j are the grid point indices and h is the grid spacing (the same in both directions). During the iteration process, however, Eq. (6.42) is not satisfied because the solution is only approximate, and can be written $\tilde{u}$. The initial value of $\tilde{u}$ might be 0 if starting from scratch, or the solution from the previous time step if the physical problem is one that involves time evolution. The residual or defect d is the error in satisfying the equation:

$$d = f - \nabla^2 \tilde{u}. \tag{6.43}$$

The goal of iterations is to reduce this residual to an acceptably small amount, at which $\tilde{u}$ is 'close enough' to the actual solution u. Using Jacobi or Gauss–Seidel iterations, d can be used to calculate a correction to $\tilde{u}$, which in this case is:

$$\tilde{u}_{ij}^{(n+1)} = \tilde{u}_{ij}^{(n)} + \omega \frac{h^2}{4} d_{ij}^{(n)}, \tag{6.44}$$

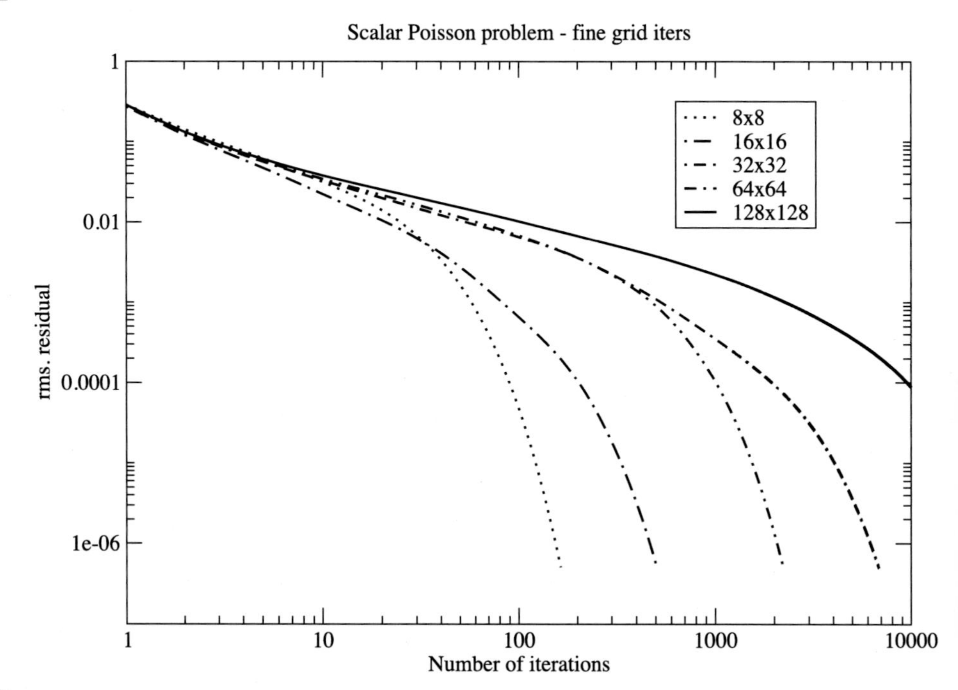

Fig. 6.1. **Residual (L_2 norm) versus iteration number for Gauss–Seidel iterations on a fixed two-dimensional grid for a finite difference Poisson solver, with resolutions varying from 8×8 to 128×128 points.**

where ω is a relaxation parameter: <1 for under-relaxation and >1 for over-relaxation, and n is the iteration number.

Figure 6.1 shows how the L_2 norm of the residual decreases with iteration for various grid sizes from 8×8 to 128×128, with a random (i.e. white noise) source field f. The number of iterations required to obtain a certain residual reduction increases rapidly with grid size and for a 128×128 grid, which is not particularly large, tens of thousands of iterations are needed to reduce d by several orders of magnitude, making this scheme impractical.

6.4.1 Two-grid cycle

Plotting the spatial distribution of the residual after various numbers of iterations (Fig. 6.2) indicates that after a few iterations, the residual becomes *smooth*. The iterations are essentially acting like diffusion, rapidly reducing small-wavelength errors but acting very slowly on long-wavelength errors. Note that this is true only for under-relaxation: over-relaxation does not smooth the residual. This observation suggests a way of speeding up convergence of the long-wavelength solution: because the residual is smooth, it can be approximated on a coarser grid (typically with twice the grid spacing). The solution to the residual can be calculated on this coarse grid, which takes much less time than calculating it on the fine grid. Note that the solution to the residual is equal to the error in the approximate

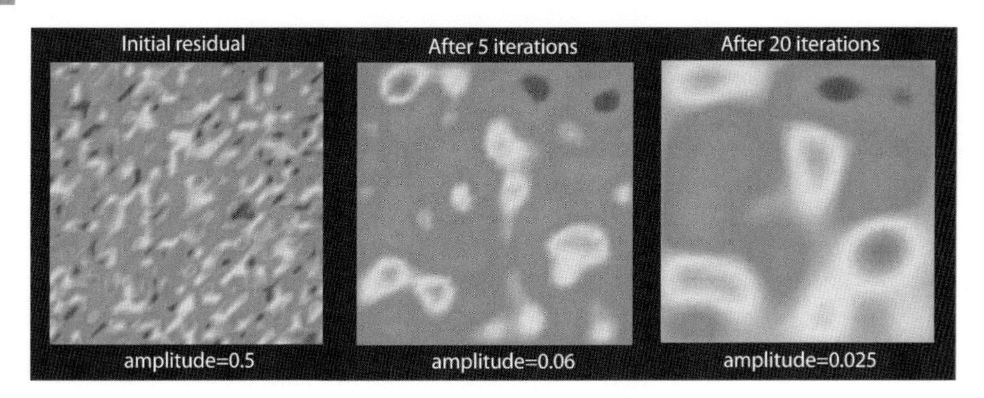

Fig. 6.2. Spatial distribution of initial residual and residual after 5 and 20 Gauss–Seidel iterations on a 32×32 grid, for a Poisson equation with random source term. The iterations smooth the residual. (In colour as Plate 1. See colour plates section.)

solution $c = u - \tilde{u}$:

$$d = f - \nabla^2 \tilde{u} = \nabla^2 (u - \tilde{u}) = \nabla^2 c, \tag{6.45}$$

and is therefore equal to the correction that must be applied to the approximate solution $\tilde{u}$. It is this error/correction that is calculated on the coarse grid, not the full solution. The two-grid cycle consists of

(i) taking a small number of iterations (e.g. two iterations) on the fine grid to smooth the residual;
(ii) restricting the fine-grid residual d to a coarser grid;
(iii) obtaining the exact solution to d, which is the correction to $\tilde{u}$, on the coarse grid; and
(iv) interpolating (prolongating) this correction onto the fine grid and adding to $\tilde{u}$.

This can now be generalised to an arbitrary linear operator L. The discretised equation on the fine grid with grid spacing h can be written as:

$$L_h u_h = f_h, \tag{6.46}$$

and the residual as

$$d_h = f_h - L\tilde{u}_h. \tag{6.47}$$

Restricting to the coarse grid with spacing H (typically $H = 2h$) leads to the coarse grid equation:

$$L_H c_H = R d_h, \tag{6.48}$$

where R is the restriction operator. The solution to this is then prolongated to the fine grid and added to the approximate solution:

$$\tilde{u}_h^{(n+1)} = \tilde{u}_h^{(n)} + P c_H, \tag{6.49}$$

after which a small number of fine-grid iterations are necessary to smooth the new approximate solution. The overall two-grid cycle must be repeated until the required level of convergence is reached.

6.4.2 Restriction, prolongation and coarse grid operators

Appropriate choice of fine-to-coarse (restriction) operator, coarse-to-fine (prolongation) operator and coarse-grid operator L_H is important. For prolongation, linear (bilinear, trilinear) interpolation is typically used. For restriction, the simplest choice is injection, i.e. taking field values at the fine-grid points that coincide with coarse-grid points and ignoring the other ones. It is, however, generally preferable to make the restriction operator the inverse (adjoint) of the prolongation operator. If the fine-grid operator L_h varies strongly from one place to another owing to variations in viscosity or diffusion coefficient, for example, then it is best to take this into account in the construction of P and R by using so-called *matrix-dependent* (or *operator-dependent*) transfer operators; for more details see Wesseling (1992).

A simple and commonly used way of constructing the coarse-grid operator L_H is to re-discretise the equation on the coarse grid, i.e. Eq. (6.46) with H instead of h. A method that often works better, particularly in the case of strongly varying coefficients on the fine grid (e.g. diffusivity or viscosity), is to construct L_H from the fine-grid operator L_h and the restriction and prolongation operators, as:

$$L_H = RL_hP, \tag{6.50}$$

which is the *Galerkin coarse grid approximation*. In essence, this gives a coarse-grid operator that is equivalent to prolongating the approximate coarse-grid solution onto the fine grid, using the fine grid operator to calculate the residual, then restricting the residual back to the coarse grid.

6.4.3 Multigrid cycle

In the two-grid cycle, the question arises how to calculate the exact coarse-grid solution, because if the coarse grid still has a lot of points then a direct method still has prohibitive memory and CPU requirements and Jacobi, Gauss–Seidel or SOR iterations still suffer from slow convergence. The best method is therefore to recursively apply the two-grid method to itself: after a few iterations on the coarse grid, the residual on that grid is smooth and can be approximated and solved for on an even coarser grid. This is done recursively, moving to grids that are progressively coarser by factors of (normally) two until a grid with only a few (e.g. four) points in each direction is reached. On this coarsest grid the solution can quickly be obtained by direct or iterative methods. The overall scheme of moving between grids can be represented on a diagram (Fig. 6.3). The simplest scheme is the V-cycle, which consists of a sweep to increasingly coarser grids, calculation of the exact solution on the coarsest grid, then a sweep to finer and finer grids. In the fine-to-coarse sweep, at each

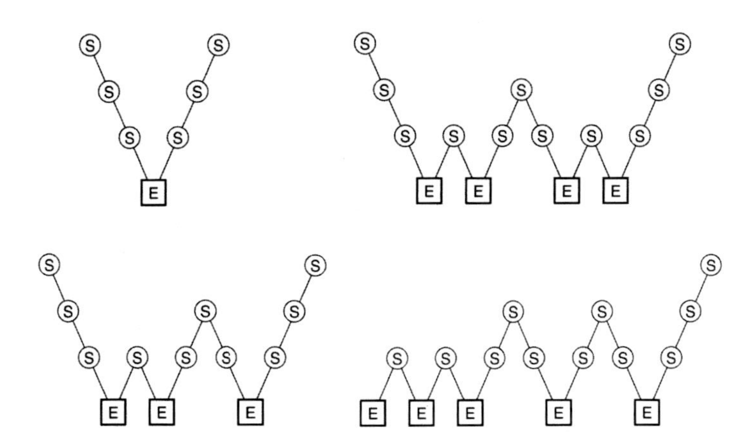

Fig. 6.3. **Different types of multigrid cycle with four grid levels: (top left) *V*-cycle, (top right) *W*-cycle, (bottom left) *F*-cycle and (bottom right) full multigrid. 'S' denotes smoothing while 'E' denotes exact coarse-grid solution. Based on figures in Press *et al.* (2007).**

level a few smoothing iterations are taken and the resulting residual is restricted to the next coarsest grid to become the right-hand side of the equation on that grid. In the coarse-to-fine sweep, the solution from each coarser grid is prolongated onto the next finest grid and used to correct the solution on that grid. To summarise, at each grid level the correction to the solution on the next finer level is calculated, i.e. the solution to the residual on that finer level. The coarser levels can also be visited in a more complex sequence, such as the F-cycle or W-cycle, which for some problems can give faster overall convergence at little additional cost, by taking more iterations at the coarser levels.

Figure 6.4 shows the L_2 norm of the residual versus number of V-cycles for different grid sizes from 8×8 to 256×256, for Poisson's equation above. Convergence is dramatically better than when using only fine-grid iterations as shown in Fig. 6.1. The rate of convergence is independent of grid size, which means that the computational effort (number of operations) is, to first order, proportional to the total number of grid points. Secondly, it converges by about an order of magnitude every two V-cycles, which means that only 10–12 cycles are required to obtain a solution to single precision accuracy. Each V-cycle takes about the same length of time as 4–5 fine-grid iterations, so for the 128×128 grid this is a speedup by a factor of about 1000.

6.4.4 Full approximation storage (for non-linear problems or grid refinement)

In the basic multigrid method given above, each coarse level deals only with the *error* from the next finer level, not the full solution to the physical equations. This is problematic in two cases.

(i) If the operator L_h is non-linear, in which case some of the assumptions made to construct the scheme above do not hold. One approach would be to linearise the operator around the current solution at each cycle, but it is also possible to treat non-linearity directly.

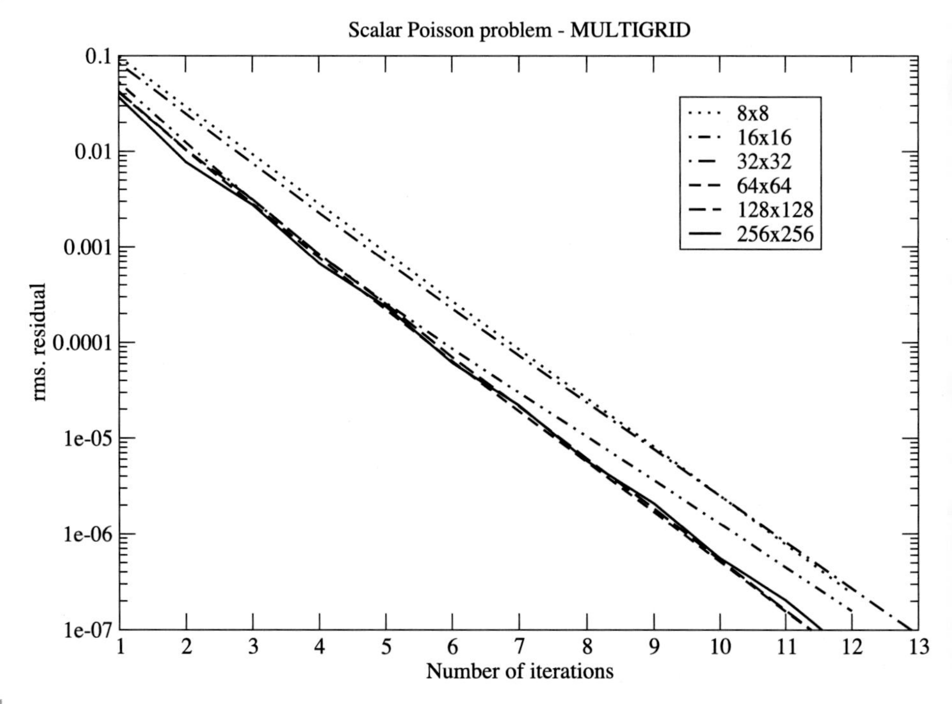

Fig. 6.4. As Fig. 6.1 but for multigrid *V*-cycles and grids from 8×8 to 256×256 points.

(ii) If the grid is refined in some areas: grid refinement can be straightforwardly implemented within the multigrid framework by going to finer levels in some regions. A 'base' fine grid exists everywhere, as do the coarser levels. Levels with finer grid spacing than the base grid exist only in regions where a finer grid is needed. The full solution then needs to be known at all the levels from the base level to the finest level, whereas coarser levels only need to deal with corrections/errors.

Fortunately, both situations can be handled using a simple extension to the above multigrid scheme, known as *full approximation storage* (FAS). In this, the basic discretised system can be written as:

$$L_h(u_h) = f_h, \tag{6.51}$$

and we seek a correction c_h to the approximate solution $\tilde{u}_h$ as follows:

$$L_h(\tilde{u}_h + c_h) - L_h(\tilde{u}_h) = f_h - L_h(\tilde{u}_h) = d_h. \tag{6.52}$$

As before, a few iterations are performed on $\tilde{u}_h$ to make d_h smooth. The difference is that now the approximate fine solution $\tilde{u}_h$ is also restricted to the coarse grid, to give a coarse-grid equation

$$L_H(u_H) = L_H(R\tilde{u}_h) + Rd_h. \tag{6.53}$$

So in the case that $\tilde{u}_h$ is already the correct solution, the coarse-grid solution u_H will simply be the restricted version of $\tilde{u}_h$. When going the other way, from coarse grid to fine grid, it is important to as before apply only the *correction* to the fine grid, not the full coarse-grid solution. That is,

$$\tilde{u}_h^{(n+1)} = \tilde{u}_h^{(n)} + P\left(\tilde{u}_H - R\tilde{u}_h^{(n)}\right). \tag{6.54}$$

In this way, the full solution is known at the coarse levels using an algorithm that is similar to the basic multigrid except with some additional storage and restriction required.

6.4.5 Full multigrid

In the basic multigrid scheme described above, V-cycles are taken until the desired level of convergence is reached. If, however, there is no initial knowledge of the solution, it is more efficient to use the *full multigrid* (FMG) algorithm. In this, the solution is first found on the coarsest grid, then interpolated to the next finest grid and improved, and so on to finer grids until the finest level is reached. At each interpolation stage, typically one or two V-cycles are taken (see Fig. 6.3).

6.4.6 Algebraic multigrid

The above discussions all refer to the *geometric multigrid*, in which the coarse-grid variables are constructed geometrically from the fine levels, then suitable prolongation, restriction and smoothing operators are sought. More robust and flexible schemes known as *algebraic multigrid* (AMG) are obtained by instead choosing coarse variables by taking into account the coefficients in the matrix $\mathbf{A}$ (Eq. (6.1)). The basic principle is to construct coarse variables by identifying, from the matrix, which fine-grid variables are closely coupled, in the sense of strongly influencing or depending on each other. This can be determined by looking at the relative size of the coefficients a_{ij}. Variables that depend on each other can be combined at the coarse level.

This approach is useful for unstructured grids, and for structured grids in which the nature of the physical problem is such that some points are more strongly coupled than others; for example, a diffusion problem with a diffusivity that varies strongly or is anisotropic or, similarly, a Stokes flow problem with strongly varying viscosity. In the case of structured grids, the AMG coarsening process can be thought of as being similar to that in the geometric multigrid, except that additional points from the fine grid are retained at the course level in areas where they are not strongly coupled to adjacent points. For example, in the case of an anisotropic diffusivity, coarsening might occur in only in one direction.

The AMG method involves two phases: set-up and solution. The set-up phase involves constructing the coarse grids and appropriate operators, while solution can use one of the standard methods discussed above. For more details about AMG the reader is referred to other sources.

6.5 Iterative methods for the Stokes equations

The Stokes equations are the basic equations describing mantle dynamics. The system of algebraic equations to be solved in this case may be represented in the form:

$$\begin{pmatrix} \mathbf{N} & \mathbf{G} \\ \mathbf{D} & 0 \end{pmatrix} \begin{pmatrix} \mathbf{u} \\ p \end{pmatrix} = \begin{pmatrix} \mathbf{f} \\ 0 \end{pmatrix}, \tag{6.55}$$

where **N**, **G** and **D** are algebraic operators representing the discretisation of the viscous term ('stiffness' operator), the pressure gradient, and the continuity equation, respectively; **u** is velocity; p is pressure; and **f** is a known vector composed of the body and boundary forces. Because the main diagonal in (6.55) contains a zero block, an application of the iterative methods is not straightforward. In this section we present approaches to numerical solution of the Stokes equations by iterative methods.

6.5.1 Distributive iterations

To apply the method of distributive iterations (Section 6.3.4), initially we define

$$\mathbf{A} = \begin{pmatrix} \mathbf{N} & \mathbf{G} \\ \mathbf{D} & 0 \end{pmatrix}. \tag{6.56}$$

We choose a distribution matrix **B** such as to represent **AB** in a block-triangular form:

$$\mathbf{AB} = \begin{pmatrix} \mathbf{Q} & 0 \\ \mathbf{R} & \mathbf{S} \end{pmatrix}. \tag{6.57}$$

Splitting $\mathbf{AB} = \mathbf{M} - \mathbf{N}$ is easily obtained by splitting **Q** and **S**, leading to simple separate updates for velocity and pressure. When using the multigrid method, smoothing analysis is simplified. A possible choice for **B** is

$$\mathbf{B} = \begin{pmatrix} \mathbf{I} & \mathbf{B}_{12} \\ 0 & \mathbf{B}_{22} \end{pmatrix}, \text{ and hence } \mathbf{AB} = \begin{pmatrix} \mathbf{N} & \mathbf{NB}_{12}+\mathbf{GB}_{22} \\ \mathbf{D} & \mathbf{DB}_{12} \end{pmatrix}. \tag{6.58}$$

Choosing **B** such that $\mathbf{NB}_{12} + \mathbf{GB}_{22} = 0$ results in the block-triangular form (6.57). Therefore we choose $\mathbf{B}_{12} = -\mathbf{N}^{-1}\mathbf{GB}_{22}$, which leads to

$$\mathbf{AB} = \begin{pmatrix} \mathbf{N} & 0 \\ \mathbf{D} & \mathbf{C} \end{pmatrix}, \qquad \mathbf{C} = -\mathbf{DN}^{-1}\mathbf{GB}_{22}, \tag{6.59}$$

with $\mathbf{B}_{22}$ still to be chosen. The main difficulty in the original formulation (6.55), namely, the zero block on the main diagonal, has disappeared. If $\mathbf{B}_{22}$ is chosen such that **C** is also an **M**-*matrix*, chances are that **AB** is an **M**-*matrix*, making the system suited for iterative solution. Various methods result from the choice of $\mathbf{B}_{22}$. We present one of the choices in the next section.

6.5.2 SIMPLE method

A method widely known in the literature as the SIMPLE method (Semi-Implicit Method for Pressure-Linked Equations) is proposed in Patankar and Spalding (1972) and discussed in detail in Patankar (1980). This is perhaps the oldest and most widely used iterative method for the Stokes equations.

The SIMPLE method is obtained by choosing $\mathbf{B}_{22} = \mathrm{I}$, so that (6.59) becomes

$$\mathbf{AB} = \begin{pmatrix} \mathbf{N} & 0 \\ \mathbf{D} & -\mathbf{DN}^{-1}\mathbf{G} \end{pmatrix}. \tag{6.60}$$

A splitting $\mathbf{AB} = \mathbf{M} - \mathbf{T}$ is defined by

$$\mathbf{M} = \begin{pmatrix} \mathbf{Q} & 0 \\ \mathbf{D} & \mathbf{R} \end{pmatrix}, \tag{6.61}$$

where $\mathbf{Q}$ and $\mathbf{R}$ are approximations to $\mathbf{N}$ and $-\mathbf{DN}^{-1}\mathbf{G}$ such that $\mathbf{Mx} = \mathbf{b}$ is easily solvable. For the distribution step in (6.55) $\mathbf{B}$ is approximated by

$$\mathbf{B} = \begin{pmatrix} \mathbf{I} & -\hat{\mathbf{N}}^{-1}\mathbf{G} \\ 0 & \mathbf{I} \end{pmatrix}, \tag{6.62}$$

where $\hat{\mathbf{N}}^{-1}$ is an easy to evaluate approximate inverse of $\mathbf{N}$. In the non-linear case one may think of $\hat{\mathbf{N}}^{-1}(\mathbf{f})$ as giving an approximate solution of $\mathbf{N}(\mathbf{u}) = \mathbf{f}$ by some iterative process.

Depending on the choice of $\hat{\mathbf{N}}^{-1}$, $\mathbf{Q}$ and $\mathbf{R}$, various variants of the SIMPLE method are obtained. $\mathbf{N}$ just represents a set of convection–diffusion equations and it is easy to use simple stationary iterative methods, thus determining $\mathbf{Q}$. In the original SIMPLE method, one chooses $\hat{\mathbf{N}} = \mathrm{diag}(\mathbf{N})$. This makes $\mathbf{D}\hat{\mathbf{N}}^{-1}\mathbf{G}$ easy to determine.

Consider now the following algorithm. Using (6.55) we have

$$\mathbf{b} - \mathbf{Ax}^{(k)} = \begin{pmatrix} \mathbf{f} \\ \mathbf{g} \end{pmatrix} - \begin{pmatrix} \mathbf{N} & \mathbf{G} \\ \mathbf{D} & 0 \end{pmatrix} \begin{pmatrix} \mathbf{u}^{(k)} \\ \mathbf{p}^{(k)} \end{pmatrix} = \begin{pmatrix} \mathbf{r}_1^{(k)} \\ \mathbf{r}_2^{(k)} \end{pmatrix}. \tag{6.63}$$

After computing the residuals $\mathbf{r}_1^{(k)}$ and $\mathbf{r}_2^{(k)}$ preliminary velocity $\delta\mathbf{u}$ and pressure $\delta\mathbf{p}$ corrections are computed by solving subsequently

$$\mathbf{Q}\delta\mathbf{u} = \mathbf{r}_1^{(k)}, \quad \mathbf{R}\delta\mathbf{p} = \mathbf{r}_2^{(k)} - \mathbf{D}\delta\mathbf{u}. \tag{6.64}$$

In the distribution step new corrections are obtained by

$$\begin{pmatrix} \delta\mathbf{u} \\ \delta\mathbf{p} \end{pmatrix} \Rightarrow \mathbf{B} \begin{pmatrix} \delta\mathbf{u} \\ \delta\mathbf{p} \end{pmatrix} = \begin{pmatrix} \delta\mathbf{u} - \hat{\mathbf{N}}\mathbf{G}\delta\mathbf{p} \\ \delta\mathbf{p} \end{pmatrix}. \tag{6.65}$$

Finally we find the velocity and pressure at next iterative step as

$$\mathbf{u}^{(k+1)} = \mathbf{u}^{(k)} + \omega_1 \delta \mathbf{u}, \quad \mathbf{p}^{(k+1)} = \mathbf{p}^{(k)} + \omega_2 \delta \mathbf{p}, \tag{6.66}$$

where ω_1 and ω_2 are relaxation parameters satisfying the following conditions: $0 < \omega_m < 1$, $m = 1, 2$. Note that the convergence will slow down upon grid refinement (Wesseling, 2001) although this method can be used as a smoother inside a multigrid algorithm (see Section 6.4 and Section 3.5.3), giving overall convergence that is almost independent of grid size.

6.5.3 Uzawa-type methods

The Uzawa-type methods solve the linear system of equations (6.55)

$$\mathbf{Nu} + \mathbf{G}p = \mathbf{f}, \tag{6.67}$$

$$\mathbf{Du} = 0, \tag{6.68}$$

by reducing the system to a set of algebraic equations regarding pressure only. Applying the operator $\mathbf{DN}^{-1}$ to Eq. (6.67) and using Eq. (6.68) to eliminate the velocity, we obtain

$$\mathbf{DN}^{-1}\mathbf{G}p = \mathbf{DN}^{-1}\mathbf{f}. \tag{6.69}$$

Each iteration requires the solution of the linear system (6.67). If an iterative method is used to solve the system, then we obtain a two-level solver with inner and outer iterations. The inner system may be solved in many ways (e.g. multigrid). Single-level methods, which approximate the Stokes problem, can also be used; however, effective pre-conditioners for this approach often rely on expensive linear system solvers.

Several geodynamical codes (e.g. CitCom, Gale) made widely available in the framework of the Computational Infrastructure for Geodynamics make use of the Uzawa algorithm to solve the Stokes problem (see http://www.geodynamics.org/cig/software/documentation/; also Cahouet and Chabard, 1988, for the Uzawa algorithm).

6.6 Alternating direction implicit method

One of common method of solving elliptic problems is to add a term containing the first time derivative to the equation and solve the resulting parabolic problem until a steady state is reached. At that point, the time derivative is zero and the solution satisfies the original elliptic equation. Considering the stability requirement, the methods for parabolic equations should be implicit in time. We consider here one of such methods called the *alternating direction implicit* (ADI) method. We refer the reader to Hageman and Young (1981) for more details on the method.

Suppose we solve two-dimensional Laplace's equation. Adding a time derivative to the equation converts it to the two-dimensional heat equation:

$$\frac{\partial \zeta}{\partial t} = \Phi \left(\frac{\partial^2 \zeta}{\partial x^2} + \frac{\partial^2 \zeta}{\partial y^2} \right). \tag{6.70}$$

If this equation is discretised by using the trapezoid rule in time and central differences are used to approximate the spatial derivatives on a uniform grid, we obtain:

$$\frac{\zeta^{(n+1)} - \zeta^{(n)}}{\Delta t} = 0.5\Phi \left[\left(\frac{\delta^2 \zeta^{(n)}}{\delta x^2} + \frac{\delta^2 \zeta^{(n)}}{\delta y^2} \right) + \left(\frac{\delta^2 \zeta^{(n+1)}}{\delta x^2} + \frac{\delta^2 \zeta^{(n+1)}}{\delta y^2} \right) \right], \tag{6.71}$$

where the following notations for the spatial finite differences are used:

$$\left(\frac{\delta^2 \zeta}{\delta x^2} \right)_{i,j} = \frac{\zeta_{i+1,j} - 2\zeta_{i,j} + \zeta_{i-1,j}}{(\Delta x)^2}, \quad \left(\frac{\delta^2 \zeta}{\delta y^2} \right)_{i,j} = \frac{\zeta_{i,j+1} - 2\zeta_{i,j} + \zeta_{i,j-1}}{(\Delta y)^2}. \tag{6.72}$$

Rearranging (6.71) we obtain

$$\begin{aligned} \left(1 - \frac{\Phi \Delta t}{2} \frac{\delta^2}{\delta x^2}\right) \left(1 - \frac{\Phi \Delta t}{2} \frac{\delta^2}{\delta y^2}\right) \zeta^{(n+1)} &= \left(1 + \frac{\Phi \Delta t}{2} \frac{\delta^2}{\delta x^2}\right) \left(1 + \frac{\Phi \Delta t}{2} \frac{\delta^2}{\delta y^2}\right) \zeta^{(n)} \\ &- \frac{(\Phi \Delta t)^2}{4} \frac{\delta^2}{\delta x^2} \left[\frac{\delta^2 (\zeta^{(n+1)} - \zeta^{(n)})}{\delta y^2} \right]. \end{aligned} \tag{6.73}$$

As $\zeta^{(n+1)} - \zeta^{(n)} \approx \Delta t \partial \zeta / \partial t$, the last term is proportional to $(\Delta t)^3$ for small Δt. Since the finite difference approximation is of second order, for small Δt, the last term is small compared with the discretisation error and can be neglected. The remaining equation can be factored into two simpler equations:

$$\left(1 - \frac{\Phi \Delta t}{2} \frac{\delta^2}{\delta x^2}\right) \zeta^* = \left(1 + \frac{\Phi \Delta t}{2} \frac{\delta^2}{\delta y^2}\right) \zeta^{(n)}, \tag{6.74}$$

$$\left(1 - \frac{\Phi \Delta t}{2} \frac{\delta^2}{\delta y^2}\right) \zeta^{(n+1)} = \left(1 + \frac{\Phi \Delta t}{2} \frac{\delta^2}{\delta x^2}\right) \zeta^*. \tag{6.75}$$

Each of these systems of equations is a set of tridiagonal equations that can be solved with one of the direct methods (see Section 6.2); this requires no iteration and is much cheaper than solving (6.71). Either (6.74) or (6.75), as a method in its own right, is only first-order accurate in time and conditionally stable but the combined method is second-order accurate and unconditionally stable. The methods based on these ideas are known as *splitting or approximate factorisation methods*.

Neglect of the third-order term, which is essential to the factorisation, is justified only when the time step is small. So, although the method is unconditionally stable, it may not be accurate in time if the time step is large. For elliptic equations, the objective is to obtain the steady-state solution as quickly as possible; this is best accomplished with the largest possible time step. However, the factorisation error becomes large when the time step is large so the method loses some of its effectiveness. In fact, there is an optimum time step, which gives the most rapid convergence. When this time step is used, the ADI method is very efficient – it converges in a number of iterations proportional to the number of points in one direction. A better strategy uses different time steps for several iterations in a cyclic fashion. This approach can make the number of iterations for convergence proportional to the square root of the number of grid points in one direction, making ADI an excellent method.

6.7 Coupled equations solving

Most problems in geodynamics require solution of coupled systems of equations, i.e. the dominant variable of each equation occurs in some of the other equations. For example, mass and heat transfer in the mantle (mantle convection) are described by the Stokes and heat balance equations where velocity (the dominant variable for the Stokes equations) and temperature (the dominant variable for the heat balance equation) enters both equations (in the case of temperature-dependent mantle viscosity).

There are two types of approaches to such problems. In the first, all variables are solved for simultaneously. In simultaneous solution methods all the equations are considered part of a single system. The discretised equations have a block-banded structure. Direct solution of these equations would be very expensive, especially when the equations are non-linear and the problem is three-dimensional. Iterative solution techniques for coupled systems are generalisations of methods for single equations. For more detail on the simultaneous solution methods, the reader is referred to Galpin and Raithby (1986), and Weiss *et al.* (1999).

When the equations are linear and tightly coupled, the simultaneous approach is best. However, the equations may be so complex and non-linear that coupled methods are difficult and expensive to use. It may then be preferable to treat each equation as if it has only a single unknown, temporarily treating the other variables as known, using the best currently available values for them. The equations are then solved in turn, repeating the cycle until all equations are satisfied. Since some terms, e.g. the coefficients and source terms that depend on the other variables change as the computation proceeds, it is inefficient to solve the equations accurately at each iteration. That being the case, direct solvers are unnecessary and iterative solvers are preferred. Iterations performed on each equation are called *inner iterations*. In order to obtain a solution that satisfies all of the equations, the coefficient matrices and source vector must be updated after each cycle and the process repeated. The cycles are called *outer iterations*.

Optimisation of this type of solution method requires careful choice of the number of inner iterations per outer iteration. It is also necessary to limit the change in each variable from one outer iteration to the next, because a change in one variable changes the coefficients in the other equations, which may slow or prevent convergence. Unfortunately, it is hard to analyse the convergence of these methods.

6.8 Non-linear equation solving

There are two types of methods for solving non-linear equations: Newton-type and global. The Newton-type methods are much faster when a good estimate of the solution is available but the global methods are guaranteed not to diverge; there is a trade-off between speed and convergence. Combinations of the two methods are also used. There is a vast literature devoted to methods for solving non-linear equations, and we present here the Newton-type methods only. For more detail on the non-linear solvers we refer the author to Lax (1954), Householder (1970), Ortega and Rheinboldt (1970), Dennis and Schnabel (1983), Allgower and Georg (1990).

The classical method for solving non-linear equations is Newton's method (sometimes called the Newton–Raphson method). Suppose that one needs to find the root of a single algebraic equation $f(x) = 0$. Newton's method linearises the function about an estimated value of x using the first two terms of the Taylor series:

$$f(x) \approx f(x_0) + f'(x_0)(x - x_0). \tag{6.76}$$

Setting the linearised function equal to zero provides new estimates of the root:

$$x_1 = x_0 - \frac{f(x_0)}{f'(x_0)}, \quad \dots, \quad x_m = x_{m-1} - \frac{f(x_{m-1})}{f'(x_{m-1})}. \tag{6.77}$$

and we continue until the change in the root $x_m - x_{m-1}$ is as small as desired. The method is equivalent to approximating the curve representing the function by its tangent at x_m. When the estimate is close enough to the root, this method converges quadratically, i.e. the error at iteration $m+1$ is proportional to the square of the error at iteration m. This means that only a few iterations are needed once the solution estimate is close to the root. For that reason, it is employed whenever it is feasible to do so.

Newton's method is easily generalised to systems of equations. A generic system of non-linear equations can be written:

$$f_i(x_1, x_2, \dots, x_n) = 0, \quad i = 1, 2, \dots, n. \tag{6.78}$$

This system can be linearised in exactly the same way as the single equation. The only difference is that now we need to use multi-variable Taylor series ($i = 1, 2, \dots, n$):

$$f_i(x_1, x_2, \dots, x_n) = f_i(x_1^{(k)}, x_2^{(k)}, \dots x_n^{(k)}) + \sum_{l=1}^{n} (x_l^{(k+1)} - x_l^{(k)}) \frac{\partial f_i(x_1^{(k)}, x_2^{(k)}, \dots x_n^{(k)})}{\partial x_l}. \tag{6.79}$$

When these equations are set to zero, we have a system of linear algebraic equations that can be solved by direct methods. The matrix of the system is the set of partial derivatives:

$$a_{il} = \frac{\partial f_i(x_1^{(k)}, x_2^{(k)}, \dots x_n^{(k)})}{\partial x_l}, \quad i, l = 1, 2, \dots, n, \tag{6.80}$$

which is called the Jacobian of the system. The system of equations can be then re-written as:

$$-f_i(x_1^{(k)}, x_2^{(k)}, \ldots x_n^{(k)}) = \sum_{l=1}^{n} a_{il}(x_l^{(k+1)} - x_l^{(k)}), \quad i = 1, 2, \ldots, n. \tag{6.81}$$

For an estimate that is close to the correct root, Newton's method for systems converges as rapidly as the method for a single equation. However, for large systems, the rapid convergence is more than offset by its principal disadvantage. For the method to be effective, the Jacobian has to be evaluated at each iteration. This presents two difficulties. The first is that, in the general case, there are n^2 elements of the Jacobian and their evaluation becomes the most expensive part of the method. The second is that a direct method of evaluating the Jacobian may not exist; many systems are such that the equations are implicit or they may be so complicated that differentiation is all but impossible. In fact, Newton's method is used quite rarely to solve the set of equations derived from geodynamical problems. It was found that the cost of generating the Jacobian and solving the system by a direct method (e.g. Gauss elimination) was so high that, even though the method does converge in just a few iterations, the overall cost is greater than that of other iterative methods.

For generic systems of non-linear equations, secant methods are much more effective. For a single equation, the secant method approximates the derivative of the function by the secant drawn between two points on the curve. This method converges more slowly than Newton's method, but as it does not require evaluation of the derivative, it may find the solution at lower overall cost and can be applied to problems in which direct evaluation of the derivative is not possible.

6.9 Convergence and iteration errors

When using iterative solvers, it is important to know when to stop the iterations. The most common procedure is based on the difference between two successive iterates; the procedure is stopped when this difference, measured by some norm, is less than a preselected value. Meanwhile, this difference may be small when the error is not small and a proper normalisation is essential. The iteration errors $\varepsilon^{(k)} = \psi - \psi^{(k)}$ can be estimated by the following criterion: $|\varepsilon^{(k)}| \approx |\delta^{(k)}|/\lambda_1 - 1$, where ψ is the exact solution, $\psi^{(k)}$ is the solution at iteration k, $\delta^{(k)}$ is the difference between solution at iterations $k+1$ and k, and λ_1 is the largest eigenvalues of the iteration matrix (Ferziger and Peric, 2002). Unfortunately, iterative methods often have complex eigenvalues. Their estimation requires an extension of the above procedure (Golub and van Loan, 1989).

Another way to terminate the iterative process is to use the reduction of the residual as a stopping criterion (Ferziger and Peric, 2002). Iteration is stopped when the residual norm has been reduced to some fraction of its original size (usually by three or four orders of magnitude). If the iteration is started from zero initial values, then the initial error is equal to the solution itself. When the residual level has fallen say three to four orders of magnitude below the initial level, the error is likely to have fallen by a comparable amount, i.e. it is of the order of 0.1% of the solution. The residual and the error usually do not fall in the same way at the beginning of iteration process; caution is also needed because, if the matrix is poorly conditioned, the error may be large even when the residual is small.

Many iterative solvers require calculation of the residual. The above approach is specifically attractive in the case of a non-linear system, as it requires no additional computation. The norm of the residual prior to the first inner iteration provides a reference for checking the convergence of inner iterations. At the same time it provides a measure of the convergence of the outer iterations. Outer iterations should not be stopped before the residual has been reduced by three to five orders of magnitude, depending on the desired accuracy.

If the order of the initial error is known, it is possible to monitor the norm of the difference between two iterates and compare it with the same quantity at the beginning of the iteration process. When the difference norm has fallen three to four orders of magnitude, the error has usually fallen by a comparable amount. Both of these methods are only approximate; however, they are better than the criterion based on the non-normalised difference between two successive iterates.

7 Numerical methods for solving ordinary and partial differential equations

7.1 Introduction

Most geodynamical processes are governed by differential equations involving more than one independent variable, and in this case the corresponding differential equations are partial differential equations (PDEs). In some cases, however, simplifying assumptions are made, which reduce the PDEs to ordinary differential equations (ODEs). An ODE is an equation stating a relationship between a function of a single independent variable and the total derivatives of this function with respect to the independent variable. We will use the variable ρ as a dependent variable. In most geodynamical problems, the independent variable is either time t or space x. If more than one independent variable exists, then partial derivatives occur, and PDEs are obtained.

The order of an ODE is the order of the highest-order derivative in the differential equation. The general first-order ODE is

$$\frac{d\rho}{dt} \equiv \rho' = f(t, \rho), \tag{7.1}$$

where $f(t, \rho)$ is called the *derivative* function. The general nth-order ODE for $\rho(t)$ has the form $a_n\rho^{(n)} + a_{n-1}\rho^{(n-1)} + \cdots + a_2\rho'' + a_1\rho' + a_0\rho = F(t)$, where $a_n \neq 0$ and the superscript i denotes ith-order differentiation ($i = n, n-1, n-2, \ldots$).

There are two different types (or classes) of ODE; they are distinguished by the type of auxiliary conditions specified. If all the auxiliary conditions are specified at the values of the independent variable and the solution is to be marched forward from that initial point, the differential equation is an *initial-value ODE*. If the auxiliary conditions are specified at two different values of the independent variable, the end points or boundaries of the domain of interest, the differential equation is a *boundary-value ODE*.

This chapter is concerned with numerical methods for solving initial-value ODEs. Equation (7.1) with the condition

$$\rho(t = t_0) = \rho_0 \tag{7.2}$$

is a classical example of an initial-value ODE.

7.2 Euler method

Suppose that an initial-value problem is given by (7.1) and (7.2). The aim is to find numerical approximate values of the unknown function ρ at points $t > t_0$, that is, at a discrete set of points $t_1 = t_0 + h$; $t_2 = t_0 + 2h$; $t_3 = t_0 + 3h$, etc. At each of these points t_n we will compute ρ_n as an approximation to $\rho(t_n)$. To derive a method of obtaining value ρ_n from its immediate predecessor, we consider the Taylor series expansion of the unknown function $\rho(t)$ about the point t_n, namely,

$$\rho(t_{n+1} = t_n + h) = \rho(t_n) + h\rho'(t_n) + \frac{h^2}{2}\rho''(t_*), \tag{7.3}$$

where the expansion is halted after the first power of h, and $t_* \in (t_n, t_{n+1})$. Equation (7.3) is exact, but cannot be used for computation, because the point t_* is unknown. Omitting the term containing the second-order derivative of the function ρ, we can write

$$\rho(t_{n+1}) \approx \rho(t_n) + h\rho'(t_n). \tag{7.4}$$

If we define ρ_n as the approximate value of $\rho(t_n)$, then we get the following computable formula for the approximate values of the unknown function

$$\rho_{n+1} = \rho_n + h\rho_n' = \rho_n + hf(t_n, \rho_n). \tag{7.5}$$

This is the *Euler method*. Equation (7.5) is a recurrence relation (or difference equation), and hence each value of ρ_n is computed from its immediate predecessor. This makes it an *explicit method*, i.e. each new value can be calculated from already-known values. Actually, the explicit Euler method is accurate to $O(h)$ and has limited usage because of the larger error that is accumulated as the process proceeds, it is unstable, unless the time step is taken to be extremely small.

7.3 Runge–Kutta methods

As mentioned in the previous section, the Euler method is not very useful in practical problems because it requires a very small step size for reasonable accuracy. Taylor's algorithm of higher order is unacceptable as a general-purpose procedure because of the need to obtain higher total derivatives of the unknown function. The Runge–Kutta methods attempt to obtain greater accuracy, and at the same time avoid the need for higher derivatives, by evaluating the right-hand side of (7.1) at selected points on each subinterval. We derive here the simplest of the Runge–Kutta methods. Consider the following recurrence relation:

$$\rho_{n+1} = \rho_n + Ak_1 + Bk_2, \tag{7.6}$$

where $k_1 = hf(t_n, \rho_n)$ and $k_2 = hf(t_n + \alpha h, \rho_n + \beta k_1)$, and A, B, α and β are constants to be determined so that (7.6) will agree with the Taylor algorithm of as high an order as possible. On expanding $\rho(t_{n+1})$ in a Taylor series through terms of order h^3, we obtain

$$\begin{aligned}\rho(t_{n+1}) &= \rho(t_n) + h\rho'(t_n) + \frac{h^2}{2}\rho''(t_n) + \frac{h^3}{6}\rho'''(t_n) + \cdots \\ &= \rho(t_n) + hf(t_n, \rho(t_n)) + \frac{h^2}{2}\left(f_t + f_\rho f\right)_n \\ &+ \frac{h^3}{6}\left(f_{tt} + 2f_{t\rho}f + f_{\rho\rho}f^2 + f_t f_\rho + f_\rho^2 f\right)_n + O(h^4),\end{aligned} \tag{7.7}$$

where the subscript n means that the function involved is to be evaluated at point (t_n, ρ_n). On the other hand, using a Taylor expansion for functions of two variables, we find that

$$\begin{aligned}\frac{k_2}{h} &= f(t_n + \alpha h, \rho_n + \beta k_1) = f(t_n, \rho_n) + \alpha h f_t + \beta k_1 f_\rho \\ &+ \frac{\alpha^2 h^2}{2} f_{tt} + \alpha h \beta k_1 f_{t\rho} + \frac{\beta^2 k_1^2}{2} f_{\rho\rho} + O(h^3),\end{aligned} \tag{7.8}$$

where all derivatives are evaluated at point (t_n, ρ_n). If we substitute (7.8) for k_2 in (7.6), we obtain upon rearrangement in powers of h that

$$\begin{aligned}\rho_{n+1} &= \rho_n + (A + B)hf + Bh^2\left(\alpha h f_t + \beta f_\rho f\right) \\ &+ Bh^3\left(\frac{\alpha^2}{2} f_{tt} + \alpha\beta f_{t\rho} f + \frac{\beta^2}{2} f^2 f_{\rho\rho}\right) + O(h^4).\end{aligned} \tag{7.9}$$

Comparing (7.7) and (7.9) we can derive the following relations:

$$A + B = 1, \qquad B\alpha = B\beta = 0.5. \tag{7.10}$$

Although we have four unknowns, we have only three equations, and hence we still have one degree of freedom in the solution of (7.10). There are many solutions to (7.10), the simplest one perhaps being $A = B = 0.5$ and $\alpha = \beta = 1$, although $A = 0$, $B = 1$ and $\alpha = \beta = 0.5$ is the most commonly used. The discretisation error of the method is of $O(h^2)$, and therefore it is called the *second-order Runge–Kutta method*. Compared with the Euler method, a larger step size can be used in computations.

Formulas of the Runge–Kutta type for any order can be derived by the method used above; however, the derivations become exceedingly complicated. One of most popular of the high-order methods is the *fourth-order Runge–Kutta method*. For initial-value problem (7.1) and (7.2) the following recurrence relation is used to compute approximations ρ_n to the unknown function $\rho(t = t_n)$:

$$\begin{aligned}\rho_{n+1} &= \rho_n + \frac{1}{6}\left(k_1 + 2k_2 + 2k_3 + k_4\right), \\ k_1 &= hf(t_n, \rho_n), \quad k_2 = hf\left(t_n + \frac{h}{2}, \rho_n + \frac{1}{2}k_1\right), \\ k_3 &= hf\left(t_n + \frac{h}{2}, \rho_n + \frac{1}{2}k_2\right), \quad k_4 = hf(t_n + h, \rho_n + k_3).\end{aligned} \tag{7.11}$$

The discretisation error of the method is of $O(h^4)$. The price we pay for the favourable discretisation error is that four function evaluations are required per step. This price may be considerable in computational time for those problems in which the function $f(t, \rho)$ is complicated. The formula (7.11) is widely used in computational geodynamics with considerable success. It has the important advantage that it requires only the value of ρ at a point $t = t_n$ to find ρ and ρ' at $t = t_{n+1}$.

7.4 Multi-step methods

The Euler and Runge–Kutta methods are called single-step methods because they use only the information from one previous point to compute the successive point. After several points have been found, it is feasible to use several prior points in the calculation. In this section we describe linear multi-step methods for the solution of differential equations. Like the Euler and Runge–Kutta methods, these are also explicit.

To derive the Euler method we truncated the Taylor series expansion of the solution at the linear term. To get a more accurate method, we could keep the quadratic term too, but the term involves a second-order derivative. Meanwhile a greater accuracy can be achieved without having to calculate higher derivatives, if a numerical integration procedure involves values of the unknown function and its derivative at more than one point.

7.4.1 The midpoint rule (leap-frog method)

We consider again the Taylor expansion of the unknown function $\rho(t)$ about the point t_n:

$$\rho(t_n + h) = \rho(t_n) + h\rho'(t_n) + h^2 \frac{\rho''(t_n)}{2} + h^3 \frac{\rho'''(t_n)}{6} + \cdots . \tag{7.12}$$

Now we rewrite (7.12) with h replaced by $-h$ to obtain

$$\rho(t_n - h) = \rho(t_n) - h\rho'(t_n) + h^2 \frac{\rho''(t_n)}{2} - h^3 \frac{\rho'''(t_n)}{6} + \cdots , \tag{7.13}$$

and then subtract (7.13) from (7.12) to get

$$\rho(t_n + h) - \rho(t_n - h) = 2h\rho'(t_n) + h^3 \frac{\rho'''(t_n)}{3} + \cdots . \tag{7.14}$$

Truncating the right side of (7.14) after the first term, we have

$$\rho_{n+1} = \rho_{n-1} + 2h\rho'_n = \rho_{n-1} + 2hf(t_n, \rho_n), \tag{7.15}$$

and this is the midpoint rule to compute the unknown function, sometimes called the leap-frog method. At first sight it seems that formula (7.15) can be used like the Euler formula (7.5), because it is a recurrence formula allowing the computation of the next value ρ_{n+1}

from two previous values ρ_n and ρ_{n-1}. The rules are quite similar, except for the fact that we cannot get started with the midpoint rule until we know the value of ρ_1 of the unknown function at the point t_1. A simple way to get the value of ρ_1 is to compute it by using the Euler method. In general, the greater accuracy of computations we design without calculations of higher derivatives, the more values of the function ρ must be known at predecessor points. To get such a formula started, several starting values should be obtained in addition to the one that is given in the statement of the initial-value problem.

7.4.2 The trapezoidal rule

We now introduce another numerical method for computing the initial-value problem (7.1) and (7.2). It is based on converting (7.1) into an integral equation and solving the integral equations using the trapezoidal approximation for the integral, instead of solving the initial-value problem.

We integrate both sides of (7.1) from t to $t + h$:

$$\rho(t+h) = \rho(t) + \int_t^{t+h} f(t, \rho(t))dt. \tag{7.16}$$

If the right-hand side of the equation is approximated by a weighted sum of values of the integrand at various points, we can get an approximate method for solving the initial-value problem.

The integral $\int_a^b f(x)dx$ can be calculated exactly as the area between the curve $y = f(x)$ and the x-axis and the lines $x = a$ and $x = b$. The trapezoidal rule states that for an approximate value of the integral we can use the area of the trapezoid whose sides are the x-axis, the lines $x = a$ and $x = b$, and the line through the points $(a, f(a))$ and $(b, f(b))$. That area is $\frac{1}{2}(f(a) + f(b))(b - a)$.

If we apply now the trapezoidal rule to the integral that appears in (7.16), we have

$$\rho(t_n + h) \approx \rho(t_n) + \frac{h}{2}\left(f(t_n, \rho(t_n)) + f(t_n + h, \rho(t_n + h))\right), \tag{7.17}$$

and using the usual abbreviation ρ_n for the computed approximate value of $\rho(t_n)$, we obtain finally

$$\rho_{n+1} = \rho_n + \frac{h}{2}\left(f(t_n, \rho_n) + f(t_{n+1}, \rho_{n+1})\right). \tag{7.18}$$

This is the *trapezoidal rule* for numerical solving of Eq. (7.1). It is classified as a semi-implicit method because the calculation of the derivative f involves both the existing value ρ_n and knowledge of the new value ρ_{n+1}. To find the next value ρ_{n+1} from the value ρ_n, an iteration process should be carried out. Initially we can guess some value for ρ_{n+1} and insert the value to calculate the entire right-hand side of (7.18). The calculated left-hand side of the equation can now be used as a new, updated value of ρ_{n+1}. If the new value agrees with the old sufficiently well, the iterations would be terminated, and the updated value can be considered as desired value of ρ_{n+1}. Otherwise, we should use the updated value on the right side of the equation just as we did previously to update ρ_{n+1}, etc.

Consider the process by which a guessed value of ρ_{n+1} is updated by using the trapezoidal formula (7.18). Suppose $\rho_{n+1}^{(m)}$ to be some guess value of ρ_{n+1} that satisfies (7.18). Then the updated value $\rho_{n+1}^{(m+1)}$ is computed from

$$\rho_{n+1}^{(m+1)} = \rho_n + \frac{h}{2}\left(f(t_n, \rho_n) + f(t_{n+1}, \rho_{n+1}^{(m)})\right). \tag{7.19}$$

To understand how rapidly the successive values of $\rho_{n+1}^{(m)}$, $m = 1, 2, 3, \ldots$ approach a limit (if at all), we rewrite (7.19) replacing m by $m - 1$:

$$\rho_{n+1}^{(m)} = \rho_n + \frac{h}{2}\left(f(t_n, \rho_n) + f(t_{n+1}, \rho_{n+1}^{(m-1)})\right), \tag{7.20}$$

and then subtract (7.20) from (7.19)

$$\begin{aligned}\rho_{n+1}^{(m+1)} - \rho_{n+1}^{(m)} &= \frac{h}{2}\left(f(t_{n+1}, \rho_{n+1}^{(m)}) + f(t_{n+1}, \rho_{n+1}^{(m-1)})\right) \\ &= \frac{h}{2}\left.\frac{\partial f}{\partial t}\right|_{(t_{n+1}, \rho_+)} \left(\rho_{n+1}^{(m)} - \rho_{n+1}^{(m-1)}\right),\end{aligned} \tag{7.21}$$

where $\rho_+ \in \left(\rho_{n+1}^{(m-1)}, \rho_{n+1}^{(m)}\right)$. According to (7.21) the iterative process will converge if h is kept small enough so that the function $\frac{h}{2}\frac{\partial f}{\partial t}$ (or the local convergence factor) is less than 1 in absolute value. If the factor is much less than 1, then the convergence will be extremely rapid.

7.5 Crank–Nicolson method

The Crank–Nicolson method is used to solve partial differential equations (e.g. the heat balance equation). It is based on central differences in space and the trapezoidal rule in time, and hence the method is semi-implicit and second-order accurate in time. For many partial differential equations (including diffusion equations) the Crank–Nicolson method is shown to be unconditionally stable. Consider the initial-value problem (7.1) and (7.2). The Crank–Nicolson scheme can be presented by Eq. (7.18) as the average of the forward Euler scheme at n and the backward Euler scheme at $n + 1$. The function f in (7.18) should be discretised spatially with a central difference.

The approximate solutions can contain spurious oscillations at large time steps. To avoid this, whenever large time steps (or high spatial resolution) are required, the less accurate, implicit backward Euler method

$$\rho_{n+1} = \rho_n + hf(t_{n+1}, \rho_{n+1}), \tag{7.22}$$

is often used because of its stability and immunity to oscillations. In the case of the 1-D heat diffusion equation

$$\frac{\partial T}{\partial t} = \kappa \frac{\partial^2 T}{\partial x^2}, \; \kappa > 0, \tag{7.23}$$

the Crank–Nicolson scheme takes the form:

$$\frac{T_i^{n+1} - T_i^n}{h} = \frac{\kappa}{2h_x^2}\left((T_{i+1}^{n+1} - 2T_i^{n+1} + T_{i-1}^{n+1}) + (T_{i+1}^n - 2T_i^n + T_{i-1}^n)\right) \tag{7.24}$$

or alternatively the form

$$-qT_{i+1}^{n+1} + (1+2q)T_i^{n+1} - qT_{i-1}^{n+1} = qT_{i+1}^n + (1-2q)T_i^n + qT_{i-1}^n, \tag{7.25}$$

where T is temperature, κ is the coefficient of heat diffusivity, h_x is the spatial step and $q = 0.5\kappa h/h_x^2$. Temperature T_i^{n+1} can be efficiently solved for by using tridiagonal matrix algorithms.

7.6 Predictor–corrector methods

In actual practice, one does not actually have to iterate (7.19) to convergence. If a good enough guess is available for the unknown value, then just one refinement by a single application of the trapezoidal formula is sufficient. The pair of formulas, one of which supplies a very good guess to the next value of ρ, and the other of which refines it to a better guess, is called a *predictor–corrector pair*, and such pairs form the basis of many of the highly accurate schemes that are used in practice.

If the trapezoidal rule is used as a corrector, for example, then a 'clever' predictor would be the midpoint rule. The reason for this will become clear if we look at both formulas together with their error terms:

$$\rho_{n+1} = \rho_{n-1} + 2h\rho_n' + \frac{h^3}{3}\rho'''(t_*),$$

$$\rho_{n+1} = \rho_n + \frac{h}{2}\left(\rho_n' + \rho_{n+1}'\right) - \frac{h^3}{12}\rho'''(t_{**}). \tag{7.26}$$

Assuming the value h to be small enough, we can regard the two values of ρ''' as being the same. The error in the trapezoidal rule is about one fourth as large as the error in the midpoint rule. The subsequent iterative refinement of that guess needs to reduce the error only by a factor of four.

When we are dealing with a predictor–corrector pair, we need to make a single refinement of the corrector if the step size is kept moderately small, that is, the step size times the local value of $\frac{\partial \rho}{\partial t}$ should be small compared with 1. For this reason, iteration to full convergence is rarely done in practice.

7.6.1 The Adams–Bashforth–Moulton method

The basic idea of the method was formulated by Bashforth and Adams (1883). Consider (7.16) in the following form:

$$\rho(t_{k+1}) = \rho(t_k) + \int_{t_k}^{t_{k+1}} f(t, \rho(t))dt. \tag{7.27}$$

The predictor uses the Lagrange polynomial approximation for $f(t, \rho(t))$ based on the points (t_k, f_k) as well as one or more previous values, depending on the order required, with second- to fourth-order schemes being in common usage. For fourth-order accuracy, previous points (t_{k-3}, f_{k-3}), (t_{k-2}, f_{k-2}) and $(t_{k-1}, f_{k-1})$are used, and the Lagrange polynomial is integrated over the interval $[t_k, t_{k+1}]$ in (7.27). This process produces the fourth-order Adams–Bashforth predictor:

$$\tilde{\rho}_{k+1} = \rho_k + \frac{h}{24}(-9f_{k-3} + 37f_{k-2} - 59f_{k-1} + 55f_k). \tag{7.28}$$

The corrector is developed similarly. The value $\tilde{\rho}_{k+1}$ computed can now be used. A second Lagrange polynomial for $f(t, \rho(t))$ is constructed, which is based on the points (t_{k-2}, f_{k-2}), (t_{k-1}, f_{k-1}), (t_k, f_k), and the new point $(t_{k+1}, f_{k+1}) = (t_{k+1}, f(t_{k+1}, \tilde{\rho}_{k+1}))$. This polynomial is then integrated over $[t_k, t_{k+1}]$ producing the Adams–Moulton corrector:

$$\rho_{k+1} = \rho_k + \frac{h}{24}(f_{k-2} - 5f_{k-1} + 19f_k + 9f_{k+1}). \tag{7.29}$$

The error terms for the numerical integration formulas used to obtain both the predictor and corrector are of the order $O(h^5)$.

The fourth-order Adams–Bashforth–Moulton method is an excellent example of a multipoint method. It has excellent stability limits, excellent accuracy, and a simple and inexpensive error estimation procedure. It is recommended as the method of choice when a multipoint method is desired.

7.7 Method of characteristics

In this section we discuss a method to solve an initial-value problem for a first-order partial differential equation (PDE). This method is based on finding the characteristic curve of the PDE. Consider the first-order PDE or the advection equation for the function $\rho(t, x)$:

$$\frac{\partial \rho}{\partial t} + u\frac{\partial \rho}{\partial x} = 0, \tag{7.30}$$

where u does not depend on t. To solve (7.30) we note that if we consider an 'observer' moving on a curve $x(t)$ then, by the chain rule, we get

$$\frac{d\rho(t,x(t))}{dt} = \frac{\partial \rho}{\partial t} + \frac{\partial \rho}{\partial x}\frac{dx}{dt}. \tag{7.31}$$

If the 'observer' is moving at a rate $\frac{dx}{dt} = u$, then by comparing (7.30) and (7.31) we find $\frac{d\rho}{dt} = 0$. Therefore (7.30) can be replaced by a set of two ODEs:

$$\frac{dx}{dt} = u, \quad \frac{d\rho}{dt} = 0. \tag{7.32}$$

These two ODEs are easy to solve. Integration of the first equation of (7.32) yields

$$x(t) = x(0) + ut, \tag{7.33}$$

and the second equation of (7.25) has a solution ρ = constant along the curve given in Eq. (7.33). The curve (7.33) is a straight line. In fact, we have a family of parallel straight lines, called *characteristics*. To obtain the general solution to (7.30) subject to the initial value

$$\rho(t=0, x(t=0)) = f(x(t=0)), \tag{7.34}$$

we note that the function ρ is constant along $x(t) = x(0) + ut$, but that constant is $f(x(0))$ from (7.34). Since $x(0) = x(t) - ut$, the general solution is then

$$\rho(t,x) = f(x(t) - ut). \tag{7.35}$$

Now we show that (7.35) is the solution to (7.30) and (7.34). First if we take $t = 0$, then (7.35) reduces to $\rho(0,x) = f(x(0) - u \cdot 0) = f(x(0))$. To check the PDE we require the first partial derivatives of ρ. Notice that f is a function of only one variable, i.e. of $x - ut$. Therefore,

$$\begin{aligned} \frac{\partial \rho}{\partial t} &= \frac{df(x-ut)}{dt} = \frac{df}{d(x-ut)}\frac{d(x-ut)}{dt} = -u\frac{df}{d(x-ut)}, \\ \frac{\partial \rho}{\partial x} &= \frac{df(x-ut)}{dx} = \frac{df}{d(x-ut)}\frac{d(x-ut)}{dx} = \frac{df}{d(x-ut)}. \end{aligned} \tag{7.36}$$

Substituting these two derivatives in (7.30) we see that the equation is satisfied.

7.8 Semi-Lagrangian method

The characteristics-based semi-Lagrangian method (Courant *et al.*, 1952; Staniforth and Coté, 1991) and the TVD method (next section) are used to compute time-dependent advection-dominated partial differential equations (e.g. advection equations for density or viscosity or temperature, the advection–diffusion heat equation).

Consider the 3-D heat advection–diffusion equation:

$$\partial T/\partial t + \mathbf{u} \cdot \nabla T = \nabla^2 T + f, \quad t \in [0, \vartheta], \quad \mathbf{x} \in \Omega, \tag{7.37}$$

where $\Omega = [0, x_1 = l_1] \times [0, x_2 = l_2] \times [0, x_3 = l_3]$ is the model domain, T is the temperature, $\mathbf{u}$ is the velocity and f is the heat source. The semi-Lagrangian method accounts for the Lagrangian nature of the advection process but, at the same time, it allows computations on a fixed grid. We rewrite the heat equation (7.37) in the following form

$$DT/Dt = \nabla^2 T + f, \quad DT/Dt = \partial T/\partial t + \mathbf{u} \cdot \nabla T. \tag{7.38}$$

The aim of such a splitting is to solve the first equation on the characteristics of the second equation. This method has been used in advection–diffusion systems owing to two useful properties of the approximations: (i) a relatively large time step may be used in a numerical simulation, and (ii) it is stable and accurate for arbitrary relations between the time and space steps (e.g. Ewing and Wang, 2001). Moreover, the implementation of this method with a high-order interpolation of the space variables yields a minimum error in the variance. In particular, such an approach is intensively used in meteorology, where the time step must be large to ensure computational efficiency (e.g. Staniforth and Coté, 1991).

Equations (7.38) are approximated by finite differences in the following form

$$\frac{T_{ijk}^{n+1} - T_d^n}{\tau} = \nabla^2 \frac{T_{ijk}^{n+1} + T_{ijk}^n}{2} + \frac{f_{ijk}^{n+1} + f_{ijk}^n}{2}, \tag{7.39}$$

$$D\mathbf{z}/Dt = \mathbf{u}(t, \mathbf{z}), \ \mathbf{z}(t_{n+1}) = \mathbf{z}_a, \tag{7.40}$$

where T_d^n is the temperature at the point $\mathbf{z}_d$. The point $\mathbf{z}_d$ is obtained by solving (7.40) backward in time with the final condition $\mathbf{z}_a$, which should coincide with the corresponding point of the regular grid ω_{ijk} at $t = t_{n+1}$. A solution to (7.40) can be obtained by solving the following system of non-linear equations by an iterative implicit method (the number of equations is equal to the number of grid points):

$$\mathbf{z}_d = \mathbf{z}_a - \mathbf{y}_k, \quad \mathbf{y}_{k+1} = \tau \mathbf{u}(t_n, \mathbf{z}_a - 0.5\mathbf{y}_k), \quad \mathbf{y}_0 = \tau \mathbf{u}(t_n, \mathbf{z}_a), \quad k = 0, 1, 2, \ldots \tag{7.41}$$

It can also be solved using the explicit predictor–corrector method

$$\mathbf{z}^* = \mathbf{z}_a - \tau \mathbf{u}(t_n, \mathbf{z}_a), \ \mathbf{z}_d = \mathbf{z}_a - \tau \mathbf{u}(t_n, \mathbf{z}^*). \tag{7.42}$$

The point $\mathbf{z}^*$ does not necessarily coincide with a grid point, and the velocity at this point can be obtained by interpolating the velocities at the adjacent grid points. The value of T_d^n at the time $t = t_n$ and at the point $\mathbf{z}_d$ can also be obtained by interpolation.

The total error of the method is estimated to be $O(\tau^2 + h^2 + \tau^s + \tau^{-1} h^{1+q})$ and is not monotonic with respect to the time step τ, where h is the spatial grid size, s is the order of integration of (7.40) backward in time, and q is the interpolation order (McDonald and

Bates 1987; Falcone and Ferretti 1998). For example, $s = 2$ for the predictor–corrector method (7.42), and $s = 4$ for the Runge–Kutta method. If cubic polynomials are used for interpolation, then $q = 3$; for linear interpolation $q = 1$.

A solution to (7.41) can be obtained in three to four iterations, if Newton's method is used to solve the set of the non-linear equations and the Courant–Friedrichs–Lewy condition $\tau \|\partial \mathbf{u}/\partial \mathbf{x}\| < 1$ is satisfied (Courant *et al.*, 1928). This condition guarantees that the trajectories of the characteristics do not intersect at one time step. The procedure of solving the characteristic equation forward and backward in time is unconditionally stable. Method (7.42) is easier to implement, but it is inferior to method (7.41) in terms of accuracy.

The three-dimensional spatial discrete operator associated with the diffusion term in (7.39) is split into one-dimensional operators as $\nabla^2 \approx \Delta_1 + \Delta_2 + \Delta_3$, and the latter operators are approximated by the central differences:

$$\Delta_1 T^n_{ijk} = \frac{T^n_{i+1jk} - 2T^n_{ijk} + T^n_{i-1jk}}{h_1^2}, \quad i = 1, 2, \ldots, n_1 - 1. \tag{7.43}$$

At the boundary grid points $i = 0$ and $i = n_1$, an approximation for Δ_1 is obtained from (7.43) with regard for prescribed boundary conditions. Expressions for Δ_2 and Δ_3 are determined similarly. The set of difference equations for the approximation of the heat equation (7.37) on a uniform rectangular grid has the form:

$$T^+_{ijk} = T(t_n, \omega_{ijk} - \tau \mathbf{u}(t_n, \mathbf{z}_d)), \tag{7.44}$$

$$T^*_{ijk} = T^+_{ijk} + 1.5\tau\Delta_1(T^*_{ijk} + T^+_{ijk}) + 1.5\tau(f^{n+1}_{ijk} + f^n_{ijk}), \tag{7.45}$$

$$T^{**}_{ijk} = T^+_{ijk} + 1.5\tau\Delta_2(T^{**}_{ijk} + T^+_{ijk}), \tag{7.46}$$

$$T^{***}_{ijk} = T^+_{ijk} + 1.5\tau\Delta_3(T^{***}_{ijk} + T^+_{ijk}), \tag{7.47}$$

$$T^{n+1}_{ijk} = (T^*_{ijk} + T^{**}_{ijk} + T^{***}_{ijk})/3. \tag{7.48}$$

In the numerical implementation of this scheme, $3(n_1 n_2 n_3)$ equations (7.41) or (7.42) and $3(n_1 n_2 + n_1 n_3 + n_2 n_3)$ independent sets of linear algebraic equations (7.45)–(7.47) with tridiagonal (diagonally dominant) matrices should be solved. To determine T^+_{ijk} the velocity and temperature should be interpolated at the point $\mathbf{z}_d$. Equations (7.45)–(7.47) can be solved independently, and hence it is straightforward to design a numerical code for multi-processor computers using the method of tridiagonal matrix factorisation (e.g. Axelsson, 1996).

7.9 Total variation diminishing methods

The concept of total variation diminishing (TVD) schemes for treating advection was introduced by Harten (1983); several individual schemes fall into this category. The idea of TVD schemes is to achieve high-order accuracy while avoiding numerical overshoots and wiggles. To introduce the method we again consider the 3-D heat advection–diffusion equation (7.37). When oscillations (e.g. owing to jumps in physical parameters or non-smoothness of the solution) arise, the numerical solution will have larger *total variation of temperature* (that is, the sum of the variations of temperature $\mathrm{TV}^n = \sum_i |T^n_{i+1jk} - T^n_{ijk}| + \sum_j |T^n_{ij+1k} - T^n_{ijk}| + \sum_k |T^n_{ijk+1} - T^n_{ijk}|$ over the whole computational domain will increase with oscillations). TVD methods are designed to yield well-resolved, non-oscillatory discontinuities by enforcing that the numerical schemes generate solutions with non-increasing total variations of temperature in time (that is $\mathrm{TV}^{n+1} \le \mathrm{TV}^n$), and thus no spurious numerical oscillations are generated (Ewing and Wang, 2001). TVD methods can treat convection problems with large temperature gradients very well, because they have high-order accuracy except in the neighbourhood of high temperature gradients, where they decrease to first-order accuracy (Wang and Hutter, 2001).

Consider initially an approximation of the advection term of (7.37):

$$\Xi_1 = u_1 \partial T/\partial x_1 \approx \left(F^+_{x_1} - F^-_{x_1}\right)/h_1, \tag{7.49}$$

$$F^+_{x_1} = 0.5 u_{1,ijk}\left(T^+_{i+1/2jk} + T^-_{i+1/2jk}\right) - 0.5\left|u_{1,ijk}\right|\left(T^+_{i+1/2jk} - T^-_{i+1/2jk}\right), \tag{7.50}$$

$$F^-_{x_1} = 0.5 u_{1,ijk}\left(T^+_{i-1/2jk} + T^-_{i-1/2jk}\right) - 0.5\left|u_{1,ijk}\right|\left(T^+_{i-1/2jk} - T^-_{i-1/2jk}\right), \tag{7.51}$$

$$T^-_{i+1/2jk} = T_{ijk} + 0.5\Upsilon(\xi_i)(T_{i+1jk} - T_{ijk}), T^+_{i+1/2jk} = T_{i+1jk} - 0.5\Upsilon(\xi_{i+1})(T_{i+2jk} - T_{i+1jk}), \tag{7.52}$$

$$T^-_{i-1/2jk} = T_{i-1jk} + 0.5\Upsilon(\xi_{i-1})(T_{ijk} - T_{i-1jk}), T^+_{i-1/2jk} = T_{ijk} - 0.5\Upsilon(\xi_i)(T_{i+1jk} - T_{ijk}), \tag{7.53}$$

$$\xi_i = \left(T_{ijk} - T_{i-1jk}\right)/\left(T_{i+1jk} - T_{ijk}\right), \quad \Upsilon(\xi) = \max\left\{0, \min\left\{1, 2\xi\right\}, \min\left\{\xi, 2\right\}\right\}, \tag{7.54}$$

where $\Upsilon(\xi)$ is a *superbee* limiter (Sweby, 1984). Expressions for Ξ_2 and Ξ_3 are determined similarly. The solution based on the implicit TVD method gives a second-order accurate solution (Wang and Hutter, 2001). Since the formula (7.54) can generate logical difficulties in the case of $T_{ijk} = T_{i-1jk} = T_{i+1jk}$, the following alternative representation of (7.54) can be used in computations:

$$\Upsilon(\xi_i)(A) = L(A,B) = \Upsilon(1/\xi_i)(B),\ A = T_{i+1jk} - T_{ijk},\ B = T_{ijk} - T_{i-1jk}, \tag{7.55}$$

$$L(A,B) = 0.5(\mathrm{sign}(A) + \mathrm{sign}(B))\max\{\min\{2|A|, |B|\}, \min\{|A|, 2|B|\}\}. \tag{7.56}$$

This representation of the limiter Υ has an explicit symmetric form compared with (7.52)–(7.54). The TVD numerical scheme was tested using known solutions to simple advection equations and also compared to another TVD numerical scheme by Samarskii and Vabishchevich (1995).

The three-dimensional spatial discrete operator associated with the diffusion term in (7.37) is split into one-dimensional operators as $\nabla^2 \approx \Delta_1 + \Delta_2 + \Delta_3$, and the latter operators are approximated by the central differences (7.43) as described in Section 7.7.

The system of difference equations for the approximation of the heat equation (7.37) on a uniform rectangular grid has the form

$$\frac{T^{*}_{ijk} - T^{n}_{ijk}}{3\tau} + \Xi_1 T^{n}_{ijk} = \Delta_1 T^{n}_{ijk}, \tag{7.57}$$

$$\frac{T^{**}_{ijk} - T^{n}_{ijk}}{3\tau} + \Xi_2 T^{n}_{ijk} = \Delta_2 T^{n}_{ijk} + f^{n}_{ijk}, \tag{7.58}$$

$$\frac{T^{***}_{ijk} - T^{n}_{ijk}}{3\tau} + \Xi_3 T^{n}_{ijk} = \Delta_3 T^{n}_{ijk}, \tag{7.59}$$

$$T^{n+1}_{ijk} = (T^{*}_{ijk} + T^{**}_{ijk} + T^{***}_{ijk})/3. \tag{7.60}$$

The total error of the numerical method is $O(\tau + h^2)$, and the iterations are stable. Considering the independence of (7.57)–(7.59), they can be solved on a parallel computer using the method of three-diagonal matrix factorisation (e.g. Axelsson, 1996).

7.10 Lagrangian methods

7.10.1 Lagrangian meshes

Some codes, particularly those designed to model the lithosphere (Section 10.2.2) use a Lagrangian mesh, in which the nodes move with the flow. Quantities stored on the nodes are thus naturally advected with no numerical diffusion. Care must, however, be taken when remeshing, which is often necessary after the mesh becomes distorted and may involve interpolation of quantities onto new nodal points. Lagrangian codes normally use a finite element discretisation so physical diffusion, if present, can be calculated straightforwardly.

7.10.2 Particle-in-cell method

The advantages of Lagrangian advection can be obtained even when using an Eulerian mesh by storing and advecting quantities on Lagrangian particles (also known as *tracers* or *markers*) that are advected with the flow (this is the basis of the *particle-in-cell method*). Tracer advection can be performed using standard Runge–Kutta or predictor–corrector methods. In geodynamic applications this approach is commonly used for composition, as discussed in Section 10.5.

It is also possible to use Lagrangian markers for a diffusive field such as temperature, with diffusion calculated on the Eulerian grid (Gerya and Yuen, 2003). This involves the following procedure. (i) Computation of temperatures on the Eulerian nodes (grid points) by locally averaging temperature values carried on tracers as

$$T_i^{node} = \sum_{j=1}^{Ntr} T_j^{tracer} S_i(\mathbf{x}_j) \Big/ \sum_{j=1}^{Ntr} S_i(\mathbf{x}_j), \tag{7.61}$$

where T_i^{node} is the temperature at node i, T_j^{tracer} is the temperature of tracer j with position $\mathbf{x}_j$, Ntr is the number of tracers, and S_i is the shape (averaging) function for node i, which in the simplest case varies linear between a value of unity at node i to zero at adjacent nodes. (ii) Computation of the change of temperature due to diffusion on the nodes by using a standard method such as explicit finite differences. (iii) Interpolation of this change in temperature onto the markers, using linear interpolation in the simplest case:

$$\Delta T_j^{tracer} = \sum_{i=1}^{Nnodes} \Delta T_i^{node} S_i(\mathbf{x}_j). \tag{7.62}$$

Interpolating the change in temperature rather than absolute temperature avoids numerical diffusion. A problem with this approach per se is that after some circulation has taken place the temperature can be quite different on tracers separated by less than one grid spacing, i.e. at the sub-grid scale, leading to a noisy temperature field. To correct this, a method of smoothing temperature variations between nearby markers at each time step was introduced by Gerya and Yuen (2003). In this method, tracer temperatures are relaxed towards a linear profile interpolated between nodes, with a time scale proportional to the diffusion time scale for the grid spacing:

$$\begin{aligned} \Delta T_j^{tracer} &= \left(\mathrm{interp}_j(T_i^{node}) - T_j^{tracer}\right)\left(1 - \exp\left[-a\frac{\Delta t}{\Delta t_{diff}}\right]\right), \\ \Delta t_{diff} &= \frac{\left(2/\Delta x^2 + 2/\Delta z^2\right)^{-1}}{\kappa}, \end{aligned} \tag{7.63}$$

where Δt is the time step, κ is the thermal diffusivity, a is a parameter between 0 and 1 that must be adjusted to give optimal results and $\mathrm{interp}_j()$ represents the interpolation operation defined in equation (7.62). On its own, this sub-grid-scale smoothing might cause some undesirable diffusion between grid points. This is subtracted in a second step. This second step involves (i) calculating the change in temperature at the nodes caused by the change in of tracer temperatures from Eq. (7.63) then (ii) interpolating the negative of this onto the tracer positions, ending up with:

$$\left(T_j^{tracer}\right)_{smoothed} = \left(T_j^{tracer}\right)_{old} + \Delta T_j^{tracer} - \mathrm{interp}_j\left(\mathrm{navg}\left(\Delta T_j^{tracer}\right)\right), \tag{7.64}$$

where navg() represents the local averaging operation defined in Eq. (7.61). This two-step smoothing procedure can also be combined with the calculation of grid-scale diffusion.

8 Data assimilation methods

8.1 Introduction

Many geodynamic problems can be described by mathematical models, i.e. by a set of partial differential equations and boundary and/or initial conditions defined in a specific domain. A mathematical model links the causal characteristics of a geodynamic process with its effects. The causal characteristics of the process include, for example, parameters of the initial and boundary conditions, coefficients of the differential equations, and geometrical parameters of a model domain. The aim of the *direct* mathematical problem is to determine the relationship between the causes and effects of the geodynamic process and hence to find a solution to the mathematical problem for a given set of parameters and coefficients. An *inverse* problem is the opposite of a direct problem. An inverse problem is considered when there is a lack of information on the causal characteristics (but information on the effects of the geodynamic process exists). Inverse problems can be subdivided into time-reverse or retrospective problems (e.g. to restore the development of a geodynamic process), coefficient problems (e.g. to determine the coefficients of the model equations and/or boundary conditions), geometrical problems (e.g. to determine the location of heat sources in a model domain or the geometry of the model boundary), and some others. In this chapter we will consider time-reverse (retrospective) problems in geodynamics.

Inverse problems are often ill-posed. *Jacques Hadamard* (1865–1963) introduced the idea of well- (and ill-) posed problems in the theory of partial differential equations (Hadamard, 1902). A mathematical model for a geophysical problem has to be well-posed in the sense that it has to have the properties of existence, uniqueness and stability of a solution to the problem. Problems for which at least one of these properties does not hold are called ill-posed. The requirement of stability is the most important one. If a problem lacks the property of stability then its solution is almost impossible to compute because computations are polluted by unavoidable errors. If the solution of a problem does not depend continuously on the initial data, then, in general, the computed solution may have nothing to do with the true solution.

The inverse (retrospective) problem of thermal convection in the mantle is an ill-posed problem, since the backward heat problem, describing both heat advection and conduction through the mantle backwards in time, possesses the properties of ill-posedness (Kirsch, 1996). In particular, the solution to the problem does not depend continuously on the initial data. This means that small changes in the present-day temperature field may result in large changes of predicted mantle temperatures in the past. Let us explain this statement in the case of the one-dimensional (1-D) diffusion equation.

Consider the following boundary-value problem for the 1-D backward diffusion equation:

$$\partial u(t,x)/\partial t = \partial^2 u(t,x)/\partial x^2,\ 0 \le x \le \pi,\ t \le 0 \tag{8.1}$$

with the following boundary and initial conditions

$$u(t,0) = 0 = u(t,\pi), \quad t \le 0, \tag{8.2}$$

$$u(0,x) = \phi_n(x), \quad 0 \le x \le \pi. \tag{8.3}$$

At the initial time we assume that the function $\phi_n(x)$ takes the following two forms:

$$\phi_n(x) = \frac{\sin((4n+1)x)}{4n+1} \tag{8.4}$$

and

$$\phi_0(x) \equiv 0. \tag{8.5}$$

Note that

$$\max_{0 \le x \le \pi} |\phi_n(x) - \phi_0(x)| \le \frac{1}{4n+1} \to 0 \quad \text{at} \quad n \to \infty. \tag{8.6}$$

The following two solutions of the problem correspond to the two chosen functions of $\phi_n(x)$, respectively:

$$u_n(t,x) = \frac{\sin((4n+1)x)}{4n+1} \exp(-(4n+1)^2 t) \quad \text{at} \quad \phi_n(x) = \phi_n \tag{8.7}$$

and

$$u_0(t,x) \equiv 0 \quad \text{at} \quad \phi_n(x) = \phi_0. \tag{8.8}$$

At $t = -1$ and $x = \pi/2$ we obtain

$$u_n(-1,\pi/2) = \frac{1}{4n+1} \exp((4n+1)^2) \quad \text{at} \quad n \to \infty. \tag{8.9}$$

At large n two closely set initial functions ϕ_n and ϕ_0 are associated with the two strongly different solutions at $t = -1$ and $x = \pi/2$. Hence, a small error in the initial data (8.6) can result in very large errors in the solution to the backward problem (8.9), and therefore the solution is unstable, and the problem is ill-posed.

Despite the fact that many inverse problems are ill-posed, there are methods for solving the problems. *Andrei Tikhonov* (1906–1993) introduced the idea of conditionally well-posed problems and the regularisation method (Tikhonov, 1963). According to Tikhonov, a class of admissible solutions to conditionally ill-posed problems should be selected to satisfy the following conditions: (i) a solution exists in this class, (ii) the solution is unique in the same class and (iii) the solution depends continuously on the input data. The Tikhonov

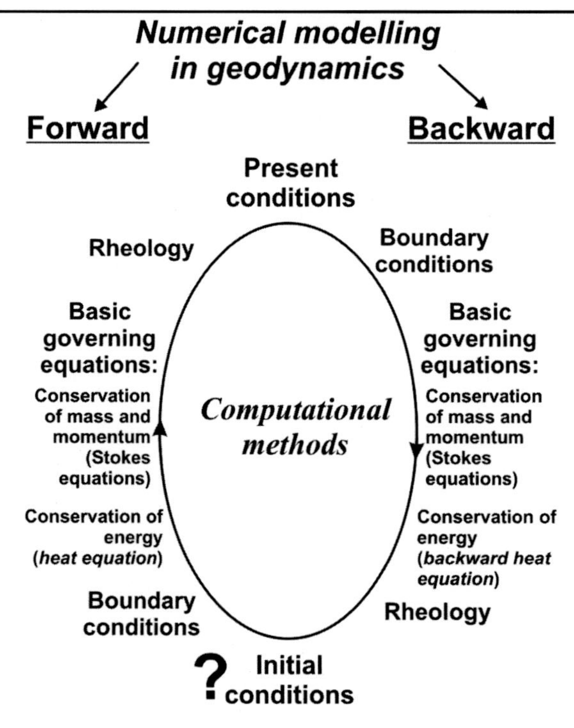

Fig. 8.1. Flowchart of forward and backward numerical modelling in geodynamics.

regularisation is essentially a trade-off between fitting the observations and reducing a norm of the solution to the mathematical model of a geophysical problem.

Forward modelling in geodynamics is associated with the solution of direct mathematical problems, and *backward modelling* with the solution of inverse (time-reverse) problems. Figure 8.1 illustrates the flow in forward and backward numerical modelling. In forward modelling one starts with unknown initial conditions, which are added to a set of governing equations, rheological law and boundary conditions to define properly the relevant mathematical problem. Once the problem is stated, a numerical model (a set of discrete equations) is solved forward in time using computational methods. The initial conditions of the numerical model vary (keeping all other model parameters unchanged) to fit model results to reality (present observations). Because the model depends on the initial conditions and they are unknown *a priori*, the task 'to fit model results to reality' becomes difficult.

Another approach is to use backward modelling. In this case present observations are employed as input conditions for the mathematical model. We shall use the term of 'input conditions' in backward modelling to distinguish it from the term of 'initial conditions' for the forward modelling, although the 'input conditions' are the initial conditions for the mathematical model in backward modelling. The aim of backward modelling in geodynamics is to find the 'initial conditions' in the geological past from the present observations and to restore mantle structures accordingly. Special methods are required to assimilate present observations to the past (Ismail-Zadeh *et al.*, 2009). In the following sections we describe the methods for data assimilation.

8.2 Data assimilation

The mantle is heated from the core and from inside owing to decay of radioactive elements. Since thermal convection in the mantle is described by heat advection and diffusion, one can ask: is it possible to tell, from the present temperature distribution estimations of the Earth, something about the Earth's temperature distribution in the geological past? Even though heat diffusion is irreversible in the physical sense, it is possible to predict accurately the heat transfer backwards in time by using data assimilation techniques without contradicting the basic thermodynamic laws (see, for example, Ismail-Zadeh *et al.*, 2004a, 2007).

To restore mantle dynamics in the geological past, data assimilation techniques can be used to constrain the initial conditions for the mantle temperature and velocity from their present observations. The initial conditions so obtained can then be used to run forward models of mantle dynamics to restore the evolution of mantle structures. *Data assimilation* can be defined as the incorporation of observations (in the present) and initial conditions (in the past) in an explicit dynamic model to provide time continuity and coupling among the physical fields (e.g. velocity, temperature). The basic principle of data assimilation is to consider the initial condition as a control variable and to optimise the initial condition in order to minimise the discrepancy between the observations and the solution of the model.

If heat diffusion is neglected, the present mantle temperature and flow can be assimilated into the past by using the backward advection (BAD). Numerical approaches to the solution of the inverse problem of the Rayleigh–Taylor instability were developed for a dynamic restoration of diapiric structures to their earlier stages (Ismail-Zadeh *et al.*, 2001b; Kaus and Podladchikov, 2001; Korotkii *et al.*, 2002; Ismail-Zadeh *et al.*, 2004b). Steinberger and O'Connell (1998) and Conrad and Gurnis (2003) modelled the mantle flow backwards in time from present-day mantle density heterogeneities inferred from seismic observations.

In sequential filtering a numerical model is computed forward in time for the interval for which observations have been made, updating the model each time where observations are available. The sequential filtering was used to compute mantle circulation models (Bunge *et al.*, 1998, 2002). Despite sequential data assimilation well adapted to mantle circulation studies, each individual observation influences the model state at later times. Information propagates from the geological past into the future, although our knowledge of the Earth's mantle at earlier times is much poorer than that at present.

The variational (VAR) data assimilation method has been pioneered by meteorologists and used very successfully to improve operational weather forecasts (see Kalnay, 2003). The data assimilation has also been widely used in oceanography (see Bennett, 1992) and in hydrological studies (see McLaughlin, 2002). The use of VAR data assimilation in models of mantle dynamics (to estimate mantle temperature and flow in the geological past) has been put forward by Bunge *et al.* (2003) and Ismail-Zadeh *et al.* (2003a, b) independently. The major differences between the two approaches are that Bunge *et al.* (2003) applied the VAR method to the coupled Stokes, continuity and heat equations (generalised inverse), whereas Ismail-Zadeh *et al.* (2003a) applied the VAR method to the heat equation only. The VAR approach by Ismail-Zadeh *et al.* (2003a) is computationally less expensive, because it does not involve the Stokes equation in the iterations between the direct and adjoint problems. Moreover, this approach admits the use of temperature-dependent viscosity.

The VAR data assimilation algorithm was employed for numerical restoration of models of present prominent mantle plumes to their past stages (Ismail-Zadeh *et al.*, 2004a; Hier-Majumder *et al.*, 2005). Effects of thermal diffusion and temperature-dependent viscosity on the evolution of mantle plumes was studied by Ismail-Zadeh *et al.* (2006) to recover the structure of mantle plumes prominent in the past from that of present plumes weakened by thermal diffusion. Liu and Gurnis (2008) simultaneously inverted mantle properties and initial conditions using the VAR data assimilation method and applied the method to reconstruct the evolution of the Farallon Plate subduction (Liu *et al.*, 2008).

The quasi-reversibility (QRV) method was introduced by Lattes and Lions (1969). The use of the QRV method implies the introduction into the backward heat equation of the additional term involving the product of a small regularisation parameter and a higher-order temperature derivative. The data assimilation in this case is based on a search of the best fit between the forecast model state and the observations by minimising the regularisation parameter. The QRV method was introduced in geodynamic modelling (Ismail-Zadeh *et al.*, 2007) and employed to assimilate data in models of mantle dynamics (Ismail-Zadeh *et al.*, 2008).

In this chapter we describe three principal techniques used to assimilate data related to geodynamics: (i) backward advection, (ii) variational (adjoint) and (iii) quasi-reversibility methods.

8.3 Backward advection (BAD) method

We consider the three-dimensional model domain $\Omega = [0, x_1 = 3h] \times [0, x_2 = 3h] \times [0, x_3 = h]$, where $\mathbf{x} = (x_1, x_2, x_3)$ are the Cartesian coordinates and h is the depth of the domain, and assume that the mantle behaves as a Newtonian incompressible fluid with a temperature-dependent viscosity and infinite Prandtl number. The mantle flow is described by heat, motion and continuity equations (Chandrasekhar, 1961). To simplify the governing equations, we make the Boussinesq approximation (Boussinesq, 1903) keeping the density constant everywhere except for buoyancy term in the equation of motion. In the Boussinesq approximation the dimensionless equations take the form:

$$\partial T/\partial t + \mathbf{u} \cdot \nabla T = \nabla^2 T, \quad \mathbf{x} \in \Omega,\ t \in (0, \vartheta), \tag{8.10}$$

$$\nabla P = \operatorname{div}[\eta \mathbf{E}] + Ra T \mathbf{e}, \quad \mathbf{E} = \{\partial u_i/\partial x_j + \partial u_j/\partial x_i\}, \quad \mathbf{e} = (0, 0, 1), \tag{8.11}$$

$$\operatorname{div}\mathbf{u} = 0, \quad t \in (0, \vartheta), \quad \mathbf{x} \in \Omega. \tag{8.12}$$

Here $T, t, \mathbf{u} = (u_1, u_2, u_3), P$ and η are dimensionless temperature, time, velocity, pressure and viscosity, respectively. The Rayleigh number is defined as $Ra = \alpha g \rho_{ref} \Delta T h^3 \eta_{ref}^{-1} \kappa^{-1}$, where α is the thermal expansivity, g is the acceleration due to gravity, ρ_{ref} and η_{ref} are the reference typical density and viscosity, respectively; ΔT is the temperature contrast between the lower and upper boundaries of the model domain; and κ is the thermal diffusivity. In Eqs. (8.10)–(8.12) length, temperature and time are normalised by h, ΔT and $h^2\kappa^{-1}$, respectively.

At the boundary Γ of the model domain Ω we set the impenetrability condition with no-slip or perfect slip conditions: $\mathbf{u} = 0$ or $\partial \mathbf{u}_\tau / \partial \mathbf{n} = 0$, $\mathbf{u} \cdot \mathbf{n} = 0$, where $\mathbf{n}$ is the outward unit normal vector at a point on the model boundary, and $\mathbf{u}_\tau$ is the projection of the velocity vector onto the tangent plane at the same point on the model boundary. We assume zero heat flux through the vertical boundaries of the box. Either temperature or heat flux are prescribed at the upper and lower boundaries of the model domain. To solve the problem forward or backward in time we assume the temperature to be known at the initial time ($t = 0$) or at the present time ($t = \vartheta$). Equations (8.10)–(8.12) together with the boundary and initial conditions describe a thermo-convective mantle flow.

The principal difficulty in solving the problem (8.10)–(8.12) backward in time is the ill-posedness of the backward heat problem and the presence of the heat diffusion term in the heat equation. The backward advection (BAD) method suggests neglecting the heat diffusion term, and the heat advection equation can then be solved backward in time. Both direct (forward in time) and inverse (backward in time) problems of the heat (density) advection are well-posed. This is because the time-dependent advection equation has the same form of characteristics for the direct and inverse velocity field (the vector velocity reverses its direction, when time is reversed). Therefore, numerical algorithms used to solve the direct problem of the gravitational instability can also be used in studies of the time-reverse problems by replacing positive time steps with negative ones.

Using the BAD method, Steinberger and O'Connell (1998) studied the motion of hotspots relative to the deep mantle. They combined the advection of plumes, which are thought to cause the hotspots on the Earth's surface, with a large-scale mantle flow field and constrained the viscosity structure of the Earth's mantle. Conrad and Gurnis (2003) modelled the history of mantle flow by using a tomographic image of the mantle beneath southern Africa as an input (initial) condition for the backward mantle advection model while reversing the direction of flow. If the resulting model of the evolution of thermal structures obtained by the BAD method is used as a starting point for a forward mantle convection model, present mantle structures can be reconstructed if the time of assimilation does not exceed 50–75 Myr.

8.4 Application of the BAD method: restoration of the evolution of salt diapirs

Salt is so buoyant and weak compared with most other rocks with which it is found that it develops distinctive structures with a wide variety of shapes and relationships with other rocks by various combinations of gravity, thermal effects and lateral forces. The crests of passive salt bodies can stay near the sedimentation surface while their surroundings are buried (downbuilt) by other sedimentary rocks (Jackson *et al.*, 1994). The profiles of downbuilt passive diapirs can simulate those of fir trees because they reflect the ratio of increase in diapir height relative to the rate of accumulation of the downbuilding sediments (Talbot, 1995) and lateral forces (Koyi, 1996). Salt movements can be triggered by faulting and driven by erosion and redeposition, differential loading, buoyancy and other geological processes. Many salt sequences are buried by overburdens sufficiently stiff to resist the buoyancy of the salt. Such salt will only be driven by differential loading into sharp-crested reactive-diapiric walls after the stiff overburden is weakened and thinned by faults (Vendeville and Jackson, 1992). Such reactive diapirs often rise up and out of the fault zone and thereafter can continue increasing in relief as by passive downbuilding of more sediment.

Active diapirs are those that lift or displace their overburdens. Although any erosion of the crests of salt structures and deposition of surrounding overburden rocks influence their growth, diapirs with significant relief have sufficient buoyancy to rise (upbuild) through stiff overburdens (Jackson *et al.*, 1994). The rapid deposition of denser and more viscous sediments over less dense and viscous salt results in the Rayleigh–Taylor instability. This leads to a gravity-driven single overturn of the salt layer with its denser but ductile overburden. Rayleigh–Taylor overturns (Ramberg, 1968) are characterised by the rise of rocksalt through overlying and younger compacting clastic sediments that are deformed as a result. The consequent salt structures evolve through a great variety of shapes. Perturbations of the interface between salt and its denser overburden result in the overburden subsiding as salt rises owing to the density inversion.

Two-dimensional (2-D) numerical models of salt diapirism were first developed by Woidt (1978) who examined how the viscosity ratio between the salt and its overburden affects the shapes and growth rate of diapirs. Schmeling (1987) demonstrated how the dominant wavelength and the geometry of gravity overturns are influenced by the initial shape of the interface between the salt and its overburden. Römer and Neugebauer (1991) presented numerical results of modelling diapiric structures in a multilayered medium. Later Poliakov *et al.* (1993a) and Naimark *et al.* (1998) developed numerical models of diapiric growth considering the effects of sedimentation and redistribution of sediments. Van Keken *et al.* (1993), Poliakov *et al.* (1993b), Daudre and Cloetingh (1994), and Poliakov *et al.* (1996) introduced non-linear rheological properties of salt and overburden into their numerical models. The authors mentioned above used various numerical methods to compute the models of salt diapirism, among them FD method, Lagrangian and Eulerian FE method and their combination.

Two-dimensional analyses of the evolution of salt structures are restricted and not suitable for examining the complicated shapes of mature diapiric patterns. Resolving the geometry of gravity overturns requires three-dimensional (3-D) numerical modelling. Ismail-Zadeh *et al.* (2000b) analysed such typical 3-D structures as deep polygonal buoyant ridges, shallow salt-stock canopies and salt walls. Kaus and Podladchikov (2001) showed how complicated 3-D diapirs developed from initial 2-D perturbations of the interface between salt and its overburden.

The increasing application of 3-D seismic exploration in oil and gas prospecting points to the need for vigorous efforts toward numerical modelling of the evolution of salt structures in three dimensions, both forward and backward in time. Most numerical models of salt diapirism involved the forward evolution of salt structures toward increasing maturity. Ismail-Zadeh *et al.* (2001b) developed a numerical approach to 2-D dynamic restoration of cross-sections across salt structures. The approach was based on solving the inverse problem of gravitational instability by the BAD method. The same method was used in

3-D cases to model Rayleigh-Taylor instability backward in time (Kaus and Podladchikov, 2001; Korotkii *et al.*, 2002; Ismail-Zadeh *et al.*, 2004b).

We consider here the advection problem (slow flow of an incompressible fluid of variable density and viscosity due to gravity) in the rectangular domain Ω. A 3-D model of the flow of salt and of the viscous deformation of the overburden of salt is described by the Stokes equations (8.11), where the term $Ra\,T$ is replaced by the term $-g\rho$, and by Eq. (8.10), where temperature T is replaced by density ρ (viscosity η) and the term on the right-hand side is omitted. Equation (8.10) in this case describes the advection of density (viscosity) with the flow. For details of the numerical model see Section 4.10.2.

Although dimensionless values and functions are used in computations, numerical results are presented in dimensional form for the reader's convenience. The time step Δt is chosen from the condition that the maximum displacement does not exceed a given small value h: $\Delta t = h/u_{\max}$, where $u_{\max}$ is the maximum value of the flow velocity. Salt diapirs in the numerical model evolve from random initial perturbations of the interface between the salt and its overburden deposited on the top of horizontal salt layer prior to the interface perturbation. Initially the evolution of salt diapirs is modelled forward in time as presented in the model example in Section 4.10.2. Figures 8.2 (a–d, a front view) and 8.3 (a–d, a top view) show the positions of the interface between salt and overburden in the model at successive times over a period of about 21 Myr.

To restore the evolution of salt diapirs predicted by the forward model through successive earlier stages, a positive time is replaced by a negative time, and the problem is solved backward in time. Such a replacement is possible, because the characteristics of the advection equations have the same form for both direct and inverse velocity fields. The final position of the interface between salt and its overburden in the forward model (Figs. 8.2d and 8.3d) is used as an initial position of the interfaces for the backward model. Figures 8.2, d–g and 8.3, d–g illustrate successive steps in the restoration of the upbuilt diapirs. Least square errors δ of the restoration are calculated by using the formula:

$$\delta(x_1,x_2) = \left(\int_0^h \left(\rho(x_1,x_2,x_3) - \tilde{\rho}(x_1,x_2,x_3)\right)^2 dx_3 \right)^{1/2}, \tag{8.13}$$

where $\rho(x_1,x_2,x_3)$ is the density at initial time, and $\tilde{\rho}(x_1,x_2,x_3)$ is the restored density (Fig. 8.3h). The maximum value δ does not exceed 120 kg m^{-3}, and the error is associated with small areas of the initial interface's perturbation.

To demonstrate the stability of the restoration results with respect to changes in the density of the overburden, the restoration procedure was tested by synthetic examples. Initially the forward model is run for 200 computational time steps (about 30 Myr). Then the density contrast ($\delta\rho$) between salt and its overburden is changed by a few per cent: namely, $\delta\rho$ was chosen to be 400, 405, 410 (the actual contrast), 415 and 420 kg m^{-3}. The evolution of the system was restored for these density contrasts. Ismail-Zadeh *et al.* (2004b) found small discrepancies (less than 0.5%) between least square errors for all these test cases. The tests show that the solution is stable to small changes in the initial conditions, and this is in agreement with the mathematical theory of well-posed problems (Tikhonov and

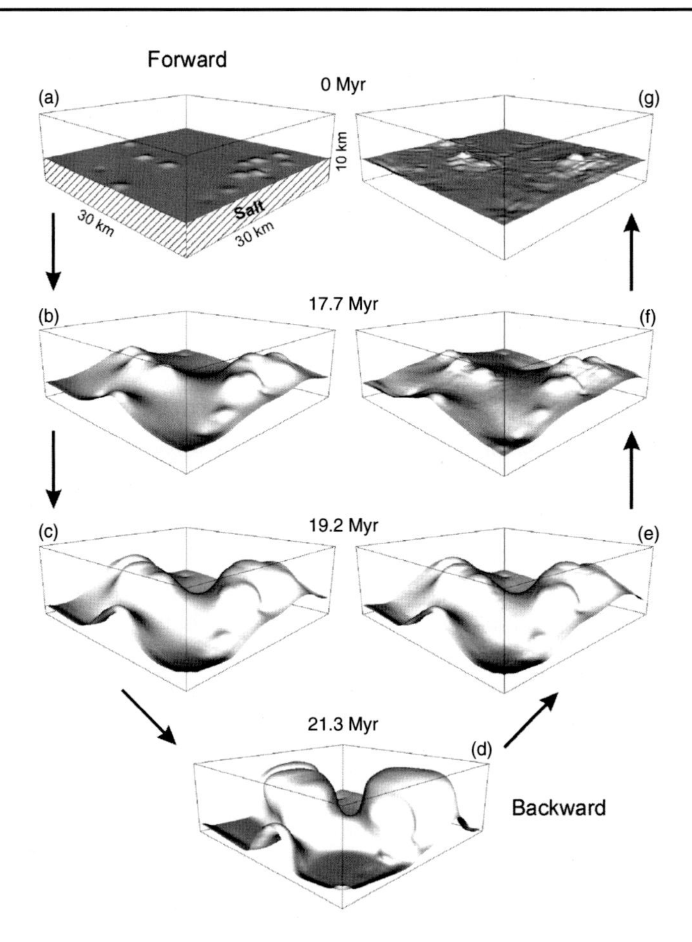

Fig. 8.2. **Evolution (front view) of salt diapirs toward increasing maturity (a)–(d) and restoration of the evolution (d)–(g). Interfaces between salt and its overburden are presented at successive times. After Ismail-Zadeh *et al.* (2004b).**

Samarskii, 1990). Meanwhile it should be mentioned that if the model is computed for a very long time and the less dense salt layer spreads uniformly into a horizontal layer near the surface, practical restoration of the layered structure becomes impossible (Ismail-Zadeh *et al.*, 2001b).

8.5 Variational (VAR) method

In this section we describe a variational approach to numerical restoration of thermo-convective mantle flow. The variational data assimilation is based on a search of the best fit between the forecast model state and the observations by minimising an objective functional (a normalised residual between the target model and observed variables) over space

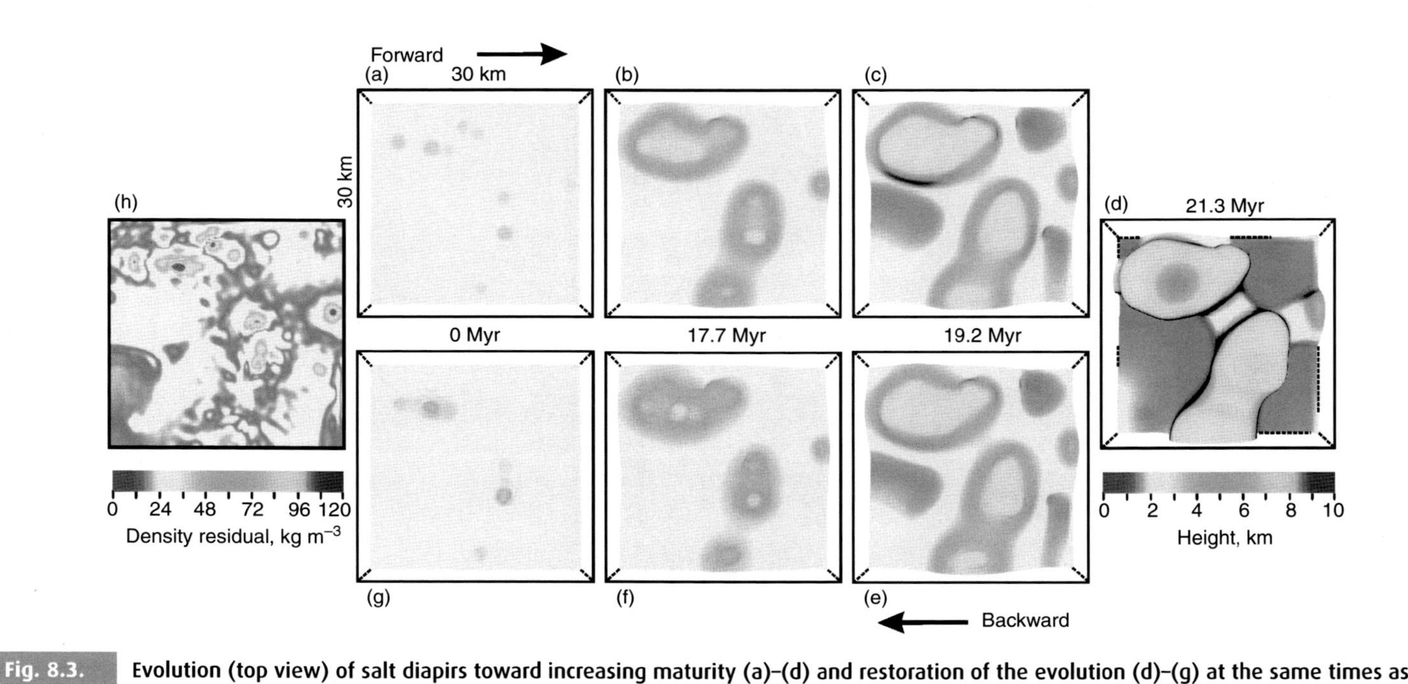

Fig. 8.3. Evolution (top view) of salt diapirs toward increasing maturity (a)–(d) and restoration of the evolution (d)–(g) at the same times as in Fig. 8.2. (h) Restoration errors. After Ismail-Zadeh *et al.* (2004b). (In colour as Plate 2. See colour plates section.)

and time. To minimise the objective functional over time, an assimilation time interval is defined and an adjoint model is typically used to find the derivatives of the objective functional with respect to the model states. The variational data assimilation is well suited for smooth problems (we discuss the problem of smoothness in Section 8.7).

The method for variational data assimilation can be formulated with a weak constraint (a generalised inverse) where errors in the model formulation are taken into account (Bunge *et al.*, 2003) or with a strong constraint where the model is assumed to be perfect except for the errors associated with the initial conditions (Ismail-Zadeh *et al.*, 2003a). Actually there are several sources of errors in forward and backward modelling of thermo-convective mantle flow, which we discuss in Section 8.12. The generalised inverse of mantle convection considers model errors, data misfit and the misfit of parameters as control variables. Unfortunately the generalised inverse presents a tremendous computational challenge and is difficult to solve in practice. Hence, Bunge *et al.* (2003) considered a simplified generalised inverse imposing a strong constraint on errors (ignoring all errors except for the initial condition errors). Therefore, the strong constraint makes the problem computationally tractable.

We consider the following objective functional at $t \in [0, \vartheta]$

$$J(\varphi) = \|T(\vartheta, \cdot; \varphi) - \chi(\cdot)\|^2, \tag{8.14}$$

where $\|\cdot\|$ denotes the norm in the space $L_2(\Omega)$ (the Hilbert space with the norm defined as $\|y\| = [\int_\Omega y^2(\mathbf{x})d\mathbf{x}]^{1/2}$). Since in what follows the dependence of solutions of the thermal boundary value problems on initial data is important, we introduce these data explicitly into the mathematical representation of temperature. Here $T(\vartheta, \cdot; \varphi)$ is the solution of the thermal boundary value problem (8.10) at the final time ϑ, which corresponds to some (unknown as yet) initial temperature distribution $\varphi(\mathbf{x})$; $\chi(\mathbf{x}) = T(\vartheta, \mathbf{x}; T_0)$ is the known temperature distribution at the final time, which corresponds to the initial temperature $T_0(\cdot)$. The functional has its unique global minimum at value $\varphi \equiv T_0$ and $J(T_0) \equiv 0$, $\nabla J(T_0) \equiv 0$ (Vasiliev, 2002).

To find the minimum of the functional we employ the gradient method ($k = 0, \ldots, j, \ldots$):

$$\varphi_{k+1} = \varphi_k - \beta_k \nabla J(\varphi_k), \quad \varphi_0 = T_*, \tag{8.15}$$

$$\beta_k = \begin{cases} J(\varphi_k)/\|\nabla J(\varphi_k)\|, & 1 \le k \le k_* \\ 1/(k+1), & k > k_* \end{cases}, \tag{8.16}$$

where T_* is an initial temperature guess. The minimisation method belongs to a class of limited-memory quasi-Newton methods (Zou *et al.*, 1993), where approximations to the inverse Hessian matrices are chosen to be the identity matrix. Equation (8.16) is used to maintain the stability of the iteration scheme (8.15). Consider that the gradient of the objective functional $\nabla J(\varphi_k)$ is computed with an error $\|\nabla J_\delta(\varphi_k) - \nabla J(\varphi_k)\| < \delta$, where $\nabla J_\delta(\varphi_k)$ is the computed value of the gradient. We introduce the function $\varphi^\infty = \varphi_0 - \sum_{k=1}^{\infty} \beta_k \nabla J(\varphi_k)$, assuming that the infinite sum exists, and the function $\varphi_\delta^\infty = \varphi_0 - \sum_{k=1}^{\infty} \beta_k \nabla J_\delta^(\varphi_k)$ as the computed value of φ^∞. For stability of the iteration method (8.15), the following inequality should be held:

$$\|\varphi_\delta^\infty - \varphi^\infty\| = \left\| \sum_{k=1}^{\infty} \beta_k (\nabla J_\delta(u_k) - \nabla J(u_k)) \right\|$$
$$\leq \sum_{k=1}^{\infty} \beta_k \|\nabla J_\delta(\varphi_k) - \nabla J(\varphi_k)\| \leq \delta \sum_{k=1}^{\infty} \beta_k.$$

The sum $\sum_{k=1}^{\infty} \beta_k$ is finite, if $\beta_k = 1/k^p, p > 1$. We use $p = 1$, but the number of iterations is limited, and therefore, the iteration method is conditionally stable, although the convergence rate of these iterations is low. Meanwhile the gradient of the objective functional $\nabla J(\varphi_k)$ decreases steadily with the number of iterations providing the convergence, although the absolute value of $J(\varphi_k)/\|\nabla J(\varphi_k)\|$ increases with the number of iterations, and it can result in instability of the iteration process (Samarskii and Vabischevich, 2004).

The minimisation algorithm requires the calculation of the gradient of the objective functional, ∇J. This can be done through the use of the *adjoint* problem for the model equations (8.10)–(8.12) with the relevant boundary and initial conditions. In the case of the heat problem, the adjoint problem can be represented in the following form:

$$\begin{aligned}
&\partial\Psi/\partial t + \mathbf{u}\cdot\nabla\Psi + \nabla^2\Psi = 0, && \mathbf{x}\in\Omega,\ t\in(0,\vartheta),\\
&\sigma_1\Psi + \sigma_2\partial\Psi/\partial\mathbf{n} = 0, && \mathbf{x}\in\Gamma,\ t\in(0,\vartheta),\\
&\Psi(\vartheta,\mathbf{x}) = 2(T(\vartheta,\mathbf{x};\varphi) - \chi(\mathbf{x})), && \mathbf{x}\in\Omega,
\end{aligned} \tag{8.17}$$

where σ_1 and σ_2 are some smooth functions or constants satisfying the condition $\sigma_1^2 + \sigma_2^2 \neq 0$. Selecting σ_1 and σ_2 we can obtain corresponding boundary conditions.

The solution to the adjoint problem (8.17) is the gradient of the objective functional (8.14). To prove the statement, we consider an increment of the functional J in the following form:

$$\begin{aligned}
J(\varphi+h) - J(\varphi) &= \int_\Omega (T(\vartheta,\mathbf{x};\varphi+h) - \chi(\mathbf{x}))^2\,d\mathbf{x} - \int_\Omega (T(\vartheta,\mathbf{x};\varphi) - \chi(\mathbf{x}))^2\,d\mathbf{x}\\
&= 2\int_\Omega (T(\vartheta,\mathbf{x};\varphi) - \chi(\mathbf{x}))\,\zeta(\vartheta,\mathbf{x})d\mathbf{x} + \int_\Omega \zeta^2(\vartheta,\mathbf{x})d\mathbf{x}\\
&= \int_\Omega \Psi(\vartheta,\mathbf{x})\zeta(\vartheta,\mathbf{x})d\mathbf{x} + \int_\Omega \zeta^2(\vartheta,\mathbf{x})d\mathbf{x}\\
&= \int_\Omega\int_0^\vartheta \frac{\partial}{\partial t}(\Psi(t,\mathbf{x})\zeta(t,\mathbf{x}))\,d\mathbf{x}dt + \int_\Omega \Psi(0,\mathbf{x})h(\mathbf{x})d\mathbf{x} + \int_\Omega \zeta^2(\vartheta,\mathbf{x})d\mathbf{x},
\end{aligned} \tag{8.18}$$

where $\Psi(t,\mathbf{x}) = 2(T(t,\mathbf{x};\vartheta) - \chi(\mathbf{x}))$; $h(\mathbf{x})$ is a small heat increment to the unknown initial temperature $\varphi(\mathbf{x})$; and $\zeta = T(t,\mathbf{x};\varphi+h) - T(t,\mathbf{x};\varphi)$ is the solution to the following forward heat problem

$$\begin{aligned} \partial\zeta/\partial t + \mathbf{u}\cdot\nabla\zeta - \nabla^2\zeta &= 0, && \mathbf{x}\in\Omega,\ t\in(0,\vartheta), \\ \sigma_1\zeta + \sigma_2\partial\zeta/\partial\mathbf{n} &= 0, && \mathbf{x}\in\Gamma,\ t\in(0,\vartheta), \\ \zeta(0,\mathbf{x}) &= h(\mathbf{x}), && \mathbf{x}\in\Omega. \end{aligned} \tag{8.19}$$

Considering the fact that $\Psi = \Psi(t,\mathbf{x})$ and $\zeta = \zeta(t,\mathbf{x})$ are the solutions to (8.17) and (8.19) respectively, and the velocity $\mathbf{u}$ satisfies (8.12) and the boundary conditions specified, we obtain

$$\begin{aligned} &\int_\Omega\int_0^\vartheta \frac{\partial}{\partial t}\left(\Psi(t,\mathbf{x})\zeta(t,\mathbf{x})\right) dtd\mathbf{x} = \int_0^\vartheta\int_\Omega \left\{\frac{\partial}{\partial t}\Psi(t,\mathbf{x})\zeta(t,\mathbf{x}) + \Psi(t,\mathbf{x})\frac{\partial\zeta(t,\mathbf{x})}{\partial t}\right\} d\mathbf{x}dt \\ &= \int_0^\vartheta\int_\Omega \zeta(t,x)\left[-\mathbf{u}\cdot\nabla\Psi - \nabla^2\Psi\right] d\mathbf{x}dt + \int_0^\vartheta\int_\Omega \Psi(t,\mathbf{x})\left[-\mathbf{u}\cdot\nabla\zeta + \nabla^2\zeta\right] d\mathbf{x}dt \\ &= \int_0^\vartheta\int_\Gamma \{\Psi\nabla\zeta\cdot\mathbf{n} - \zeta\nabla\Psi\cdot\mathbf{n}\} d\Gamma dt + \int_0^\vartheta\int_\Omega \{\nabla\Psi\cdot\nabla\zeta - \nabla\zeta\cdot\nabla\Psi\} d\mathbf{x}dt \\ &+ \int_0^\vartheta\int_\Omega \{\zeta\Psi\nabla\cdot\mathbf{u} + \Psi\mathbf{u}\cdot\nabla\zeta - \Psi\mathbf{u}\cdot\nabla\zeta\}\, dxdt - 2\int_0^\vartheta\int_\Gamma \zeta\Psi\mathbf{u}\cdot\mathbf{n}\, d\Gamma dt = 0. \end{aligned} \tag{8.20}$$

Hence

$$J(\varphi+h) - J(\varphi) = \int_\Omega \Psi(0,\mathbf{x})h(\mathbf{x})d\mathbf{x} + \int_\Omega \zeta^2(\vartheta,\mathbf{x})d\mathbf{x} = \int_\Omega \Psi(0,\mathbf{x})h(\mathbf{x})d\mathbf{x} + o(\|h\|). \tag{8.21}$$

The gradient is derived by using the Gateaux derivative of the objective functional. Therefore, we obtain that the gradient of the functional is represented as

$$\nabla J(\varphi) = \Psi(0,\cdot). \tag{8.22}$$

Thus, the solution of the backward heat problem is reduced to solutions of series of forward problems, which are known to be well-posed (Tikhonov and Samarskii, 1990). The algorithm can be used to solve the problem over any subinterval of time in $[0,\vartheta]$.

We note that information on the properties of the Hessian matrix $(\nabla^2 J)$ is important in many aspects of minimisation problems (Daescu and Navon, 2003). To obtain sufficient conditions for the existence of the minimum of the problem, the Hessian matrix must be positive definite at T_0 (optimal initial temperature). However, an explicit evaluation of the Hessian matrix in many cases is prohibitive owing to the number of variables.

We now describe the algorithm for numerical solution of the inverse problem of mantle convection, that is, the numerical algorithm to solve (8.10)–(8.12) backward in time using the VAR method. A uniform partition of the time axis is defined at points $t_n = \vartheta - \delta t\, n$, where δt is the time step, and n successively takes integer values from 0 to some natural number $m = \vartheta/\delta t$. At each subinterval of time $[t_n+_1, t_n]$, the search of the temperature T and flow velocity $\mathbf{u}$ at $t = t_n+_1$ consists of the following basic steps.

Step 1. Given the temperature $T = T(t_n, \mathbf{x})$ at $t = t_n$ solve a set of linear algebraic equations derived from (8.11) and (8.12) with the appropriate boundary conditions in order to determine the velocity $\mathbf{u}$.

Step 2. The 'advective' temperature $T_{adv} = T_{adv}(t_{n+1}, \mathbf{x})$ is determined by solving the advection heat equation backward in time, neglecting the diffusion term in Eq. (8.10). This can be done by replacing positive time steps by negative ones (see Section 8.4). Given the temperature $T = T_{adv}$ at $t = t_{n+1}$ steps 1 and 2 are then repeated to find the velocity $\mathbf{u}_{adv} = \mathbf{u}(t_{n+1}, \mathbf{x}; T_{adv})$.

Step 3. The heat equation (8.10) is solved with appropriate boundary conditions and initial condition $\varphi_k(\mathbf{x}) = T_{adv}(t_{n+1}, \mathbf{x})$, $k = 0, 1, 2, \ldots, m, \ldots$ forward in time using velocity $\mathbf{u}_{adv}$ in order to find $T(t_n, \mathbf{x}; \varphi_k)$.

Step 4. The adjoint equation of (8.17) is then solved backward in time with appropriate boundary conditions and initial condition $\Psi(t_n, \mathbf{x}) = 2(T(t_n, \mathbf{x}; \varphi_k) - \chi(\mathbf{x}))$ using velocity $\mathbf{u}$ in order to determine $\nabla J(\varphi_k) = \Psi(t_{n+1}, \mathbf{x}; \varphi_k)$.

Step 5. The coefficient β_k is determined from (8.16), and the temperature is updated (i.e. φ_{k+1} is determined) from (8.15).

Steps 3 to 5 are repeated until

$$\delta\varphi_n = J(\varphi_n) + \|\nabla J(\varphi_n)\|^2 < \varepsilon, \tag{8.23}$$

where ε is a small constant. Temperature φ_k is then considered to be the approximation to the target value of the initial temperature $T(t_{n+1}, \mathbf{x})$. And finally, step 1 is used to determine the flow velocity $\mathbf{u}(t_{n+1}, \mathbf{x}; T(t_{n+1}, \mathbf{x}))$. Step 2 introduces a pre-conditioner to accelerate the convergence of temperature iterations in steps 3 to 5 at high Rayleigh number. At low *Ra*, step 2 is omitted and $\mathbf{u}_{adv}$ is replaced by $\mathbf{u}$. After these algorithmic steps, we obtain temperature $T = T(t_n, \mathbf{x})$ and flow velocity $\mathbf{u} = \mathbf{u}(t_n, \mathbf{x})$ corresponding to $t = t_n$, $n = 0, \ldots, m$. Based on the obtained results, we can use interpolation to reconstruct, when required, the entire process on the time interval $[0, \vartheta]$ in more detail.

Thus, at each subinterval of time we apply the VAR method to the heat equation only, iterate the direct and conjugate problems for the heat equation in order to find temperature, and determine backward flow from the Stokes and continuity equations twice (for 'advective' and 'true' temperatures). Compared to the VAR approach by Bunge *et al.* (2003), the described numerical approach is computationally less expensive, because we do not involve the Stokes equation in the iterations between the direct and conjugate problems (the numerical solution of the Stokes equation is the most time consuming calculation).

8.6 Application of the VAR method: restoration of mantle plume evolution

A plume is hot, narrow mantle upwelling that is invoked to explain hotspot volcanism. In a temperature-dependent viscosity fluid such as the mantle, a plume is characterised by a mushroom-shaped head and a thin tail. Upon impinging under a moving lithosphere, such a mantle upwelling should therefore produce a large amount of melt and successive massive eruption, followed by smaller but long-lived hot-spot activity fed from the plume tail (Morgan, 1972; Richards *et al.*, 1989; Sleep, 1990). Meanwhile, slowly rising plumes (a buoyancy flux of less than 10^3 kg s^{-1}) coming from the core–mantle boundary should have cooled so much that they would not melt beneath old lithosphere (Albers and Christensen, 1996).

Mantle plumes evolve in three distinguishing stages: (i) *immature*, i.e. an origin and initial rise of the plumes; (ii) *mature*, i.e. plume–lithosphere interaction, gravity spreading of plume head and development of overhangs beneath the bottom of the lithosphere, and partial melting of the plume material (see Ribe and Christensen, 1994; Moore *et al.*, 1998); and (iii) *overmature*, i.e. slowing-down of the plume rise and fading of the mantle plumes due to thermal diffusion (Davaille and Vatteville, 2005; Ismail-Zadeh *et al.*, 2006). The ascent and evolution of mantle plumes depend on the properties of the source region (that is, the thermal boundary layer) and the viscosity and thermal diffusivity of the ambient mantle. The properties of the source region determine the temperature and viscosity of the mantle plumes. Structure, flow rate and heat flux of the plumes are controlled by the properties of the mantle through which the plumes rise. While properties of the lower mantle (e.g. viscosity, thermal conductivity) are relatively constant during about 150 Myr lifetime of most plumes, source region properties can vary substantially with time as the thermal basal boundary layer feeding the plume is depleted of hot material (Schubert *et al.*, 2001). Complete local depletion of this boundary layer cuts the plume off from its source.

A mantle plume is a well-established structure in computer modelling and laboratory experiments. Numerical experiments on dynamics of mantle plumes (Trompert and Hansen, 1998a,b; Zhong, 2005) showed that the number of plumes increases and the rising plumes become thinner with an increase in Rayleigh number. Disconnected thermal plume structures appear in thermal convection at *Ra* greater than 10^7 (Hansen *et al.*, 1990; Malevsky *et al.*, 1992). At high *Ra* (in the hard turbulence regime) thermal plumes are torn off the boundary layer by the large-scale circulation or by non-linear interactions between plumes (Malevsky and Yuen, 1993). Plume tails can also be disconnected when the plumes are tilted by plate scale flow (see Olson and Singer, 1985; Steinberger and O'Connell, 1998). Ismail-Zadeh *et al.* (2006) presented an alternative explanation for the disconnected mantle plume heads and tails that is based on thermal diffusion of mantle plumes.

A dimensionless temperature-dependent viscosity law (Busse *et al.*, 1993) is employed in the models discussed in this chapter

$$\eta(T) = \exp\left(\frac{M}{T+G} - \frac{M}{0.5+G}\right), \tag{8.24}$$

where $M = [225/\ln(r)] - 0.25\ln(r)$, $G = 15/\ln(r) - 0.5$, and r is the viscosity ratio between the upper and lower boundaries of the model domain. The temperature-dependent viscosity profile has its minimum at the core–mantle boundary. A more realistic viscosity profile (Forte and Mitrovica, 2001) will influence the evolution of mantle plumes, though it will not influence the restoration of the plumes.

The model domain is divided into $37 \times 37 \times 29$ rectangular finite elements to approximate the vector velocity potential by tricubic splines, and a uniform grid $112 \times 112 \times 88$ is employed for approximation of temperature, velocity and viscosity. Temperature in the heat equation (8.10) is approximated by finite differences and determined by the semi-Lagrangian method (see Section 7.8). A numerical solution to the Stokes and incompressibility equations (8.11) and (8.12) is based on the introduction of a two-component vector velocity potential and on the application of the Eulerian finite-element method with a tricubic-spline basis for computing the potential (Section 4.9 and 4.10). Such a procedure results in a set of linear algebraic equations with a symmetric positive-definite banded matrix. We solve the set of equations by the conjugate gradient method (Section 6.3.3).

8.6.1 Forward modelling

Here the evolution of mature mantle plumes is modelled initially forward in time. With $\alpha = 3\times10^{-5}\,\mathrm{K}^{-1}$, $\rho_{ref} = 4000\,\mathrm{kg\,m}^{-3}$, $\Delta T = 3000\,\mathrm{K}$, $h = 2800\,\mathrm{km}$, $\eta_{ref} = 8\times10^{22}\,\mathrm{Pa\,s}$, and $\kappa = 10^{-6}\,\mathrm{m}^{-2}\,\mathrm{s}^{-1}$, the initial Rayleigh number is $Ra = 9.5 \times 10^5$. While plumes evolve in the convecting heterogeneous mantle, at the initial time it is assumed that the plumes develop in a laterally homogeneous temperature field, and hence the initial mantle temperature is considered to increase linearly with depth.

Mantle plumes are generated by random temperature perturbations at the top of the thermal source layer associated with the core–mantle boundary (Fig. 8.4a). The mantle material in the basal source layer flows horizontally toward the plumes. The reduced viscosity in this basal layer promotes the flow of the material to the plumes. Vertical upwelling of hot mantle material is concentrated in low viscosity conduits near the centrelines of the emerging plumes (Fig. 8.4b,c). The plumes move upward through the model domain, gradually forming structures with well-developed heads and tails. Colder material overlying the source layer (e.g. portions of lithospheric slabs subducted to the core–mantle boundary) replaces hot material at the locations where the source material is fed into mantle plumes. Some time is required to recover the volume of source material depleted due to plume feeding (Howard, 1966). Because the volume of upwelling material is comparable to the volume of the thermal source layer feeding the mantle plumes, hot material could eventually be exhausted, and mantle plumes would be starved thereafter.

The plumes diminish in size with time (Fig. 8.4d), and the plume tails disappear before the plume heads (Fig. 8.4e,f). We note that Fig. 8.4 presents a hot isothermal surface of the plumes. If colder isotherms are considered, the disappearance of the isotherms will occur later. But anyhow, hot or cold isotherms are plotted, plume tails will vanish before their heads. Results of recent laboratory experiments (Davaille and Vatteville, 2005) support

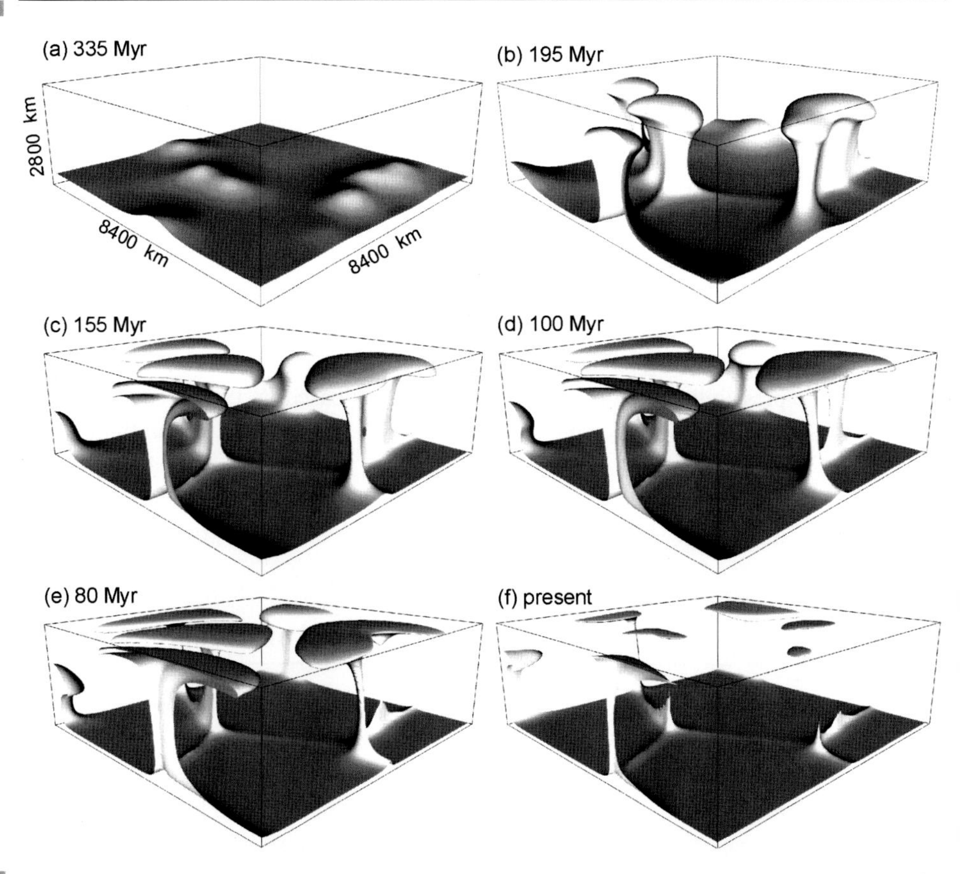

Fig. 8.4. **Mantle plumes in the forward modelling at successive diffusion times: from 335 Myr ago (a) to the 'present' state of the plumes (f). The plumes are represented here and in Figs. 8.5 and 8.6 by isothermal surfaces at 3000 K. After Ismail-Zadeh *et al.* (2006).**

strongly the numerical findings that plumes start disappearing from the bottom up and fade away by thermal diffusion.

At different stages in the plume decay one sees quite isolated plume heads, plume heads with short tails, and plumes with nearly pinched off tails. Different amounts of time are required for different mantle plumes to vanish into the ambient mantle, the required time depending on the geometry of the plume tails. Temperature loss is greater for sheet-like tails than for cylindrical tails. The tails of the cylindrical plumes (e.g. Fig. 8.4c, in the left part of the model domain) are still detectable after about 155 Myr. However, at this time the sheet-like tail of the large plume in the right part of the model domain (Fig. 8.4c) is already invisible and only its head is preserved in the uppermost mantle (Fig. 8.4f). Two-dimensional numerical experiments of steady-state convection (Leitch *et al.*, 1996) reveal a significant change in the centreline temperature of sheet-like plume tails compared with the cylindrical plume tail due to heat conduction in the horizontal direction.

The numerical results may have important implications for the interpretation of seismic tomographic images of mantle plumes. Finite-frequency seismic tomography images

(Montelli *et al.*, 2004) show that a number of plumes extend to mid-mantle depths but are not visible below these depths. From a seismological point of view, the absence of the plume tails could be explained as a combination of several factors (Romanowicz and Gung, 2002): elastic velocities are sensitive to composition as well as temperature; the effect of temperature on velocities decreases with increasing pressure (Karato, 1993); and wavefront healing effects make it difficult to accurately image low velocity bodies (Nolet and Dahlen, 2000). The 'disappearance' of the plume tails can hence be explained as the effects of poor tomographic resolution at deeper levels. Apart from this, the numerical results demonstrate the plausibility of finding a great diversity in the morphology of seismically imaged mantle plumes, including plume heads without tails and plumes with tails that are detached from their sources.

8.6.2 Backward modelling

To restore the prominent state of the plumes (Fig. 8.4d) in the past from their 'present' weak state (Fig. 8.4f), the VAR method can be employed. Figure 8.5 illustrates the restored states of the plumes (middle panel) and the temperature residuals δT (right panel) between the temperature $T(\mathbf{x})$ predicted by the forward model and the temperature $\tilde{T}(\mathbf{x})$ reconstructed to the same age:

$$\delta T(x_1, x_2) = \left[\int_0^h \left(T(x_1, x_2, x_3) - \tilde{T}(x_1, x_2, x_3) \right)^2 dx_3 \right]^{1/2}. \tag{8.25}$$

To study the effect of thermal diffusion on the restoration of mantle plumes, several experiments on mantle plume restoration were run for various Rayleigh number Ra (typically less than the initial Ra) and viscosity ratio r. Figure 8.6 presents the case of $r = 200$ and $Ra = 9.5 \times 10^3$ and shows several stages in the diffusive decay of the mantle plumes.

The dimensional temperature residuals are within a few degrees for the initial restoration period (Figs. 8.5i and 8.6h). The computations show that the errors (temperature residuals) get larger the farther the restorations move backward in time (e.g. $\delta T \approx 300$ K at the restoration time of more than 300 Myr, $r = 200$, and $Ra = 9.5 \times 10^3$). Compared with the case of $Ra = 9.5 \times 10^5$, one can see that the residuals become larger as the Rayleigh number decreases or thermal diffusion increases and viscosity ratio increases.

The quality of the restoration depends on the dimensionless Péclet number $Pe = hu_{\max}\kappa^{-1}$, where $u_{\max}$ is the maximum flow velocity. According to the numerical experiments, the Péclet number corresponding to the temperature residual $\delta T = 600$ K is $Pe = 10$; Pe should not be less than about 10 for a high quality plume restoration.

8.6.3 Performance of the numerical algorithm

Here we analyse the performance of the VAR data assimilation algorithm for various Ra and r. The performance of the algorithm is evaluated in terms of the number of iterations

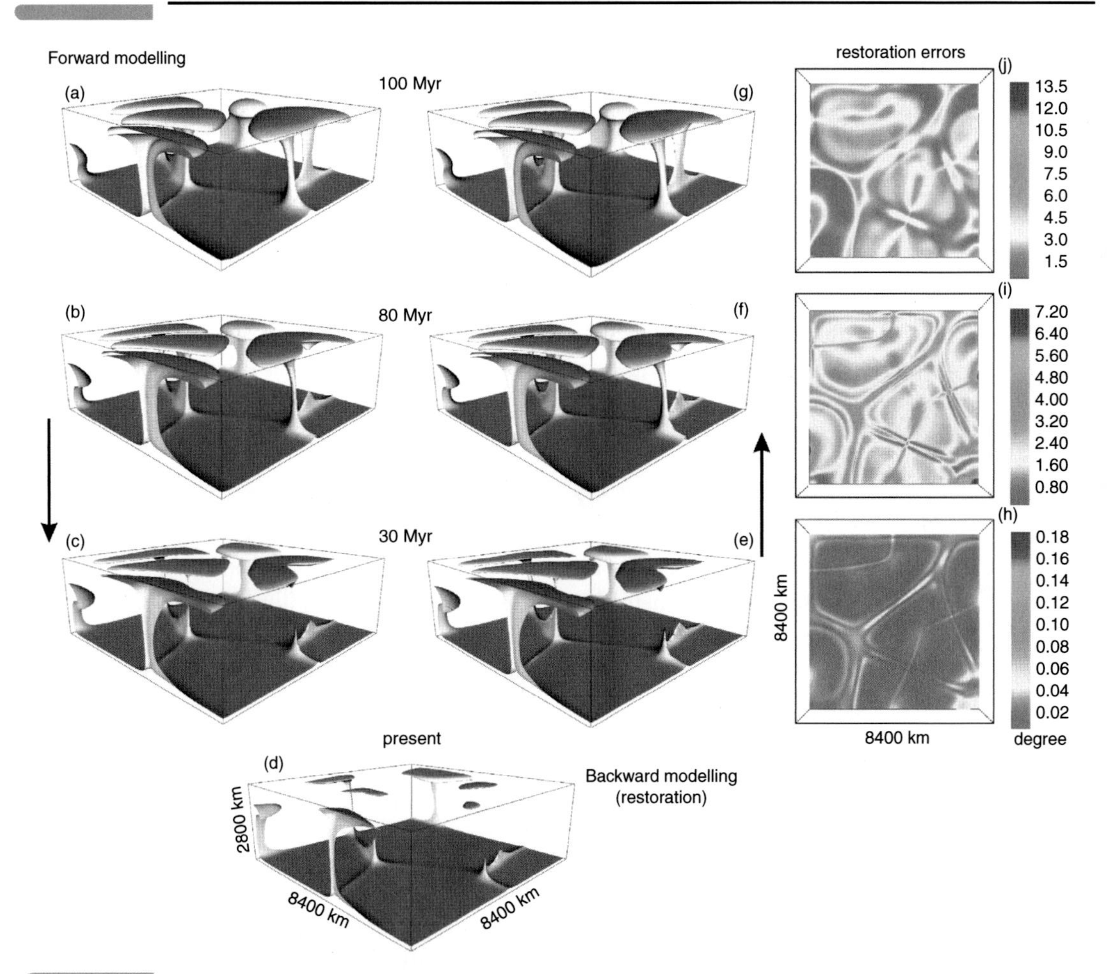

Fig. 8.5. **Mantle plume diffusion ($r = 20$ and $Ra = 9.5 \times 10^5$) in the forward modelling at successive diffusion times: from 100 Myr ago to the 'present' state of the plumes (left panel, a–d). Restored mantle plumes in the backward modelling (central panel, e–g) and restoration errors (right panel, h–j). After Ismail-Zadeh *et al.* (2006). (In colour as Plate 3. See colour plates section.)**

n required to achieve a prescribed relative reduction of $\delta\varphi_n$ (inequality (8.23)). Figure 8.7 presents the evolution of the objective functional $J(\varphi_n)$ and the norm of the gradient of the objective functional $\|\nabla J(\varphi_n)\|$ versus the number of iterations at time about 0.5θ. For other time steps we observe a similar evolution of J and $\|\nabla J\|$.

Both the objective functional and the norm of its gradient show a quite rapid decrease after about seven iterations for $Ra = 9.5 \times 10^5$ and $r = 20$ (curves *1*). The same rapid convergence as a function of adjoint iterations is observed in the Bunge *et al.* (2003) case. As Ra decreases and thermal diffusion increases (curves *2–4*) the performance of the algorithm becomes poor: more iterations are needed to achieve the prescribed ε. All curves illustrate that the first four to seven iterations contribute mainly to the reduction of $\delta\varphi_n$. The convergence drops after a relatively small number of iterations. The curves approach the horizontal line with an increase in the number of iterations, because β_k tends to zero with a large number of iterations (see Eq. (8.6)). The increase of $\|\nabla J\|$ at $k = 2$ is associated with uncertainty of this gradient at $k = 1$.

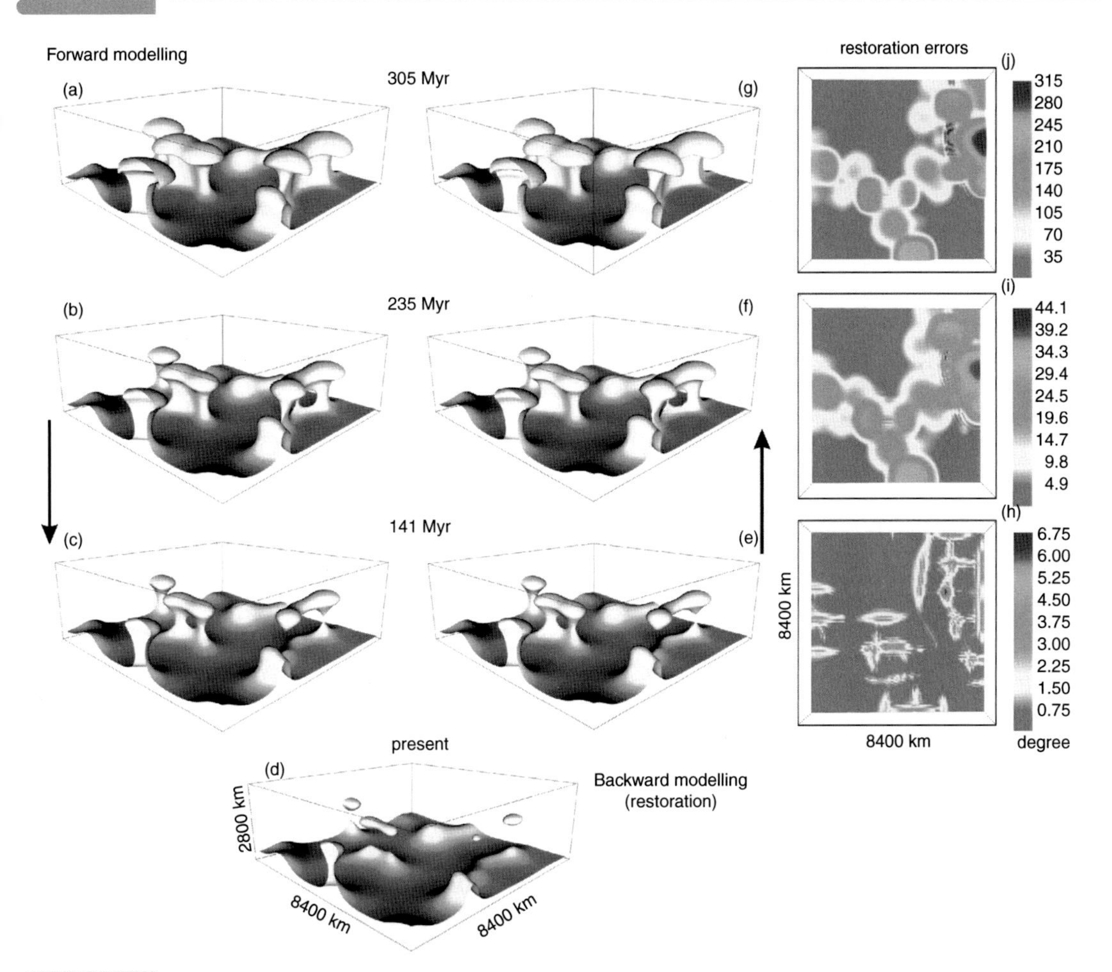

Fig. 8.6. **Mantle plume diffusion ($r = 200$ and $Ra = 9.5 \times 10^3$) in the forward modelling at successive diffusion times: from 305 Myr ago to the 'present' state of the plumes (left panel, a–d). Restored mantle plumes in the backward modelling (central panel, e–g) and restoration errors (right panel, h–j). After Ismail-Zadeh *et al.* (2006). (In colour as Plate 4. See colour plates section.)**

Implementation of minimisation algorithms requires the evaluation of both the objective functional and its gradient. Each evaluation of the objective functional requires an integration of the model equation (8.10) with the appropriate boundary and initial conditions, whereas the gradient is obtained through the backward integration of the adjoint equations (8.17). The performance analysis shows that the CPU time required to evaluate the gradient J is about the CPU time required to evaluate the objective functional itself, and this is because the direct and adjoint heat problems are described by the same equations.

Despite its simplicity, the minimisation algorithm used in this study provides for a rapid convergence and good quality of optimisation at high Rayleigh numbers (low thermal diffusion). The convergence rate and the quality of optimisation become worse with the decreasing Rayleigh number. The use of the limited-memory quasi-Newton algorithm L-BFGS (Liu and Nocedal, 1989) might provide for a better convergence rate and quality of optimisation (Zou *et al.*, 1993). Meanwhile, we note that although an improvement of the convergence rate by using another minimisation algorithm (e.g. L-BFGS) will reduce

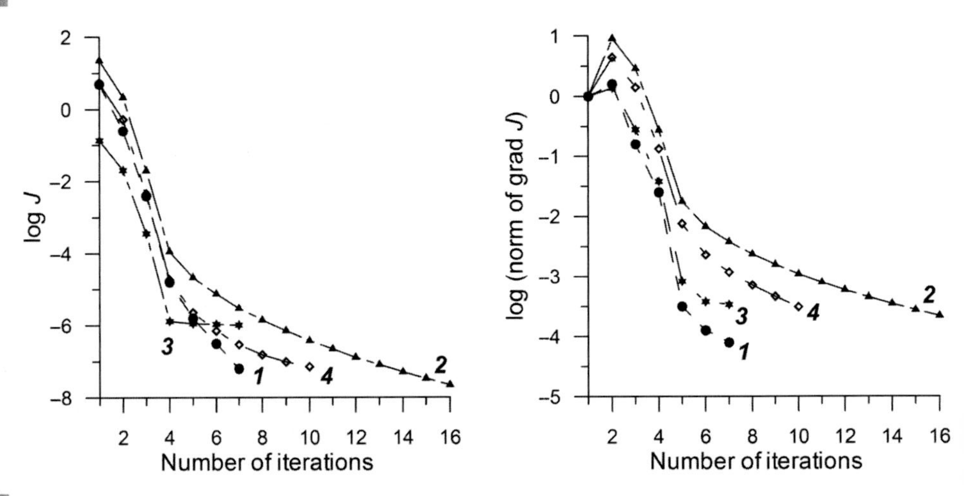

Fig. 8.7. **Relative reductions of the objective functional J (left panel) and the norm of the gradient of J (right panel) as functions of the number of iterations. Curves: 1, $r = 20$, $Ra = 9.5 \times 10^5$; 2, $r = 20$, $Ra = 9.5 \times 10^2$; 3, $r = 200$, $Ra = 9.5 \times 10^3$; 4, $r = 200$, $Ra = 9.5 \times 10^2$. After Ismail-Zadeh *et al.* (2006).**

the computational expense associated with the solving of the problem under question, this reduction would be not significant, because the large portion (about 70%) of the computer time is spent to solve the 3-D Stokes equations.

8.7 Challenges in VAR data assimilation

Although the VAR data assimilation technique described above can theoretically be applied to many problems of mantle and lithosphere dynamics, practical implementation of the technique for modelling of real geodynamic processes backward in time (to restore the temperature and flow pattern in the past) is not a simple task. The mathematical model of mantle dynamics described by a set of equations (8.10)–(8.12) is simple, and many complications are omitted. A viscosity increase from the upper to the lower mantle is not included in the model, although it is suggested by studies of the geoid (Ricard *et al.*, 1993), post-glacial rebound (Mitrovica, 1996), and joint inversion of convection and glacial isostatic adjustment data (Mitrovica and Forte, 2004). The adiabatic heating/cooling term in the heat equation can provide more realistic distribution of temperature in the mantle, especially near the thermal boundary layer. The numerical models presented in Section 8.6 do not include phase transformations (Liu *et al.*, 1991; Honda *et al.*, 1993a,b; Harder and Christensen, 1996), although the phase changes can influence the evolution of mantle plumes retarding/accelerating their ascent. The coefficient of thermal expansion (see Chopelas and Boehler, 1989; Hansen *et al.*, 1991; 1993) and the coefficient of thermal conductivity (Hofmeister, 1999) are not constant in the mantle and vary with depth and temperature. Moreover, if the findings of Badro *et al.* (2004) of a significant increase in the radiative thermal conductivity at high pressure are relevant to the lower mantle, plume tails should diffuse away even faster than the studied models predict. To consider these complications in the VAR data assimilation, the adjoint equations should be derived each time when the set of the equations is changed. The cost to be paid is in software development since an adjoint model has to be developed.

8.7.1 Smoothness of observational data

The solution $T(\vartheta, \cdot; \varphi)$ of the heat equation (8.10) with appropriate boundary and initial conditions is a sufficiently smooth function and belongs to space $L_2(\Omega)$. The present temperature χ_δ derived from the seismic tomography is a representation of the exact temperature χ of the Earth and so it must also belong to this space and hence be rather smooth; otherwise, the objective functional J cannot be defined. Therefore, before any assimilation of the present temperature data can be attempted, the data must be smoothed. The smoothing of the present temperature improves the convergence of the iterations.

8.7.2 Smoothness of the target temperature

If mantle temperature in the geological past was not a smooth function of space variables, recovery of this temperature by using the VAR method is not effective because the iterations converge very slowly to the target temperature. Here we explain the problem of recovering the initial temperature on the basis of three one-dimensional model tasks: restoration of a smooth, piece-wise smooth and discontinuous target function. We note that the temperature in the Earth's mantle is not a discontinuous function but its shape can be close to a step function.

The dynamics of a physical system is assumed to be described by the Burgers equation $u_t + uu_x = u_{xx}$, $0 \le t \le 1$, $0 \le x \le 2\pi$ with the boundary conditions $u(t, 0) = 0$, $u(t, 2\pi) = 0$, $0 \le t \le 1$ and the condition $u_\theta = u(1, x; u_0)$, $0 \le x \le 2\pi$ at $t = 1$, where the variable u can denote temperature. The problem is to recover the function $u_0 = u_0(x)$, $0 \le x \le 2\pi$ at $t = 0$ (the state in the past) from the function $u_\theta = u_\theta(x)$, $0 \le x \le 2\pi$ at $t = 1$ (its present state). The finite difference approximations and the variational method are applied to the Burgers equation with the appropriate boundary and initial conditions.

Task 1. Consider the sufficiently smooth function $u_0 = \sin(x)$, $0 \le x \le 2\pi$. The functions u_0 and u_θ are shown in Fig. 8.8a. Figures 8.8b and c illustrate the iterations φ_k using the iterative scheme similar to Eq. (8.15) for $k = 0, 4, 6$ and the residual $r_6(x) = u_0(x) - \varphi_6(x)$, $0 \le x \le 2\pi$ respectively. We see that iterations converge rather rapid for the sufficiently smooth target function.

Task 2. Now consider the continuous piece-wise smooth function $u_0 = 3x/(2\pi)$, $0 \le x \le 2\pi/3$ and $u_0 = 3/2 - 3x/(2\pi)$, $2\pi/3 \le x \le 2\pi$. Figure 8.8 presents (d) the functions u_0 and u_θ, (e) the successive approximations φ_k for $k = 0, 4, 1000$, and (f) the residual $r_{1000}(x) = u_0(x) - \varphi_{1000}(x)$, $0 \le x \le 2\pi$, respectively. This example shows that a large number of iterations is required to reach the target function.

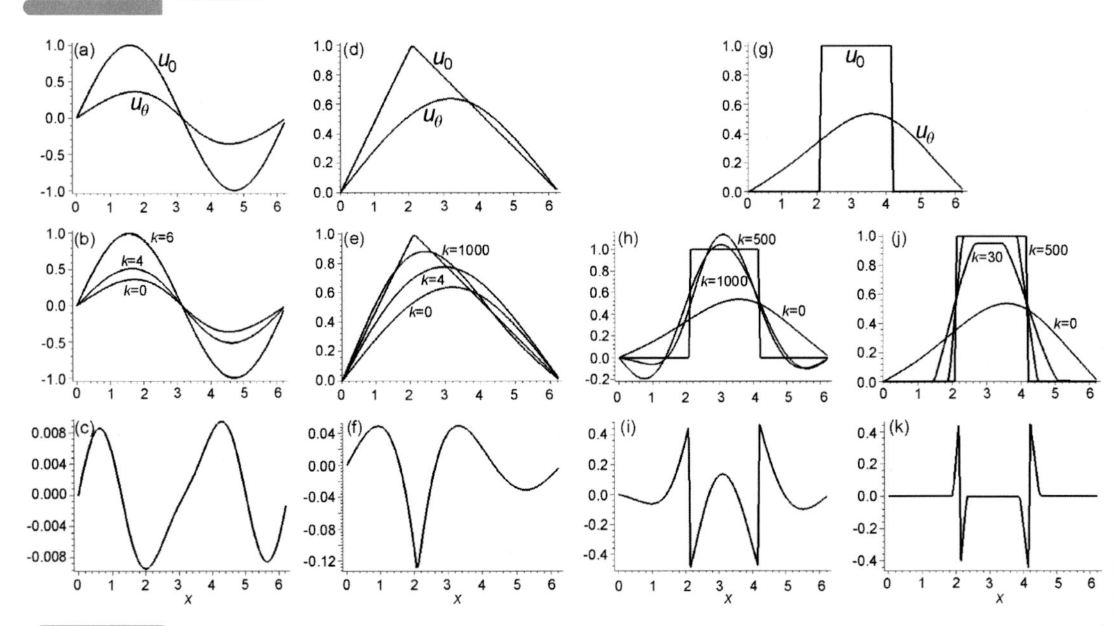

Fig. 8.8. **Recovering function u_0 from the smooth guess function u_θ. The sufficiently smooth u_0 (a–c); continuous piece-wise smooth function u_0 (d–f); and discontinuous function u_0 (g–k). Plots of u_0 and u_θ are presented at (a), (d) and (g); successive approximations to u_0 at (b), (e), (h) and (j); and the residual functions at (c), (f), (i) and (k). After Ismail-Zadeh *et al.* (2006).**

Task 3. Consider the discontinuous function u_0, which takes 1 at $2\pi/3 \le x \le 4\pi/3$ and 0 in other points of the closed interval $0 \le x \le 2\pi$. Figure 8.8 presents (g) the functions u_0 and u_θ, (h) the successive approximations φ_k for $k = 0, 500, 1000$, and (e) the residual $r_{1000}(x) = u_0(x) - \varphi_{1000}(x)$, $0 \le x \le 2\pi$, respectively. We see that convergence to the target temperature is very poor.

To improve the convergence to the target function, a modification of the variational method based on *a priori* information about a desired solution can be used (Korotkii and Tsepelev, 2003). Figure 8.8 (j) shows the successive approximations $\tilde{\varphi}_k$ for $k = 0, 30, 500$, and (k) the residual $\tilde{r}_{500}(x) = u_0(x) - \tilde{\varphi}_{500}(x)$, $0 \le x \le 2\pi$, respectively. The approximations $\tilde{\varphi}_k$ based on the method of gradient projection (Vasiliev, 2002) converge to the target solution better than approximations generated by Eq. (8.5).

8.7.3 Numerical noise

If the initial temperature guess φ_0 is a smooth function, all successive temperature iterations φ_k in scheme (8.15) should be smooth functions too, because the gradient of the objective functional ∇J is a smooth function since it is the solution to the adjoint problem (8.17). However, the temperature iterations φ_k are polluted by small perturbations (errors), which are inherent in any numerical experiment (Section 8.12). These perturbations can grow with time. Samarskii *et al.* (1997) applied a VAR method to a 1-D backward heat diffusion problem and showed that the solution to this problem becomes noisy if the initial temperature guess is slightly perturbed, and the amplitude of this noise increases with the initial perturbations of the temperature guess. To reduce the noise they used a special filter and illustrated the efficiency of the filter. This filter is based on the replacement of iterations (8.15) by the following iterative scheme:

$$\mathbf{B}(\varphi_{k+1} - \varphi_k) = -\beta_k \nabla J(\varphi_k), \tag{8.26}$$

where $\mathbf{B}y = y - \nabla^2 y$. Unfortunately, employment of this filter increases the number of iterations to obtain the target temperature and it becomes quite expensive computationally, especially when the model is three-dimensional. Another way to reduce the noise is to employ high-order adjoint (Alekseev and Navon, 2001) or regularisation (Tikhonov, 1963; Lattes and Lions, 1969; Samarskii and Vabischevich, 2004) techniques.

8.8 Quasi-reversibility (QRV) method

The principal idea of the quasi-reversibility (QRV) method is based on the transformation of an ill-posed problem into a well-posed problem (Lattes and Lions, 1969). In the case of the backward heat equation, this implies an introduction of an additional term into the equation, which involves the product of a small regularisation parameter and higher-order temperature derivative. The additional term should be sufficiently small compared to other terms of the heat equation and allow for simple additional boundary conditions. The data assimilation in this case is based on a search of the best fit between the forecast model state and the observations by minimising the regularisation parameter. The QRV method is proven to be well suited for smooth and non-smooth input data (Lattes and Lions, 1969; Samarskii and Vabishchevich, 2004).

To explain the transformation of the problem, we follow Ismail-Zadeh *et al.* (2007) and consider the following boundary-value problem for the one-dimensional heat conduction problem

$$\frac{\partial T(t,x)}{\partial t} = \frac{\partial^2 T(t,x)}{\partial x^2}, \quad 0 \le x \le \pi, \quad 0 \le t \le t^*, \tag{8.27}$$

$$T(t,x=0) = T(t,x=\pi) = 0, \quad 0 \le t \le t^*, \tag{8.28}$$

$$T(t=0,x) = \frac{1}{4n+1} \sin((4n+1)x), \quad 0 \le x \le \pi. \tag{8.29}$$

The analytical solution to (8.27)–(8.29) can be obtained in the following form

$$T(t,x) = \frac{1}{4n+1} \exp(-(4n+1)^2 t) \sin((4n+1)x). \tag{8.30}$$

Figure 8.9 presents the solution (solid curves) for time interval $0 \le t \le t^* = 0.14$ and $n = 1$.

It is known that the backward heat conduction problem is ill-posed (e.g. Kirsch, 1996). To transform the problem into a well-posed problem, we introduce a term in Eq. (8.27) involving

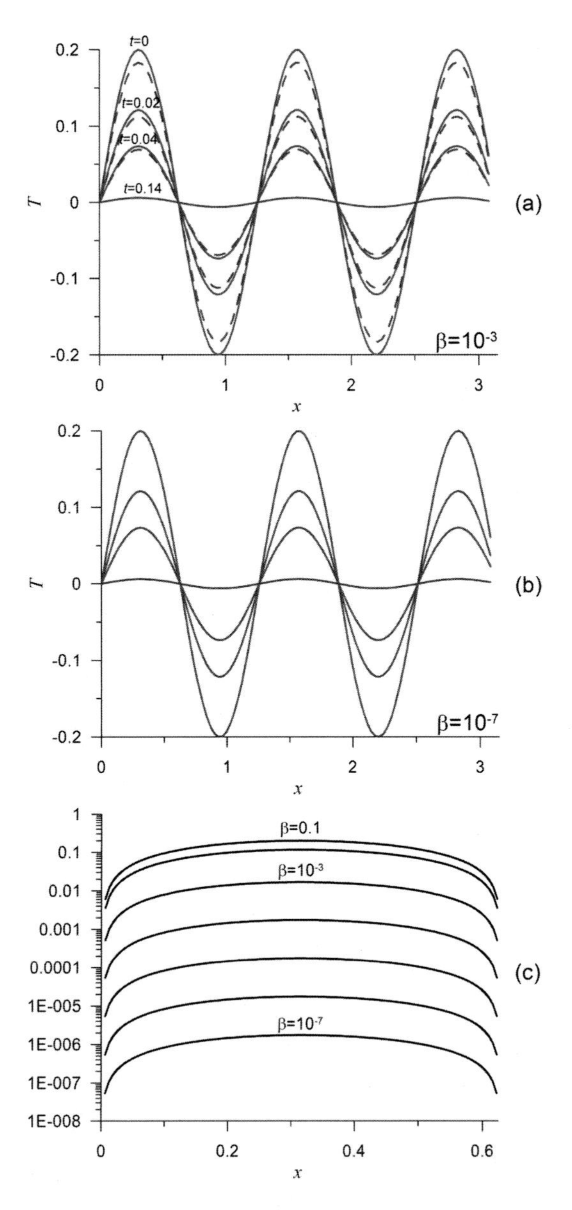

Fig. 8.9. Comparison of the exact solutions to the heat conduction problem (red solid curves; a and b) and to the regularised backward heat conduction problem (a: $\beta = 10^{-3}$ and b: $\beta = 10^{-7}$; blue dashed curves). The temperature residual between two solutions is presented in panel c at various values of the regularisation parameter β. After Ismail-Zadeh *et al.* (2007). (In colour as Plate 5. See colour plates section.)

the product of a small parameter $\beta > 0$ and the higher-order temperature derivative:

$$\frac{\partial T_\beta(t,x)}{\partial t} = \frac{\partial^2 T_\beta(t,x)}{\partial x^2} - \beta \frac{\partial^4}{\partial x^4}\left(\frac{\partial T_\beta(t,x)}{\partial t}\right), \quad 0 \le x \le \pi, \quad 0 \le t \le t^*, \tag{8.31}$$

$$T_\beta(t, x=0) = T_\beta(t, x=\pi) = 0, \quad 0 \le t \le t^*, \tag{8.32}$$

$$\frac{\partial^2 T_\beta(t, x=0)}{\partial x^2} = \frac{\partial^2 T_\beta(t, x=\pi)}{\partial x^2} = 0, \quad 0 \le t \le t^*, \tag{8.33}$$

$$T_\beta(t=t^*, x) = \frac{1}{4n+1} \exp(-(4n+1)^2 t^*) \sin((4n+1)x), \quad 0 \le x \le \pi. \tag{8.34}$$

Here the initial condition is assumed to be the solution (8.30) to the heat conduction problem (8.27)–(8.29) at $t = t^*$. The subscript β at T_β is used to emphasise the dependence of the solution to problem (8.31)–(8.34) on the regularisation parameter. The analytical solution to the regularised backward heat conduction problem (8.31)–(8.34) is represented as:

$$T_\beta(t, x) = A_n \exp\left(\frac{-(4n+1)^2 t}{1+\beta(4n+1)^4}\right) \sin((4n+1)x),$$

$$A_n = \frac{1}{4n+1} \exp(-(4n+1)^2 t^*) \exp^{-1}\left(\frac{-(4n+1)^2 t^*}{1+\beta(4n+1)^4}\right), \tag{8.35}$$

and the solution approaches the initial condition for the problem (8.27)–(8.29) at $t = 0$ and $\beta \to 0$. Figure 8.9a,b illustrates the solution to the regularised problem at two values of β (dashed curves) and $n = 1$. The temperature residual (Fig. 8.9c) indicates that the solution (8.35) approaches the solution (8.30) with $\beta \to 0$.

Samarskii and Vabischevich (2004) estimated the stability of the solution to problem (8.31)–(8.33) with respect to the initial condition expressed in the form $T_\beta(t = t^*, x) = T_\beta^*$:

$$\|T_\beta(t,x)\| + \beta \|\partial T_\beta(t,x)/\partial x\| \le C\left(\left\|T_\beta^*\right\| + \beta \left\|\partial T_\beta^*/\partial x\right\|\right) \exp\left[(t^* - t)\beta^{-1/2}\right],$$

where C is a constant, and showed that the natural logarithm of errors will increase in direct proportion to time and inversely to the root square of the regularisation parameter.

Any regularisation has its advantages and disadvantages. A regularising operator is used in a mathematical problem to (i) accelerate a convergence; (ii) fulfil the physical laws (e.g. maximum principal, conversation of energy, etc.) in discrete equations; (iii) suppress a noise in input data and in numerical computations; and (iv) take into account *a priori* information about an unknown solution and hence to improve a quality of computations. The major drawback of regularisation is that the accuracy of the solution to a regularised problem is always lower than that to a non-regularised problem.

We should mention that the transformation to the regularised backward heat problem is not only a mathematical approach to solving ill-posed backward heat problems, but has some physical meaning: it can be explained on the basis of the concept of relaxing heat flux for heat conduction (Vernotte, 1958). The classical Fourier heat conduction theory provides the infinite velocity of heat propagation in a region. The instantaneous heat propagation is unrealistic, because the heat is a result of the vibration of atoms and the vibration propagates in a finite speed (Morse and Feshbach, 1953). To accommodate the finite velocity of heat propagation, a modified heat flux model was proposed by Vernotte (1958) and Cattaneo (1958).

The modified Fourier constitutive equation (sometimes called the Riemann law of heat conduction) is expressed as $\vec{Q} = -k\nabla T - \tau\, \partial\vec{Q}/\partial t$, where $\vec{Q}$ is the heat flux, and k is the coefficient of thermal conductivity. The thermal relaxation time $\tau = k/\left(\rho c_p v^2\right)$ is usually recognised to be a small parameter (Yu *et al.*, 2004), where ρ is the density, c_p is the specific heat, and v is the heat propagation velocity. The situation for $\tau \to 0$ leads to instantaneous diffusion at infinite propagation speed, which coincides with the classical thermal diffusion theory. The heat conduction equation $\partial T/\partial t = \nabla^2 T + \tau\, \partial^2 T/\partial t^2$ based on non-Fourier heat flux can be considered as a regularised heat equation. If the Fourier law is modified further by an addition of the second derivative of heat flux, e.g. $\vec{Q} = -k\nabla T + \beta \frac{\partial^2 \vec{Q}}{\partial t^2}$, where small β is the relaxation parameter of heat flux (Bubnov, 1976, 1981), the heat conduction equation can be transformed into a higher-order regularised heat equation similar to Eq. (8.31).

8.8.1 The QRV method for restoration of thermo-convective flow

For convenience, we present a set of equations (8.10)–(8.12) with the relevant boundary and initial conditions as two mathematical problems. Namely, we consider the *boundary-value problem for the flow velocity* (it includes the Stokes equation, the incompressibility equation subject to appropriate boundary conditions)

$$\nabla P = \operatorname{div}\left(\eta(T)\mathbf{E}\right) + RaT\mathbf{e}, \qquad \mathbf{x} \in \Omega, \tag{8.36}$$

$$\operatorname{div}\mathbf{u} = 0, \qquad \mathbf{x} \in \Omega, \tag{8.37}$$

$$\mathbf{u}\cdot\mathbf{n} = 0, \quad \partial \mathbf{u}_\tau/\partial \mathbf{n} = 0, \qquad \mathbf{x} \in \partial\Omega, \tag{8.38}$$

where $\mathbf{u}_\tau$ is the projection of the velocity vector onto the tangent plane at the same point on the model boundary, and *the initial-boundary-value problem for temperature* (it includes the heat equation subject to appropriate boundary and initial conditions)

$$\partial T/\partial t + \mathbf{u}\cdot\nabla T = \nabla^2 T + f, \quad t \in [0, \vartheta], \quad \mathbf{x} \in \Omega, \tag{8.39}$$

$$\sigma_1 T + \sigma_2 \partial T/\partial \mathbf{n} = T_*, \qquad t \in [0, \vartheta], \quad \mathbf{x} \in \partial\Omega, \tag{8.40}$$

$$T(0, \mathbf{x}) = T_0(\mathbf{x}), \qquad \mathbf{x} \in \Omega, \tag{8.41}$$

where T_* is the given temperature.

The direct problem of thermo-convective flow can be formulated as follows: find the velocity $\mathbf{u} = \mathbf{u}(t, \mathbf{x})$, the pressure $P = P(t, \mathbf{x})$, and the temperature $T = T(t, \mathbf{x})$ satisfying boundary value problem (8.36)–(8.38) and initial-boundary-value problem (8.39)–(8.41). We can formulate the inverse problem in this case as follows: find the velocity, pressure, and temperature satisfying boundary-value problem (8.36)–(8.38) and the final-boundary value problem that includes Eqs. (8.39) and (8.40) and the final condition:

$$T(\vartheta, \mathbf{x}) = T_\vartheta(\mathbf{x}), \quad \mathbf{x} \in \Omega, \tag{8.42}$$

where T_ϑ is the temperature at time $t = \vartheta$.

To solve the inverse problem by the QRV method, Ismail-Zadeh *et al.* (2007) considered the following regularised backward heat problem to define temperature in the past from the known temperature $T_\vartheta(\mathbf{x})$ at present time $t = \vartheta$:

$$\partial T_\beta/\partial t - \mathbf{u}_\beta \cdot \nabla T_\beta = \nabla^2 T_\beta + f - \beta\Lambda(\partial T_\beta/\partial t), \quad t \in [0, \vartheta], \quad \mathbf{x} \in \Omega, \tag{8.43}$$

$$\sigma_1 T_\beta + \sigma_2 \partial T_\beta/\partial \mathbf{n} = T_*, \quad t \in (0, \vartheta), \quad \mathbf{x} \in \partial\Omega, \tag{8.44}$$

$$\sigma_1 \partial^2 T_\beta/\partial \mathbf{n}^2 + \sigma_2 \partial^3 T_\beta/\partial \mathbf{n}^3 = 0, \quad t \in (0, \vartheta), \quad \mathbf{x} \in \partial\Omega, \tag{8.45}$$

$$T_\beta(\vartheta, \mathbf{x}) = T_\vartheta(\mathbf{x}), \quad \mathbf{x} \in \Omega, \tag{8.46}$$

where $\Lambda(T) = \partial^4 T/\partial x_1^4 + \partial^4 T/\partial x_2^4 + \partial^4 T/\partial x_3^4$, and the boundary value problem to determine the fluid flow:

$$\nabla P_\beta = -\mathrm{div}\left[\eta(T_\beta)\mathbf{E}(\mathbf{u}_\beta)\right] + Ra T_\beta \mathbf{e}, \quad \mathbf{x} \in \Omega, \tag{8.47}$$

$$\mathrm{div}\mathbf{u}_\beta = 0, \quad \mathbf{x} \in \Omega, \tag{8.48}$$

$$\mathbf{u}_\beta \cdot \mathbf{n} = 0 \text{ (and/or } \partial(\mathbf{u}_\beta)_\tau/\partial \mathbf{n} = 0), \quad \mathbf{x} \in \partial\Omega, \tag{8.49}$$

where the sign of the velocity field is changed ($\mathbf{u}_\beta$ by $-\mathbf{u}_\beta$) in Eqs. (8.43) and (8.47) to simplify the application of the total variation diminishing (TVD) method (see Section 7.9) for solving (8.43)–(8.46). Hereinafter we refer to temperature T_ϑ as the input temperature for the problem (8.43)–(8.49). The core of the transformation of the heat equation is the addition of a high-order differential expression $\Lambda(\partial T_\beta/\partial t)$ multiplied by a small parameter $\beta > 0$. Note that Eq. (8.45) is added to the boundary conditions to properly define the regularised backward heat problem. The solution to the regularised backward heat problem is stable for $\beta > 0$, and the approximate solution to (8.43)–(8.49) converges to the solution of (8.36)–(8.40), and (8.42) in some spaces, where the conditions of well-posedness are met (Samarskii and Vabischevich, 2004). Thus, the inverse problem of thermo-convective mantle flow is reduced to determination of the velocity $\mathbf{u}_\beta = \mathbf{u}_\beta(t, \mathbf{x})$, the pressure $P_\beta = P_\beta(t, \mathbf{x})$, and the temperature $T_\beta = T_\beta(t, \mathbf{x})$ satisfying (8.43)–(8.49).

8.8.2 Optimisation problem

A maximum of the following functional is sought with respect to the regularisation parameter β:

$$\delta - \left\| T(t = \vartheta, \cdot; T_{\beta_k}(t = 0, \cdot)) - \varphi(\cdot) \right\| \to \max_k, \tag{8.50}$$

$$\beta_k = \beta_0 q^{k-1}, \quad k = 1, 2, \ldots, \mathfrak{R}, \tag{8.51}$$

where sign $\|\cdot\|$ denotes the norm in the space $L_2(\Omega)$. Since in what follows the dependence of solutions on initial temperature data is important, we introduce these data explicitly into the mathematical representation of temperature. Here $T_k = T_{\beta_k}(t = 0, \cdot)$ is the solution to the regularised backward heat problem (8.43)–(8.45) at $t = 0$; $T(t = \vartheta, \cdot; T_k)$ is the solution to the heat problem (8.39)–(8.41) at the initial condition $T(t = 0, \cdot) = T_k$ at time $t = \vartheta$; φ is the known temperature at $t = \vartheta$ (the input data on the present temperature);

small parameters $\beta_0 > 0$ and $0 < q < 1$ are defined below; and $\delta > 0$ is a given accuracy. When q tends to unity, the computational cost becomes large; and when q tends to zero, the optimal solution can be missed.

The prescribed accuracy δ is composed from the accuracy of the initial data and the accuracy of computations. When the input noise decreases and the accuracy of computations increases, the regularisation parameter is expected to decrease. However, estimates of the initial data errors are usually inaccurate. Estimates of the computation accuracy are not always known, and when they are available, the estimates are coarse. In practical computations, it is more convenient to minimise the following functional with respect to (8.51)

$$\left\| T_{\beta_{k+1}}(t = 0, \cdot) - T_{\beta_k}(t = 0, \cdot) \right\| \to \min_k, \tag{8.52}$$

where misfit between temperatures obtained at two adjacent iterations must be compared. To implement the minimisation of temperature residual (8.50), the inverse problem (8.43)–(8.49) must be solved on the entire time interval as well as the direct problem (8.36)–(8.41) on the same time interval. This at least doubles the amount of computations. The minimisation of functional (8.52) has a lower computational cost, but it does not rely on *a priori* information.

8.8.3 Numerical algorithm for QRV data assimilation

In this section we describe the numerical algorithm for solving the inverse problem of thermo-convective mantle flow using the QRV method. We consider a uniform temporal partition $t_n = \vartheta - \delta t\, n$ (as defined in Section 8.5) and prescribe some values to parameters β_0, q and $\Re$ (e.g. $\beta_0 = 10^{-3}$, $q = 0.1$ and $\Re = 10$). According to (8.51) a sequence of the values of the regularisation parameter $\{\beta_k\}$ is defined. For each value $\beta = \beta_k$ model temperature and velocity are determined in the following way.

Step 1. Given the temperature $T_\beta = T_\beta(t, \cdot)$ at $t = t_n$, the velocity $\mathbf{u}_\beta = \mathbf{u}_\beta(t_n, \cdot)$ is found by solving problem (8.47)–(8.49). This velocity is assumed to be constant on the time interval $[t_n+_1, t_n]$.

Step 2. Given the velocity $\mathbf{u}_\beta = \mathbf{u}_\beta(t_n, \cdot)$, the new temperature $T_\beta = T_\beta(t, \cdot)$ at $t = t_{n+1}$ is found on the time interval $[t_{n+1}, t_n]$ subject to the final condition $T_\beta = T_\beta(t_n, \cdot)$ by solving the regularised problem (8.43)–(8.46) backward in time.

Step 3. Upon the completion of steps 1 and 2 for all $n = 0, 1, \ldots, m$, the temperature $T_\beta = T_\beta(t_n, \cdot)$ and the velocity $\mathbf{u}_\beta = \mathbf{u}_\beta(t_n, \cdot)$ are obtained at each $t = t_n$. Based on the computed solution we can find the temperature and flow velocity at each point of time interval $[0, \vartheta]$ using interpolation.

Step 4a. The direct problem (8.39)–(8.41) is solved assuming that the initial temperature is given as $T_\beta = T_\beta(t = 0, \cdot)$, and the temperature residual (8.50) is found. If the residual does not exceed the predefined accuracy, the calculations are terminated, and the results obtained at step 3 are considered as the final ones. Otherwise, parameters β_0, q and $\Re$ entering Eq. (8.51) are modified, and the calculations are continued from step 1 for new set $\{\beta_k\}$.

Step 4b. The functional (8.52) is calculated. If the residual between the solutions obtained for two adjacent regularisation parameters satisfies a predefined criterion (the criterion should be defined by a user, because no *a priori* data are used at this step), the calculation is terminated, and the results obtained at step 3 are considered as the final ones. Otherwise, parameters β_0, q and $\Re$ entering Eq. (8.51) are modified, and the calculations are continued from step 1 for new set $\{\beta_k\}$.

In a particular implementation, either step 4a or step 4b is used to terminate the computation. This algorithm allows (i) organising a certain number of independent computational modules for various values of the regularised parameter β_k that find the solution to the regularised problem using steps 1–3 and (ii) determining *a posteriori* an acceptable result according to step 4a or step 4b.

8.9 Application of the QRV method: restoration of mantle plume evolution

To compare the numerical results obtained by the QRV method with that obtained by the VAR and BAD methods described in this chapter, we develop the same forward model for mantle plume evolution as presented in Section 8.6. Figure 8.10 (panels a–d) illustrates the evolution of mantle plumes in the forward model. The state of the plumes at the 'present' time (Fig. 8.10d) obtained by solving the direct problem was used as the input temperature for the inverse problem (an assimilation of the 'present' temperature to the past). Note that this initial state (input temperature) is given with an error introduced by the numerical algorithm used to solve the direct problem. Figure 8.10 illustrates the states of the plumes restored by the QRV method (panels e–g) and the residual δT (see Eq. (8.26) and panel h) between the initial temperature for the forward model (Fig. 8.10a) and the temperature $\tilde{T}(\mathbf{x})$ assimilated to the same age (Fig. 8.10g). To check the stability of the algorithm, a forward model of the restored plumes is computed using the solution to the inverse problem at the time of 265 Myr ago (Fig. 8.10g) as the initial state for the forward model. The result of this run is shown in Fig. 8.10i.

To compare the accuracy of the data assimilation methods, a restoration model from the 'present' time (Fig. 8.10d) to the time of 265 Myr ago was developed using the BAD method. Figure 8.10 shows the BAD model results (panels e1–g1) together with the temperature residual (panel h1) between the initial temperature (panel a) and the temperature assimilated to the same age (panel g1). The VAR method was not used to assimilate data within the time interval of more than 100 Myr (for $Ra \approx 10^6$), because proper filtering of the increasing noise is required to smooth the input data and solution (Section 8.7).

Figure 8.11a presents the residual $J_1(\beta) = \|T_0(\cdot) - T_\beta(t = t_0, \cdot; T_\vartheta)\|$ between the initial temperature T_0 at $t_0 = 265$ Myr ago and the restored temperature (to the same time) obtained by solving the inverse problem with the input temperature T_ϑ. The optimal accuracy is attained at $\beta^* = \arg\min\{J_1(\beta) : \beta = \beta_k, k = 1, 2, \ldots, 10\} \approx 10^{-7}$ in the case of $r = 20$, and at $\beta^* \approx 10^{-6}$ and $\beta^* \approx 10^{-5.5}$ in the cases of the viscosity ratio $r = 200$ and $r = 1000$,

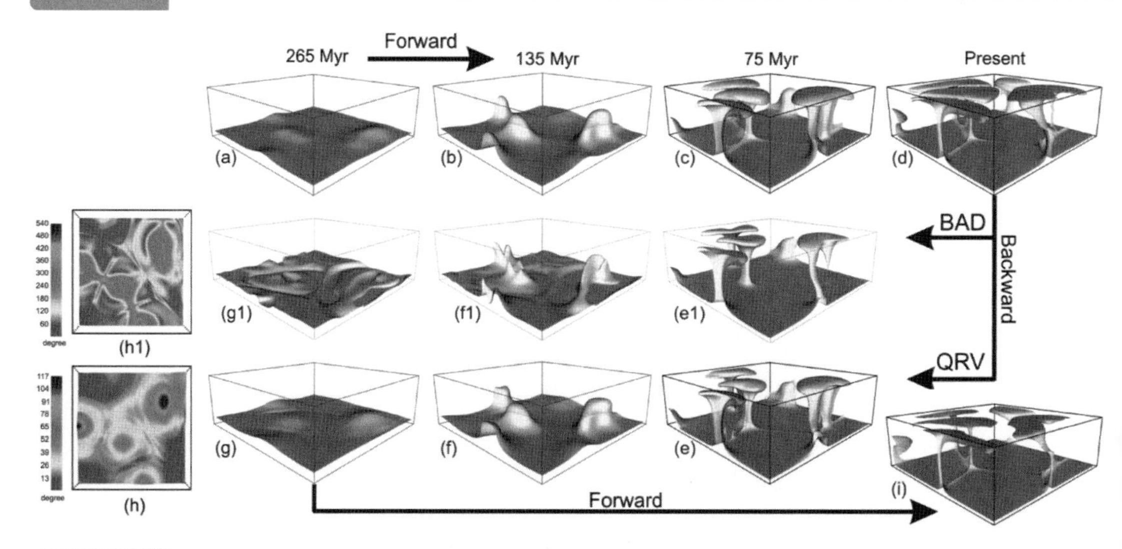

Fig. 8.10. Model of mantle plume evolution forward in time at successive times: (a–d) from 265 Myr ago to the present state of the plumes ($r = 20$). Assimilation of the mantle temperature and flow from the present state back to the geological past using the QRV (d–g; $\beta = 10^{-7}$) and BAD (d, e1–g1) methods. Verification of the QRV assimilation accuracy: forward model of the plume evolution starting from the initial (restored) state of the plumes (g) to their present state (i). Temperature residuals between the initial temperature for the forward model and the temperature assimilated to the same age using the QRV and BAD methods are presented in panels (h) and (h1), respectively. After Ismail-Zadeh *et al.* (2007). (In colour as Plate 6. See colour plates section.)

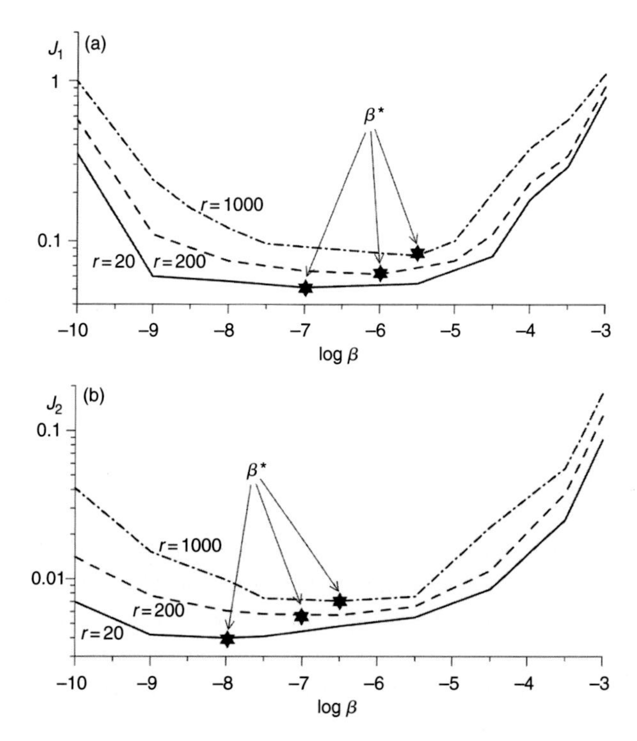

Fig. 8.11. Temperature misfit (a) J_1 and (b) J_2 as functions of the regularisation parameter β. The minimum of the temperature misfit is achieved at β^*, an optimal regularisation parameter. Solid curves: $r = 20$; dashed curves: $r = 200$; and dash-dotted curves: $r = 1000$. After Ismail-Zadeh *et al.* (2007).

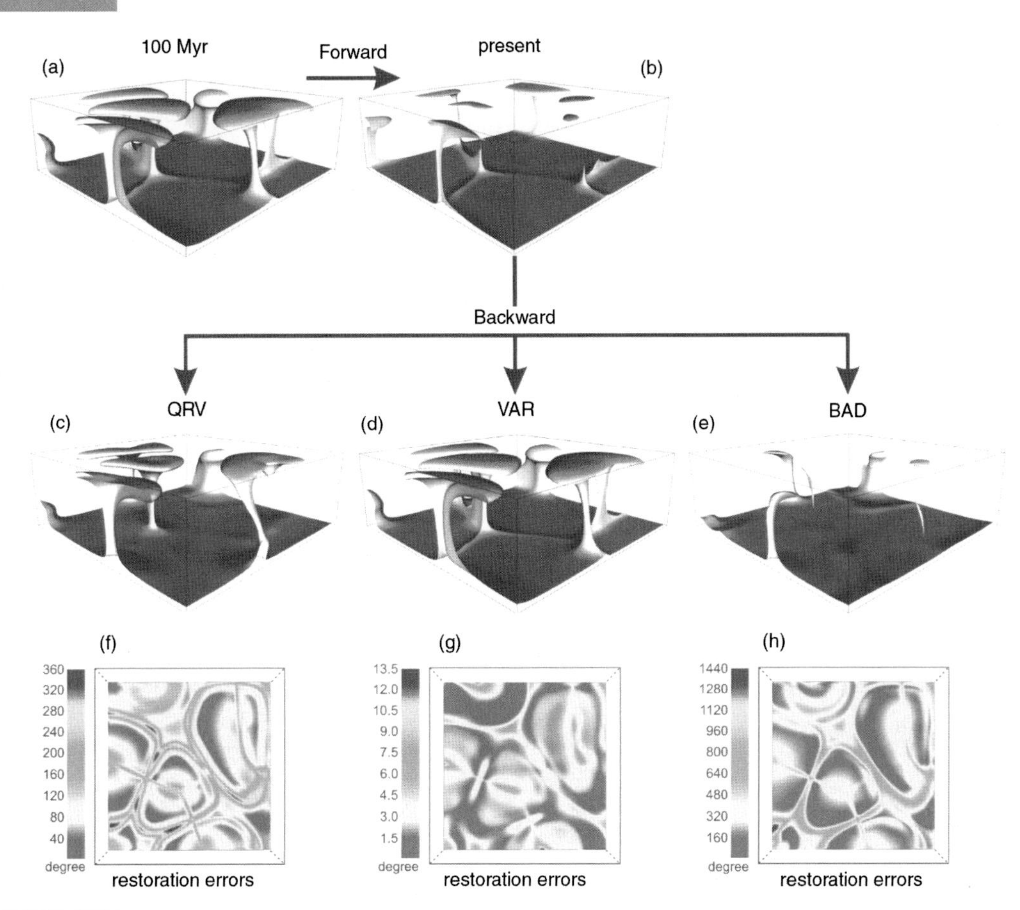

Fig. 8.12. **Model of mantle plume diffusion forward in time (a and b; $r = 20$). Assimilation of the mantle temperature and flow to the time of 100 Myrs ago and temperature residuals between the present temperature model (b) and the temperature assimilated to the same age, using the QRV (c and f; $\beta = 10^{-7}$), VAR (d and g), and BAD (e and h) methods, respectively. After Ismail-Zadeh *et al.* (2007). (In colour as Plate 7. See colour plates section.)**

respectively. Figure 8.11b illustrates the residual $J_2(\beta) = \|T_\beta(t_0, \cdot; T_\vartheta) - T_{\widehat{\beta}}(t_0, \cdot; T_\vartheta)\|$ between the reconstructed temperature at $t_0 = 265$ Myr ago obtained for various values of β in the range $10^{-9} \le \beta \le 10^{-3}$ and $\widehat{\beta} = \beta/2$. These results show the choice of the optimal value of the regularisation parameter using step 4b of the numerical algorithm for the QRV data assimilation (Section 8.8.3). In the case of $r = 20$ the parameter $\beta^* = \arg\min\{J_2(\beta) : \beta = \beta_k, k = 1, 2, \ldots, 12\} \approx 10^{-8}$ provides the optimal accuracy for the solution; in the cases of $r = 200$ and $r = 1000$ the optimal accuracy is achieved at $\beta^* \approx 10^{-7}$ and $\beta^* \approx 10^{-6.5}$, respectively. Comparison of the temperature residuals for three values of the viscosity ratio r indicates that the residuals become larger as the viscosity ratio increases. The numerical experiments show that the algorithm for solving the inverse problem performs well when the regularisation parameter is in the range $10^{-8} \le \beta \le 10^{-6}$. For greater values, the solution of the inverse problem retains the stability but is less accurate. For $\beta < 10^{-9}$ the numerical procedure becomes unstable, and the computations must be stopped.

To compare how the techniques for data assimilation can restore the prominent state of the thermal plumes in the past from their 'present' weak state, a forward model was initially developed from the prominent state of the plumes (Fig. 8.12a) to their diffusive state in 100 Myr (Fig. 8.12b) using $50 \times 50 \times 50$ finite rectangular elements to approximate the vector velocity potential and a finite difference grid $148 \times 148 \times 148$ for approximation of temperature, velocity and viscosity. All other parameters of the model are the same.

The VAR method (Fig. 8.12d, g) provides the best performance for the diffused plume restoration. The BAD method (Fig. 8.12e, h) cannot restore the diffused parts of the plumes, because temperature is only advected backward in time. The QRV method (Fig. 8.12c, f) restores the diffused thermal plumes, meanwhile the restoration results are not so perfect as in the case of VAR method (compare temperature residuals in Fig. 8.12, panels f and g). Although the accuracy of the QRV data assimilation is lower compared with the VAR data assimilation, the QRV method does not require any additional smoothing of the input data and filtering of temperature noise as the VAR method does.

8.10 Application of the QRV method: restoration of descending lithosphere evolution

8.10.1 The Vrancea seismicity and the relic descending slab

Repeated large intermediate-depth earthquakes in the southeastern (SE-) Carpathians (the Vrancea region) cause destruction in Bucharest, the capital city of Romania, and shake central and eastern European cities several hundred kilometres away from the hypocentres of the events. The earthquake-prone Vrancea region (Fig. 8.13) is bounded to the north and north-east by the Eastern European platform (EEP), to the east by the Scythian platform (SCP), to the south-east by the North Dobrogea orogen (DOB), to the south and south-west by the Moesian platform (MOP), and to the north-west by the Transylvanian basin (TRB). The epicentres of the sub-crustal earthquakes in the Vrancea region are concentrated within a very small seismogenic volume about 70×30 km^2 in planform and between depths of about 70 and 180 km. Below this depth the seismicity ends abruptly: one seismic event at 220 km depth is an exception (Oncescu and Bonjer, 1997).

The 1940 $M_W = 7.7$ earthquake gave rise to the development of a number of geodynamic models for this region. McKenzie (1972) suggested that this seismicity is associated with a relic slab sinking in the mantle and now overlain by continental crust. The 1977 large earthquake and later the 1986 and 1990 earthquakes again raised questions about the nature of the earthquakes. A seismic gap at depths of 40–70 km beneath Vrancea led to the assumption that the lithospheric slab had already detached from the continental crust (Fuchs *et al.*, 1979). Oncescu (1984) proposed that the intermediate-depth events are generated in a zone that separates the sinking slab from the neighbouring immobile part of the lithosphere rather than in the sinking slab itself. Linzer (1996) explained the nearly vertical position of the Vrancea slab as the final rollback stage of a small fragment of oceanic lithosphere. Various types of slab detachment or delamination (see, for example, Girbacea and Frisch, 1998; Wortel and Spakman, 2000; Gvirtzman, 2002; Sperner *et al.*, 2005) have been

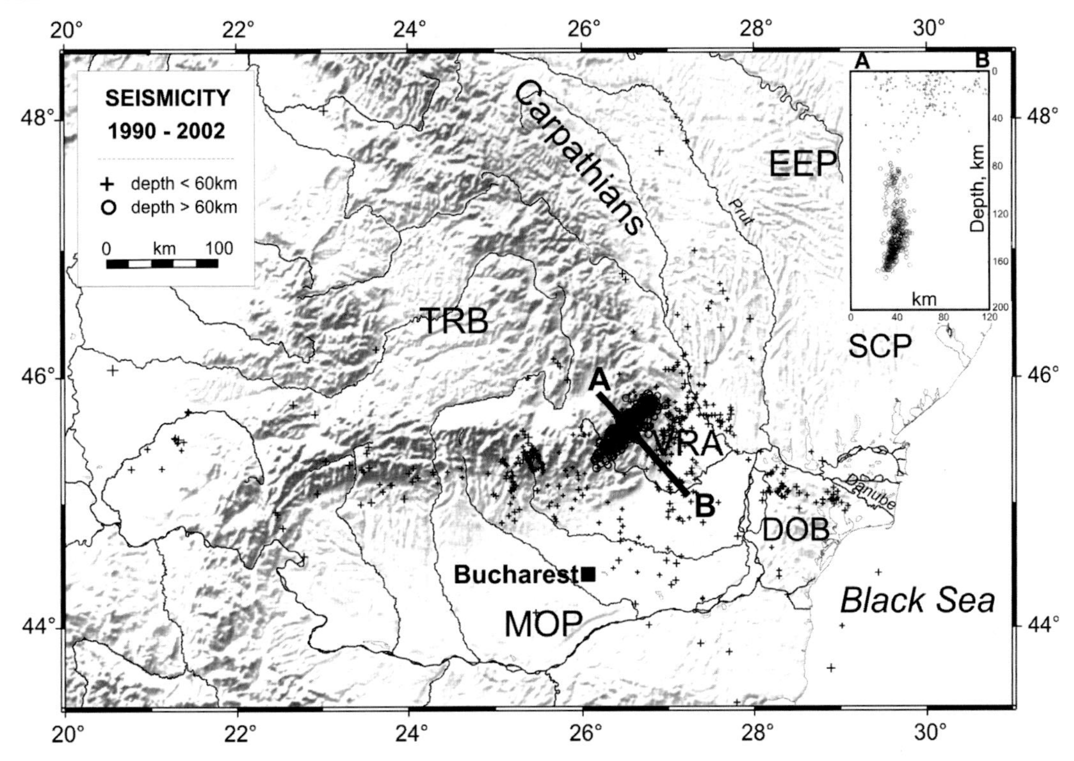

Fig. 8.13. Topography map of the SE-Carpathians and epicentres of Vrancea earthquakes (magnitude ≥3). The upper right panel presents hypocentres of the same earthquakes projected onto the NW–SE vertical plane AB. DOB, Dobrogea orogen; EEP, Eastern European platform; MOP, Moesian platform; SCP, Scythian platform; TRB, Transylvanian basin; and VRA, Vrancea. After Ismail-Zadeh *et al.* (2008).

proposed to explain the present-day seismic images of the descending slab. Cloetingh *et al.* (2004) argued in favour of the complex configuration of the underthrusted lithosphere and its thermo-mechanical age as primary factors in the behaviour of the descending slab after continental collision. The origin of the descending lithosphere in the region, i.e. whether the Vrancea slab is oceanic or continental, is still under debate. Pana and Erdmer (1996) and Pana and Morris (1999) argued that because there is no geological evidence of Miocene oceanic crust in the eastern Carpathians, the descending lithosphere is likely to be thinned continental or transitional lithosphere.

The Neogene to Late Miocene (*c* 11 Myr) evolution of the Carpathian region is mainly driven by the north-eastward, later eastward and south-eastward roll-back or slab retreat (Royden, 1988; Sperner *et al.*, 2001) into a Carpathians embayment, consisting of the last remnants of an oceanic or thinned continental domain attached to the European continent (see Balla, 1987; Csontos *et al.*, 1992). When the mechanically strong East-European and Scythian platforms started to enter the subduction zone, the buoyancy forces of the thick continental crust exceeded the slab pull forces and convergence stopped after only a short period of continental thrusting (Tarapoanca *et al.*, 2004; Sperner *et al.*, 2005). Continental convergence in the SE-Carpathians ceased about 11 Myr (Jiricek, 1979; Csontos *et al.*,

1992), and after that the lithospheric slab descended beneath the Vrancea region due to gravity. The hydrostatic buoyancy forces promote the sinking of the slab, but viscous and frictional forces resist the descent. The combination of these forces produces shear stresses at intermediate depths that are high enough to cause earthquakes (Ismail-Zadeh *et al.*, 2000a, 2005b).

In this section we present a quantitative model of the thermal evolution of the descending slab in the SE-Carpathians suggested by Ismail-Zadeh *et al.* (2008). The model is based on assimilation of present crust/mantle temperature and flow in the geological past using the QRV method. Mantle thermal structures are restored and analysed in the context of modern regional geodynamics.

8.10.2 Temperature model

Temperature is a key physical parameter controlling the density and rheology of the Earth's material and hence crustal and mantle dynamics. Besides direct measurements of temperature in boreholes in the shallow portion of the crust, there are no direct measurements of deep crustal and mantle temperatures, and therefore the temperatures must be estimated indirectly from seismic wave anomalies, geochemical data and surface heat flow observations.

Ismail-Zadeh *et al.* (2005a, 2008) developed a model of the present crustal and mantle temperature beneath the SE-Carpathians by using the most recent high-resolution seismic tomography image (map of the anomalies of P-wave velocities) of the lithosphere and asthenosphere in the region (Martin *et al.*, 2005, 2006). The tomography image shows a high velocity body beneath the Vrancea region and the Moesian platform interpreted as the subducted lithospheric slab (Martin *et al.*, 2006). The seismic tomographic model of the region consists of eight horizontal layers of different thickness (15 km up to 70 km) starting from the depth of 35 km and extending down to a depth of 440 km. Each layer of about $1000 \times 1000\ \text{km}^2$ is subdivided horizontally into $16 \times 16\ \text{km}^2$ blocks. To restrict numerical errors in our data assimilation we smooth the velocity anomaly data between the blocks and the layers using a spline interpolation. Ismail-Zadeh *et al.* (2005a) converted seismic wave velocity anomalies into temperature considering the effects of mantle composition, anelasticity, and partial melting on seismic velocities. The temperature in the crust is constrained by measurements of surface heat flux corrected for palaeoclimate changes and for the effects of sedimentation (Demetrescu *et al.*, 2001).

Depth slices of the present temperature model are illustrated in Fig. 8.14. The pattern of resulting mantle temperature anomalies (predicted temperature minus background temperature) is similar to the pattern of observed P-wave velocity anomalies (Martin *et al.*, 2006), but not an exact copy because of the non-linear inversion of the seismic anomalies to temperature. The low temperatures are associated with the high-velocity body beneath the Vrancea region (VRA) and the East European platform (EEP) and are already visible at depths of 50 km. The slab image becomes clear at 70–110 km depth as a NE–SW oriented cold anomaly. With increasing depth (110–200 km depth) the thermal image of the slab broadens in NW–SE direction. The orientation of the cold body changes from NE–SW to N–S below the depth of 200 km. The slab extends down to 280–320 km depth beneath

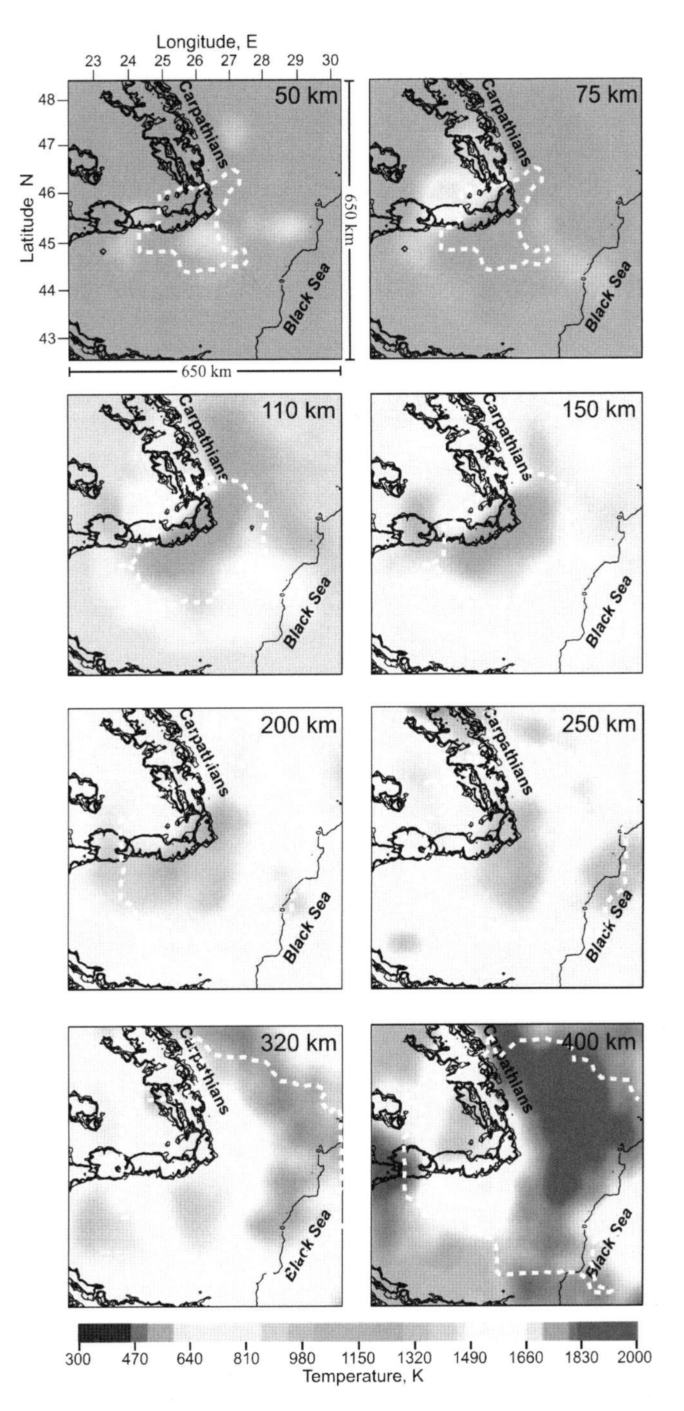

Fig. 8.14. Present temperature model as the result of the inversion of the *P*-wave velocity model. Theoretically well-resolved regions are bounded by dashed line (see text and Martin *et al.*, 2006). Each slice presents a part of the horizontal section of the model domain Ω corresponding to $[x_1 = 177.5\text{ km}, x_1 = 825.5\text{ km}] \times [x_2 = 177.5\text{ km}, x_2 = 825.5\text{ km}]$, and the isolines present the surface topography (also in Figs. 8.15 and 8.17). After Ismail-Zadeh *et al.* (2008). (In colour as Plate 8. See colour plates section.)

the Vrancea region itself. A cold anomaly beneath the Transylvanian Basin is estimated at depths of 370–440 km. According to Wortel and Spakman (2000) and Martin *et al.* (2006) this cold material can be interpreted as a remnant of subducted lithosphere detached during the Miocene along the Carpathian Arc and residing within the upper mantle transition zone. High temperatures are predicted beneath the Transylvanian Basin (TRB) at about 70–110 km depth. Two other high temperature regions are found at 110–150 km depth below the Moesian platform (MOP) and deeper than 200 km under the EEP and the Dobrogea orogen (DOB), which might be correlated with the regional lithosphere/asthenosphere boundary.

8.10.3 QRV data assimilation

To minimise boundary effects, the studied region (650 $\times$ 650 km^2 and 440 km deep, see Fig. 8.14) has been bordered horizontally by a 200 km area and extended vertically to the depth of 670 km. Therefore, a rectangular domain $\Omega = [0,\ l_1 = 1050\,\text{km}] \times [0, l_2 = 1050\,\text{km}] \times [0, h = 670\,\text{km}]$ is considered for assimilation of present temperature and mantle flow beneath the SE-Carpathians.

Our ability to reverse mantle flow is limited by our knowledge of past movements in the region, which are well constrained in only some cases. In reality, the Earth's crust and lithospheric mantle are driven by mantle convection and the gravitational pull of dense descending slabs. However, when a numerical model is constructed for a particular region, external lateral forces can influence the regional crustal and uppermost mantle movements. Yet in order to make useful predictions that can be tested geologically, a time-dependent numerical model should include the history of surface motions. Since this is not currently achievable in a dynamical way, it is necessary to prescribe surface motions by using velocity boundary conditions.

The simulations are performed backward in time for a period of 22 Myr. Perfect slip conditions are assumed at the vertical and lower boundaries of the model domain. For the first 11 Myr (starting from the present time), when the rates of continental convergence were insignificant (Jiricek, 1979; Csontos *et al.*, 1992), no velocity is imposed at the surface, and the conditions at the upper boundary are free slip. The north-westward velocity is imposed in the portion of the upper model boundary (Fig. 8.15a) for the time interval from 11 Myr to 16 Myr and the westward velocity in the same portion of the boundary (Fig. 8.15b) for the interval from 16 Myr to 22 Myr. The velocities are consistent with the direction and rates of the regional convergence in the Early and Middle Miocene (Morley, 1996; Fügenschuh and Schmid, 2005; Sperner *et al.*, 2005). The effect of the surface loading due to the Carpathian Mountains is not considered, because this loading would have insignificant influence on the dynamics of the region (as was shown in two-dimensional models of the Vrancea slab evolution; Ismail-Zadeh *et al.*, 2005b).

The heat flux through the vertical boundaries of the model domain Ω is set to zero. The upper and lower boundaries are assumed to be isothermal surfaces. The present temperature above 440 km depth is derived from the seismic velocity anomalies and heat flow data. The adiabatic geotherm for potential temperature 1750 K (Katsura *et al.*, 2004) was used to define the present temperature below 440 km (where seismic tomography data are not

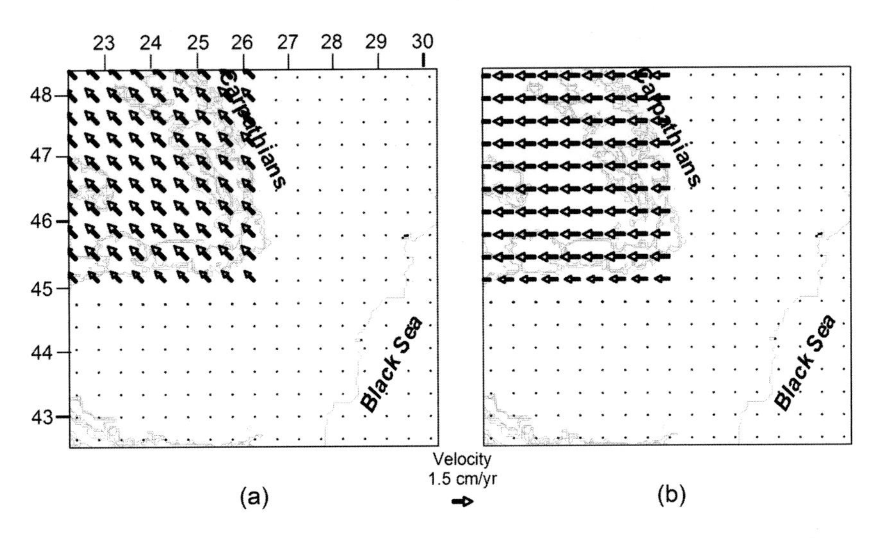

Fig. 8.15. **Surface velocity imposed on the part of the upper boundary of the model domain (see the caption of Fig. 8.14) in data assimilation modelling for the time interval from 11 Myr to 16 Myr ago (a) and for that from 16 Myr to 22 Myr ago (b). After Ismail-Zadeh *et al.* (2008).**

available). Equations (8.36)–(8.49) with the specified boundary and initial conditions are solved numerically.

To estimate the accuracy of the results of data assimilation, the temperature and mantle flow restored to the time of 22 Myr ago were employed as the initial condition for a model of the slab evolution forward in time; the model was run to the present; and the temperature residual (the difference between the present temperature and that predicted by the forward model with the restored temperature as an initial temperature distribution) was analysed subsequently. The maximum temperature residual does not exceed 50 K.

A sensitivity analysis was performed to understand how stable is the numerical solution to small perturbations of input (present) temperatures. The model of the present temperature (Section 8.10.2) has been perturbed randomly by 0.5% to 2% and then assimilated to the past to find the initial temperature. A misfit between the initial temperatures related to the perturbed and unperturbed present temperature is rather small (2% to 4%), which proves that the solution is stable. The numerical models, with a spatial resolution of 7 km × 7 km × 5 km, were run on parallel computers. The accuracy of the numerical solutions has been verified by several tests, including grid and total mass changes (Ismail-Zadeh *et al.*, 2001a).

8.10.4 What the past tells us

We discuss here the results of assimilation of the present temperature model beneath the SE-Carpathians into Miocene times. Although there is some evidence that the lithospheric slab was already partly subducted some 75 Myr ago (Sandulescu, 1988), the assimilation interval was restricted to the Miocene, because the pre-Miocene evolution of the descending slab, as well as the regional horizontal movements, are poorly known. Incorporation of insufficiently accurate data into the assimilation model could result in incorrect scenarios of mantle and lithosphere dynamics in the region. Moreover, to restore the history of pre-Miocene slab subduction, a high-resolution seismic tomography image of the deeper mantle is required (the present image is restricted to the depth of 440 km).

Early Miocene subduction beneath the Carpathian arc and the subsequent gentle continental collision transported cold and dense lithospheric material into the hotter mantle. Figure 8.16 presents the 3-D thermal image of the slab and pattern of contemporary flow induced by the descending slab. Note that the direction of the flow is reversed, because we solve the problem backward in time: cold slabs move upward during the numerical modelling. The 3-D flow is rather complicated: toroidal (in horizontal planes) flow at depths between about 100 km and 200 km coexists with poloidal (in vertical planes) flow.

The relatively cold (blue to dark green) region seen at depths of 40 km to 230 km (Fig. 8.17b) can be interpreted as the earlier evolutionary stages of the lithospheric slab. The slab is poorly visible at shallow depth in the model of the present temperature (Fig. 8.17a). Since active subduction of the lithospheric slab in the region ended in Late Miocene times and earlier rates of convergence were low before it, Ismail-Zadeh *et al.* (2006) argue that the cold slab, descending slowly at these depths, has been warmed up, and its thermal shape has faded due to heat diffusion. Thermal conduction in the shallow Earth (where viscosity is high) plays a significant part in heat transfer compared to thermal convection. The deeper we look in the region, the larger are the effects of thermal advection compared to conduction: the lithosphere has moved upwards to the place where it had been in Miocene times. Below 230 km depth the thermal roots of the cold slab are clearly visible in the present temperature model (Figs. 8.14, 8.16 and 8.17a), but they are almost invisible in Fig. 8.17b and in Fig. 8.18 of the models of the assimilated temperature, because the slab did not reach these depths in Miocene times.

The geometry of the restored slab clearly shows two parts of the sinking body (Figs. 8.17b and 8.18). The NW–SE oriented part of the body is located in the vicinity of the boundary between the EEP and Scythian platform (SCP) and may be a relic of cold lithosphere that has travelled eastward. Another part has a NE–SW orientation and is associated with the present descending slab. An interesting geometrical feature of the restored slab is its curvature beneath the SE-Carpathians. In Miocene times the slab had a concave surface confirming the curvature of the Carpathian arc down to depths of about 60 km. At greater depths the slab changed its shape to that of a convex surface and split into two parts at a depth of about 200 km. Although such a change in slab curvature is visible neither in the model of the present temperature nor in the seismic tomography image, most likely because of slab warming and heat diffusion, we suggest that the convex shape of the slab is likely to be preserved at the present time. Ismail-Zadeh *et al.* (2008) proposed that this change in the geometry of the descending slab can cause stress localisation due to slab bending and subsequent stress release resulting in earthquakes, which occur at depths of 70–180 km in the region.

Moreover, the north–south (NS)-oriented cold material visible at the depths of 230 km to 320 km (Figs. 8.14 and 8.17a) does not appear as a separate (from the NE–SW-oriented slab) body in the models of Miocene time. Instead, it looks more like two differently oriented branches of the SW-end of the slab at 60–130 km depth (visible in Figs. 8.17b and 8.18).

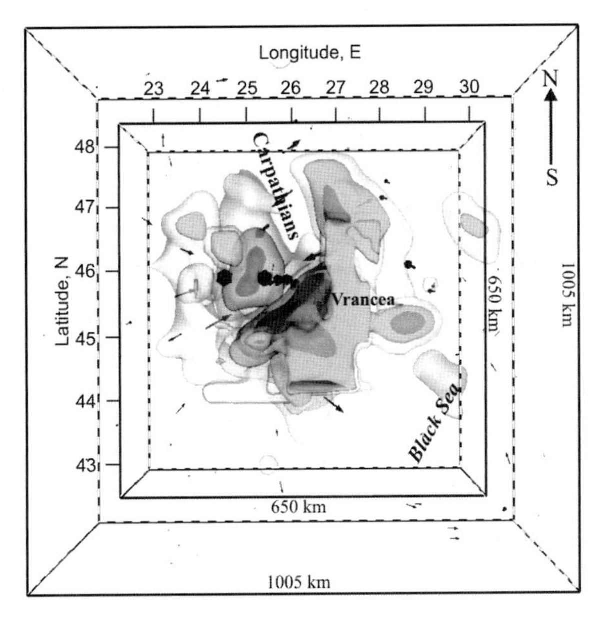

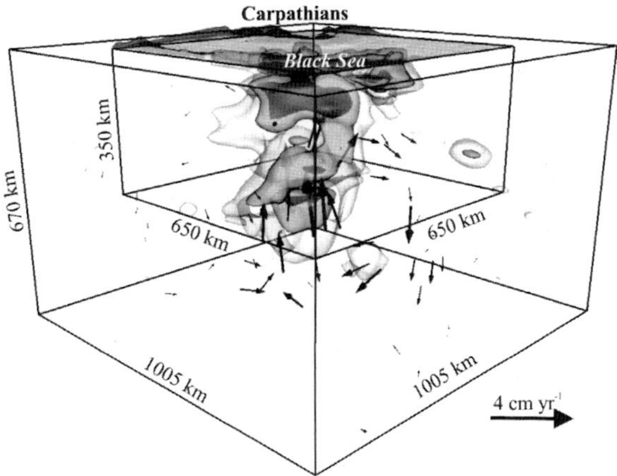

Fig. 8.16. A 3-D thermal shape of the Vrancea slab and contemporary flow induced by the descending slab beneath the SE-Carpathians. Upper panel: top view. Lower panel: side view from the SE toward NW. Arrows illustrate the direction and magnitude of the flow. The marked sub-domain of the model domain presents the region around the Vrancea shown in Fig. 8.17 (in horizontal slices) and in Fig. 8.18. The surfaces marked by blue, dark cyan and light cyan illustrate the surfaces of 0.07, 0.14 and 0.21 temperature anomaly δT, respectively, where $\delta T = (T_{hav} - T)/T_{hav}$ and T_{hav} is the horizontally averaged temperature. The top surface presents the topography, and the red star marks the location of the intermediate-depth earthquakes. After Ismail-Zadeh *et al.* (2008). (In colour as Plate 9. See colour plates section.)

Therefore, the results of the assimilation of the present temperature model to Miocene time provide a plausible explanation for the change in the spatial orientation of the slab from NE–SW to NS beneath 200 km observed in the seismic tomography image (Martin *et al.*, 2006).

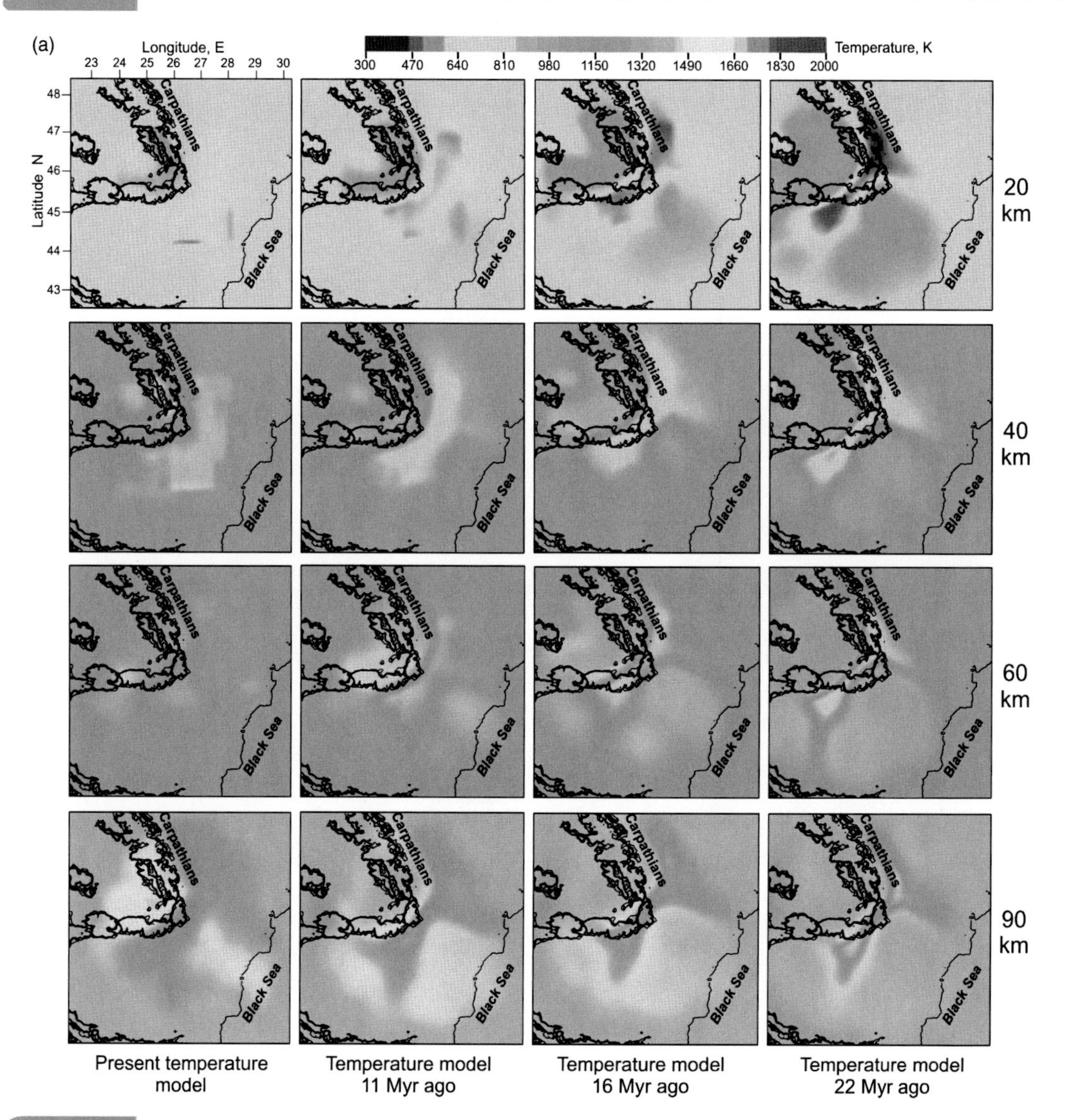

Fig. 8.17. **Thermal evolution of the crust and mantle beneath the SE-Carpathians. Horizontal sections of temperature obtained by the assimilation of the present temperature to the Miocene times. After Ismail-Zadeh *et al.* (2008). (In colour as Plate 10. See colour plates section.)**

The slab bending might be related to a complex interaction between two parts of the sinking body and the surrounding mantle. The sinking body displaces the mantle, which, in its turn, forces the slab to deform due to corner (toroidal) flows different within each of two sub-regions (to NW and to SE from the present descending slab). Also, the curvature of the descending slab can be influenced by slab heterogeneities due to variations in its thickness and viscosity (Cloetingh *et al.*, 2004; Morra *et al.*, 2006).

Martin *et al.* (2006) interpret the negative velocity anomalies NW of the present slab at depths between 70 km and 110 km (see the relevant temperature slices in Figs. 8.14 and 8.17a) as a shallow asthenospheric upwelling associated with possible slab rollback. Also, they mention partial melting as an additional contribution to the reduction of seismic

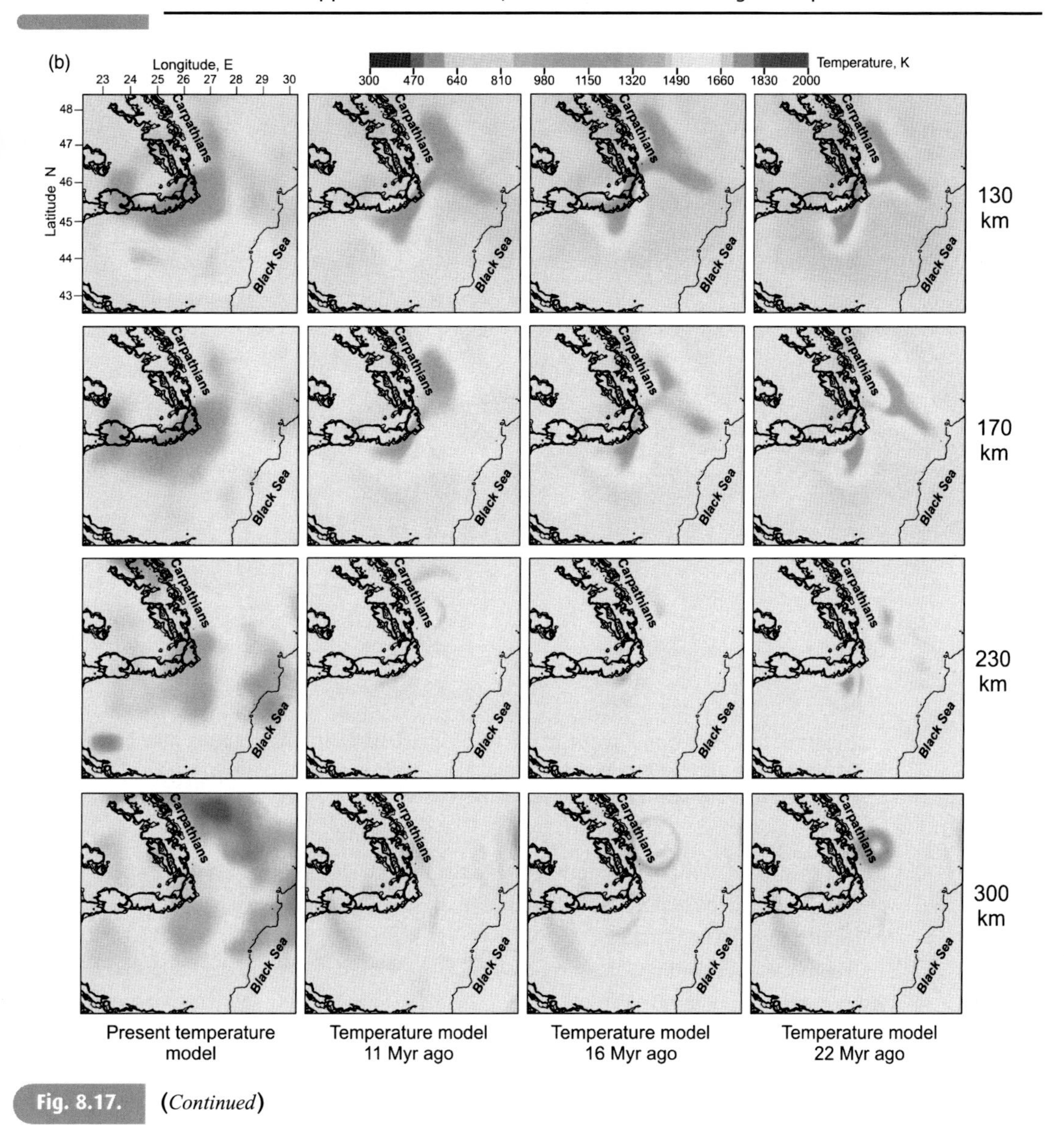

Fig. 8.17. (*Continued*)

velocities at these depths. The results of our assimilation show that the descending slab is surrounded by a border of hotter rocks at depths down to about 250 km. The rocks could be heated owing to partial melting as a result of slab dehydration. Although the effects of slab dehydration or partial melting were not considered in the modelling, the numerical results support the hypothesis of dehydration of the descending lithosphere and its partial melting as the source of reduction of seismic velocities at these depths and probably deeper (see temperature slices at the depths of 130–220 km). Alternatively, the hot anomalies beneath the Transylvanian basin and partly beneath the Moesian platform could be dragged down by the descending slab since the Miocene times, and therefore, the slab was surrounded by the hotter rocks. Using numerical experiments, Honda *et al.* (2007) showed recently how the lithospheric plate subducting beneath the Honshu Island in Japan dragged down a hot anomaly adjacent to the plate. Some areas of high temperature at depths below 280 km

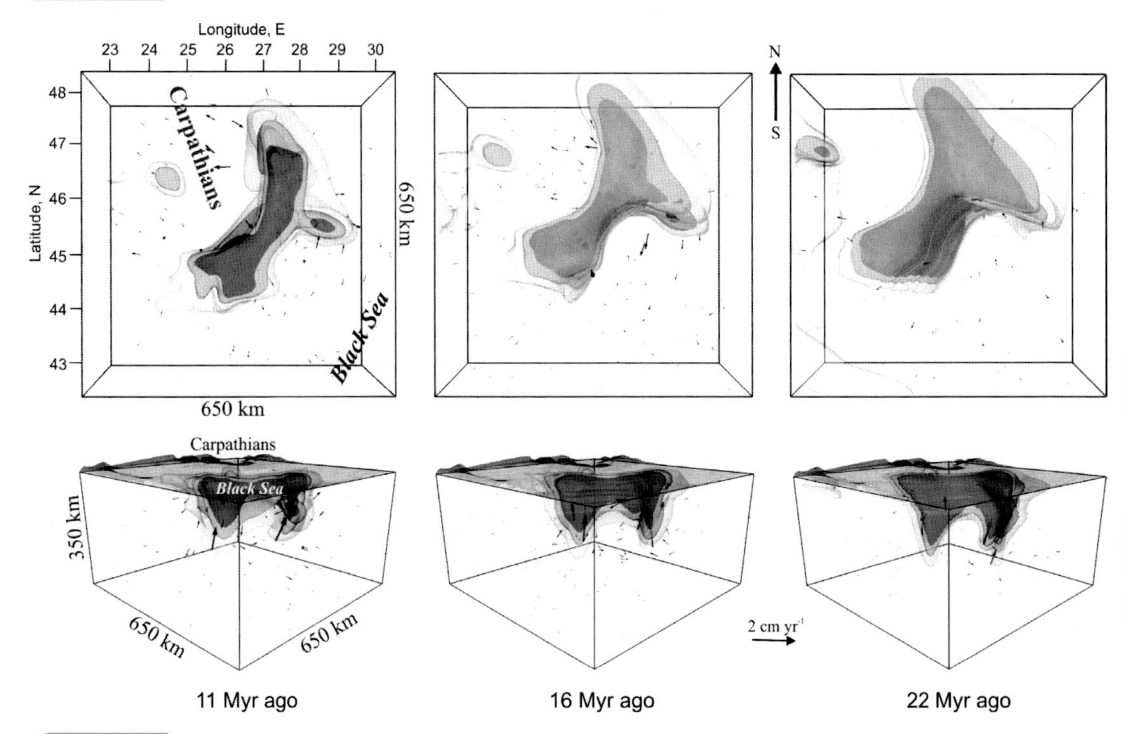

Fig. 8.18. **Snapshots of the 3-D thermal shape of the Vrancea slab and pattern of mantle flow beneath the SE-Carpathians in the Miocene times. See Fig. 8.16 for other notations. After Ismail-Zadeh *et al.* (2008). (In colour as Plate 11. See colour plates section.)**

can be associated with mantle upwelling in the region. High-temperature anomalies are not clearly visible in the restored temperatures at these depths, because the upwelling was likely not active in Miocene times.

The numerical results were compared with that obtained by the backward advection of temperature (using the BAD method). Figure 8.19 (dashed curve) shows that the maximum temperature residual is about 360 K. The neglect of heat diffusion leads to an inaccurate restoration of mantle temperature, especially in the areas of low temperature and high viscosity. The similar results for the BAD data assimilation have been obtained in the synthetic case study (see Fig. 8.12e and h).

8.10.5 Limitations and uncertainties

There is a major physical limitation of the restoration of mantle structures. If a thermal feature created, let us say, several hundred million years ago has completely diffused away by the present, it is impossible to restore the feature, which was more prominent in the past. The time to which a present thermal structure in the upper mantle can be restored should be restricted by the characteristic thermal diffusion time, the time when the temperatures of the evolved structure and the ambient mantle are nearly indistinguishable (Ismail-Zadeh *et al.*, 2004a). The time (t) for restoration of seismic thermal structures depends on depth

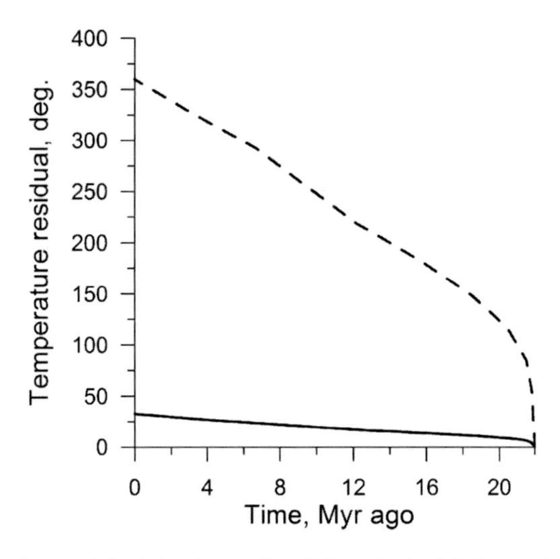

Fig. 8.19. **Temperature misfit in the model of the descending lithospheric slab beneath the southeastern Carpathians. The misfit is defined as an integral difference between the temperature assimilated to any time $t \in$ [present, 22 Myr ago] and that predicted by the forward model (8.21)–(8.26) to the same time assuming the assimilated temperature 22 Myr ago as the initial condition for the forward model. Solid and dashed curves present the misfits for the cases of temperature assimilation using the QRV and BAD methods, respectively.**

(d) of seismic tomography images and can be roughly estimated as $t = d/v$, where v is the average vertical velocity of mantle flow. For example, the time for restoration of the Vrancea slab evolution in the studied models should be less than about 80 Myr, considering $d = 400$ km and $v \approx 0.5$ cm yr^{-1}.

Other sources of uncertainty in the modelling of mantle temperature in the SE-Carpathians come from the choice of mantle composition (Nitoi *et al.*, 2002; Seghedi *et al.*, 2004; Szabó *et al.*, 2004), the seismic attenuation model (Popa *et al.*, 2005; Weidle *et al.*, 2007), and poor knowledge of the presence of water at mantle depths. The drop of electrical resistivity below 1 Ω m (Stanica and Stanica, 1993) can be an indicator of the presence of fluids (due to dehydration of mantle rocks) below the SE-Carpathians; however, the information is very limited and cannot be used in quantitative modelling.

Viscosity is an important physical parameter in numerical modelling of mantle dynamics, because it influences the stress state and results in strengthening or weakening of Earth's material. Though it is the least-known physical parameter of the model, the viscosity of the Vrancea slab was constrained by observations of the regional strain rates (Ismail-Zadeh *et al.*, 2005a).

The geometry of the mantle structures changes with time, diminishing the degree of surface curvature of the structures. Like Ricci flow, which tends to diffuse regions of high curvature into ones of lower curvature (Hamilton, 1982; Perelman, 2002), heat conduction smoothes the complex thermal surfaces of mantle bodies with time. Present seismic tomography images of mantle structures do not allow definition of the sharp shapes of these structures. Assimilation of mantle temperature and flow to the geological past instead provides a quantitative tool to restore thermal shapes of prominent structures in the past from their diffusive shapes at present. High-resolution experiments on seismic wave attenuation, improved knowledge of crustal and mantle mineral composition, accurate GPS measurements of regional movements, and precise geological palaeoreconstructions of crustal movements will assist to refine the present models and our knowledge of the regional thermal evolutions. The basic knowledge we have gained from the case studies is the dynamics of the Earth's interior in the past, which could result in its present dynamics.

8.11 Comparison of data assimilation methods

We compare the VAR, QRV and BAD methods in terms of solution stability, convergence, and accuracy, time interval for data assimilation, analytical and algorithmic works, and computer performance (see Tables 8.1–8.3). The VAR data assimilation assumes that the direct and adjoint problems are constructed and solved iteratively forward in time. The structure of the adjoint problem is identical to the structure of the original problem, which considerably simplifies the numerical implementation. However, the VAR method imposes some requirements for the mathematical model (i.e. a derivation of the adjoint problem). Moreover, for an efficient numerical implementation of the VAR method, the error level of the computations must be adjusted to the parameters of the algorithm, and this complicates computations.

The QRV method allows employing sophisticated mathematical models (because it does not require derivation of an adjoint problem as in the VAR data assimilation) and hence expands the scope for applications in geodynamics (e.g. thermo-chemical convection, phase transformations in the mantle). It does not require that the desired accuracy of computations be directly related to the parameters of the numerical algorithm. However, the regularising operators usually used in the QRV method enhance the order of the system of differential equations to be solved.

The BAD is the simplest method for data assimilation in models of mantle dynamics, because it does not require any additional work (neither analytical nor computational). The major difference between the BAD method and two other methods (VAR and QRV methods) is that the BAD method is by design expected to work (and hence can be used) *only* in advection-dominated heat flow. In the regions of high temperature/low mantle viscosity, where heat is transferred mainly by convective flow, the use of the BAD method is justified, and the results of numerical reconstructions can be considered to be satisfactory. Otherwise, in the regions of conduction-dominated heat flow (due to either high mantle viscosity or high conductivity of mantle rocks), the use of the BAD method cannot guarantee any similarity of reconstructed structures. If mantle structures are diffused significantly, the remaining features of the structures can be only backward advected with the flow.

The comparison between the data assimilation methods is summarised in Table 8.2 in terms of a quality of numerical results. The quality of the results is defined here as a relative (not absolute) measure of their accuracy. The results are good, satisfactory or poor compared with other methods for data assimilation considered in this study. The numerical results of the reconstructions for both synthetic and geophysical case studies show the comparison quantitatively.

Table 8.1. Comparison of methods for data assimilation in models of mantle dynamics

	QRV method	VAR method	BAD method
Method	Solving the regularised backward heat problem with respect to parameter β	Iterative sequential solving of the direct and adjoint heat problems	Solving of heat advection equation backward in time
Solution's stability	Stable for parameter β to numerical errors (see text; also in[1]) and conditionally stable for parameter β to arbitrarily assigned initial conditions (numerically[2])	Conditionally stable to numerical errors depending on the number of iterations (theoretically[3]) and unstable to arbitrarily assigned initial conditions (numerically[4])	Stable theoretically and numerically
Solution's convergence	Numerical solution to the regularised backward heat problem converges to the solution of the backward heat problem in the special class of admissible solutions[5]	Numerical solution converges to the exact solution in the Hilbert space[6]	Not applied
Solution's accuracy[7]	Acceptable accuracy for both synthetic and geophysical data	High accuracy for synthetic data	Low accuracy for both synthetic and geophysical data in conduction-dominated mantle flow
Time interval for data assimilation[8]	Limited by the characteristic thermal diffusion time	Limited by the characteristic thermal diffusion time and the accuracy of the numerical solution	No specific time limitation; depends on mantle flow intensity
Analytical work	Choice of the regularising operator	Derivation of the adjoint problem	No additional analytical work
Algorithmic work	New solver for the regularised equation should be developed	No new solver should be developed	Solver for the advection equation is to be used

[1]Lattes and Lions, 1969; [2]see Fig. 8.11 and relevant text; [3]Ismail-Zadeh *et al.*, 2004a; [4]Ismail-Zadeh *et al.*, 2006; [5]Tikhonov and Arsenin, 1977; [6]Tikhonov and Samarskii, 1990; [7]see Table 8.2; [8]see text for details.

Table 8.2. Quality of the numerical results obtained by different methods for data assimilation

	Synthetic data		*Geophysical data*	
Quality	*Advection-dominated regime*	*Diffusion-dominated region*	*Advection-dominated regime*	*Diffusion-dominated region*
Good	VAR	VAR	—	—
Satisfactory	QRV, BAD	QRV	QRV, BAD	QRV
Poor	—	BAD	—	BAD

Table 8.3. Performance of data assimilation methods

	CPU time (circa, in s)		
Method	Solving the Stokes problem using $50 \times 50 \times 50$ finite elements	Solving the backward heat problem using $148 \times 148 \times 148$ finite difference mesh	Total
BAD	180	2.5	182.5
QRV	100 to 180	3	103 to 183
VAR	360	$1.5\,n$	$360 + 1.5\,n$

The time interval for the VAR data assimilation depends strongly on smoothness of the input data and the solution. The time interval for the BAD data assimilation depends on the intensity of mantle convection: it is short for conduction-dominated heat transfer and becomes longer for advection-dominated heat flow. In the absence of thermal diffusion the backwards advection of a low-density fluid in the gravity field will finally yield a uniformly stratified, inverted density structure, where the low-density fluid overlain by a dense fluid spreads across the lower boundary of the model domain to form a horizontal layer. Once the layer is formed, information about the evolution of the low-density fluid will be lost, and hence any forward modelling will be useless, because no information on initial conditions will be available (Ismail-Zadeh *et al.* 2001b; Kaus and Podladchikov 2001).

The QRV method can provide stable results within the characteristic thermal diffusion time interval. However, the length of the time interval for QRV data assimilation depends on several factors. Let us explain this by the example of heat conduction equation (8.27). Assume that the solution to the backward heat conduction equation with the boundary conditions (8.28) and the initial condition $T(t = t^*, x) = T^*(x)$ satisfies the inequality $\|\partial^4 T/\partial x^4\| \le L_d$ at any time t. This strong additional requirement can be considered as the requirement of sufficient smoothness of the solution and initial data. Considering the regularised backward heat conduction equation (8.31) with the boundary conditions (8.32)–(8.33) and the input temperature $T_\beta(t = t^*, x) = T^*_\beta(x)$ and assuming that $\|T^*_\beta - T^*\| \le \delta$, Samarskii and Vabishchevich (2004) estimated the temperature misfit between the solution $T(t,x)$ to the backward heat conduction problem and the solution $T_\beta(t,x)$ to the regularised backward heat conduction equation:

$$\left\| T(t,x) - T_\beta(t,x) \right\| \leq \tilde{C}\delta \exp[\beta^{-1/2}(t^* - t)] + \beta L_d t, \quad 0 \leq t \leq t^*, \tag{8.53}$$

where constant $\tilde{C}$ is determined from the *a priori* known parameters of the backward heat conduction problem. For the given regularisation parameter β, errors in the input data δ, and smoothness parameter L_d, it is possible to evaluate the time interval $0 \leq t \leq t^*$ of data assimilation for which the temperature misfit would not exceed a prescribed value.

Computer performance of the data assimilation methods can be estimated by a comparison of CPU times for solving the inverse problem of thermal convection. Table 8.3 lists the CPU times required to perform one time-step computations on 16 processors. The CPU time for the case of the QRV method is presented for a given regularisation parameter β; in general, the total CPU time increases by a factor of $\mathfrak{R}$, where $\mathfrak{R}$ is the number of runs required to determine the optimal regularisation parameter β^*. The numerical solution of the Stokes problem (by the conjugate gradient method) is the most time consuming calculation: it takes about 180 s to reach a high accuracy in computations of the velocity potential. The reduction in the CPU time for the QRV method is attained by employing the velocity potential computed at β_i as an initial guess function for the conjugate gradient method to compute the vector potential at β_{i+1}. An application of the VAR method requires to compute the Stokes problem twice to determine the 'advected' and 'true' velocities (Ismail-Zadeh *et al.*, 2004a). The CPU time required to compute the backward heat problem using the TVD solver (Section 7.9) is about 3 s in the case of the QRV method and 2.5 s in the case of the BAD method. For the VAR case, the CPU time required to solve the direct and adjoint heat problems by the semi-Lagrangian method (Section 7.8) is $1.5 \times n$, where n is the number of iterations in the gradient method (Eq. (8.15)) used to minimise the cost functional (Eq. (8.14)).

8.12 Errors in forward and backward modelling

A numerical model has three kinds of variables: state variables, input variables and parameters. *State variables* describe the physical properties of the medium (velocity, pressure, temperature) and depend on time and space. *Input variables* have to be provided to the model (initial or boundary conditions), most of the time these variables are not directly measured but they can be estimated through data assimilation. Most models contain also a set of *parameters* (e.g. viscosity, thermal diffusivity), which have to be tuned to adjust the model to the observations. All the variables can be polluted by errors.

There are three kinds of systematic errors in numerical modelling of geodynamical problems: model, discretisation and iteration errors. *Model errors* are associated with the idealisation of the Earth's dynamics by a set of conservation equations governing the dynamics. The model errors are defined as the difference between the actual Earth dynamics and the exact solution of the mathematical model. *Discretisation errors* are defined as the difference between the exact solution of the conservation equations and the exact solution of the algebraic system of equations obtained by discretising these equations. And *iteration errors* are defined as the difference between the iterative and exact solutions of the algebraic system of equations. It is important to be aware of the existence of these errors, and even more to try to distinguish one from another.

Apart from the errors associated with the numerical modelling, another two components of errors are essential when mantle temperature data are assimilated into the past: (i) data misfit associated with the uncertainties in the present temperature distribution in the Earth's mantle and (ii) errors associated with the uncertainties in initial and boundary conditions. Since there are no direct measurements of mantle temperatures, the temperatures can be estimated indirectly from either seismic wave (and their anomalies), geochemical analysis or through the extrapolation of surface heat flow observations. Many models of mantle temperature are based on the conversion of seismic tomography data into temperature. Meanwhile, a seismic tomography image of the Earth's mantle is a model indeed and incorporates its own model errors. Another source of uncertainty comes from the choice of mantle compositions in the modelling of mantle temperature from the seismic velocities. Therefore, if the present mantle temperature models are biased, information on temperature can be improperly propagated to the geological past.

The temperature at the lower boundary of the model domain used in forward and backward numerical modelling is, of course, an approximation to the real temperature, which is unknown and may change over time at this boundary. Hence, errors associated with the knowledge of the temperature (or heat flux) evolution at the core–mantle boundary are another essential component of errors, which can be propagated into the past during the data assimilation.

In numerical modelling sensitivity analysis assists in understanding the stability of the model solution to small perturbations in input variables or parameters. For instance, if we consider mantle temperature in the past as a solution to the backward model, what will be its variation if there is some perturbation on the inputs of the model (e.g. present temperature data)? The gradient of the objective functional with respect to input parameters in variational data assimilation gives the first-order sensitivity coefficients. The second-order adjoint sensitivity analysis presents some challenge associated with cumbersome computations of the product of the Hessian matrix of the objective functional with some vector (Le Dimet *et al.*, 2002), and hence it is omitted in our study. Hier-Majumder *et al.* (2006) performed first-order sensitivity analysis for two-dimensional problems of thermo-convective flow in the mantle. See Cacuci (2003) and Cacuci *et al.* (2005) for more detail on sensitivity and uncertainty analysis.

9 Parallel computing

9.1 Introduction

This chapter introduces only the basics of parallel computing and does not intend to cover all aspects of this topic. The major challenge in writing this chapter was to keep up with the progress in computer science, which, compared with mathematics and computational methods, is a rapidly evolving discipline, and many current approaches to parallel computing may become almost useless in about a decade or so. Meanwhile many geodynamic models cannot be solved today on a single processor because of memory or time restrictions, and hence parallel computers should be employed to run the models. Researchers dealing with computational geodynamics should know at least the basics of parallel coding and computing, and that motivated us to write this chapter. We discuss here the principal differences between sequential and parallel computing, shared and distributed memory, introduce a domain decomposition approach, message passing and MPI, analyse the cost of parallel processing, and present simple examples of codes for parallel computing. We refer the reader to the books and journals on parallel computing where the topic is described in much detail (see, for example, Lipovski and Malek, 1987; Crichlow, 1988; Fox *et al.*, 1988; Gibbons and Rytter, 1988; Fountain, 1994; Foster, 1995; Hord, 1999; Roosta, 2000; Snyder and Lin, 2008; Barney, 2009; Elsevier's *Parallel Computing Journal* and *Journal of Parallel and Distributed Computing*, World Scientific's *International Journal of High Speed Computing*).

9.2 Parallel versus sequential processing

The complexity of geodynamic problems and resulting mathematical models (three-dimensional time-dependent problems, the use of large observational data sets, visualisation of numerical results, etc.) demands the use of multi-processor computers to reduce time for computations by distributing or sharing data among processors. Parallel processing is the use of multiple processors to execute different parts of the same computation simultaneously. The main goal of parallel processing is to reduce the time that a researcher has to spend waiting for the numerical solution (that is, the own time of the researcher).

Imagine a student having to prepare a set of mineral samples for laboratory analysis: a typical solution would be to distinguish initially the minerals by their type (e.g. silicate and non-silicate minerals) and then to separate them by the number of defects within each type of the mineral samples. If there were two students doing this, they could split the set of mineral samples between themselves and both could follow the above strategy, combining their partial solutions at the end; or one of two students could sort by mineral type, and the other by defects within each type, both of them working simultaneously. Both scenarios are examples of the application of parallel processing to a particular task, and the reason for doing so is very simple: to reduce the amount of time before achieving a solution.

The above analogy can be used to distinguish the power and the weakness of the parallel approach. As the number of people involved in a particular task (e.g. the number of students to assist in the selection of minerals) increases, a characteristic speed-up curve can be observed. Such a curve demonstrates (i) how beneficial is the increase in the number of people involved in the task and (ii) when any further increase will not reduce significantly the time the people spend on the work. Consider, for example, how little it would help to have 30–40 students crowding around a table, each responsible for putting one particular mineral into its proper place in the table. This is exactly what is meant by the proverb 'Too many cooks spoil the broth'.

It should be pointed out that reducing the time to solution is not the only goal of the parallel processing. Notice that running programs (codes) costs money, and different ways of achieving the same solution could have significantly different costs. Remember that running a code in parallel across a large number of workstation-type computers or PC cluster could cost considerably less than submitting it to a large, mainframe-style supercomputer. All processors to be used in computation of a particular task should be located close to each other in terms of communication. The more communication latency incurred by a particular numerical task, the longer the task will run, and the more the user of the task might be charged.

As researchers become increasingly computationally sophisticated, the complexity of the problems they tackle increases proportionally (researchers are sometimes trying to bite off more than their computers can chew). One of the first resources to get exhausted is local memory. The amount of memory available on a single system is rarely going to be sufficient for the computational and data storage needs encountered during runs of numerical codes. This situation is greatly simplified by having access to the aggregate memory made available by distributed computing environments. Working storage (main memory) requirements can be spread around the various processors engaged in the cooperative computation, and long-term storage (tape and disk) can be accommodated at different locations. Any limited resource can be considered as the object of optimisation, if it is deemed to be the most important quantity to conserve. In most cases involving large-scale computation, however, user time is considered to be the most valuable resource to be conserved.

Accepting reduction of user time as the fundamental goal, why necessarily focus on parallel programming and processing as the means to this end? Is there not another approach that can also yield fast turnarounds? Actually such an approach exists. For example, make the single-processor design larger (e.g. increase the amount of memory it can directly address), more powerful and faster. However, there are several fundamental limitations in developing this approach: limits of communication speed (the most common strategies for increasing speed involve faster processors) and limits to miniaturisation (even though there are efforts directed at atomic-level component structures). Moreover, it is increasingly expensive to make a single processor faster. Therefore, the main road in the development of *computational science* is to put all single processors in parallel. A recent development in this field is to integrate several processors into a single chip, i.e. dual-core, quad-core, etc.

There are a number of bottlenecks typically encountered in the transition from serial processing to parallel processing. Some geoscientists who developed their serial codes a long time ago are understandably reluctant to have to learn a new way of designing their codes and then properly rewrite the codes. An efficient parallel algorithm often has little similarity to an efficient serial algorithm. The very first task in the conversion effort is to step way back from the existing serial application and to examine the question: can this task be effectively and efficiently performed in parallel, and, if so, how best can that be accomplished?

Very often, an existing serial code has to be almost completely ignored, and the parallel version written virtually from scratch. This can be a major commitment of resources, and for some old codes the projected return from such an investment is often considered to be insufficient to warrant the effort. However, once the decision has been made to move from serial to parallel, the real work of code conversion can very often be helped along by application of the growing number of automatic tools, well seasoned by the manual use of hard-learned rules of thumb.

9.3 Terminology of parallel processing

Parallel processing has own lexicon of terms and phrases. The following are some of the more commonly encountered terms listed here in the assumption that the reader does not know any of the terms.

- *Task* is a logically discrete section of computational work.
- *Parallel task*s are the tasks whose computations are independent of each other, so that all such tasks can be performed simultaneously with correct results.
- *Serial execution* is an execution of a computer code sequentially, one statement at a time.
- *Parallelisable problem* is a problem that can be divided into parallel tasks. This may require changes in the code and/or the underlying algorithm.

The simplest example of a parallelisable problem is to calculate the multiplication of two matrices $\mathbf{A} = \{a_{ij}\}$ and $\mathbf{B} = \{b_{ij}\}$ of size $n \times n$: each row of one matrix $(a_{i1}, a_{i2}, \ldots, a_{in})$ can be multiplied by each column of the other matrix $(b_{1j}, b_{2j}, \ldots, b_{nj})^T$, independently and simultaneously. As an example of a non-parallelisable problem, we can consider the calculation of the Fibonacci series (1, 1, 2, 3, 5, 8, 13, 21, ...): this series is calculated using the formula: $F(k+2) = F(k+1) + F(k)$, where $F(1) = F(2) = 1$. We notice that the calculation of the $k+2$ value uses those of both $k+1$ and k and these three terms of the series cannot be calculated independently, therefore, in parallel. A non-parallelisable problem, such as the calculation of the Fibonacci series, would entail dependent calculations rather than independent ones.

There are two basic ways to partition computational work among parallel tasks: *data* and *functional parallelism*. If each task performs the same series of calculations, but applies them to different data, the computational work is referred to as data parallelism. Considering the same example of matrix multiplication (as above) using n processors, we note that each processor does the exact same operations, but works on different parts of the matrices. If each task performs different calculations, i.e. carries out different functions of the overall problem, then the work is called a functional parallelism. This can be done on the same data or different data. For example, m processors can compute m different functions related to a rheological law of the mantle. Data parallelism is used more intensively in computations of geodynamic problems than functional parallelism.

Observed *speedup* (S) of a code that is parallelised, can be estimated by using the ratio between the *time of serial execution* (t_s) and the *time of parallel execution* (t_p): $S = t_s/t_p$. A typical relationship between speedup and number of nodes is presented in Fig. 9.1. The estimation of the observed speedup, one of the most widely used indicators of parallelisability, is intuitively satisfying as well as potentially misleading. A well-parallelised code usually runs in a fraction of the time that it takes the serial version. Meanwhile, serial and parallel codes are different, they perform different tasks, and the algorithms may be entirely distinct, and therefore, the comparison of t_s and t_p is not a rightful business unless the same version of the code is used to measure both t_s and t_p. Still, a good job of parallelisation should be evident in the amount of user time saved; what is debatable is the converse: if it is not evident that a lot of time has been saved, is it because the problem itself is not parallelisable, or because the parallelisation simply was not done well? An alternative way to determine a speedup is to compare the time of a serial job and a parallel job with the number of processors set to one. The difference in this case may be considerable if significant changes to the algorithm were necessary to enable parallelisation. Linear speedup is rarely seen (Fig. 9.1), because of a cost associated with using more nodes. The best parallel code should give the time of communication to be not more than 10% of the total.

The temporal coordination of parallel tasks is referred to as *synchronisation*. It involves waiting until two or more tasks reach a specified point (a sync point) before continuing any of the tasks. Synchronisation is needed to coordinate information exchange among tasks. In the previous example calculating the multiplication of two matrices: all of the conformations had to be completed before the resultant could be found. Synchronisation can consume time because processors sit idle waiting for tasks on other processors to complete. Synchronisation can be a major factor in decreasing parallel speedup, because the time spent waiting could have been spent in useful calculation.

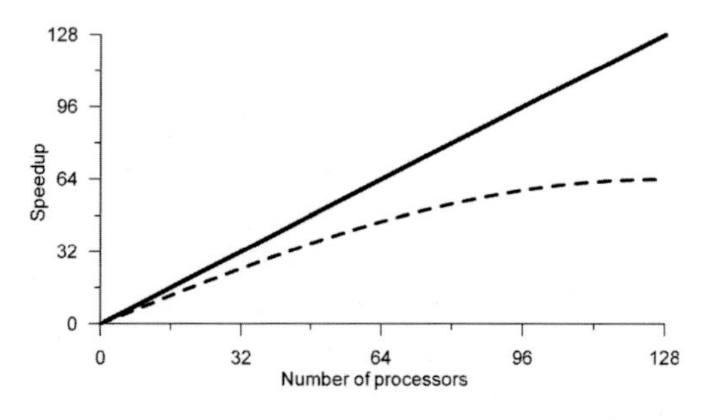

Fig. 9.1. **Efficiency of parallel performance – a speedup curve. The solid and dashed lines represent the ideal and actual speedups, respectively.**

The amount of time required to coordinate parallel tasks is called *parallel overhead*, as opposed to doing useful work. Parallelisation does not come free, and one of the most insidious costs is the time and cycles put into making sure that all of those separate tasks are doing what they are supposed to be doing. Things that are simply taken for granted in serial execution, or that do not apply, take on special significance when there are many tasks instead of just one; the three most commonly encountered coordination tasks are (i) time to start a task, (ii) time to terminate a task and (iii) synchronisation time. To start a task, a user needs to identify the task, to locate a processor in order to run it, to load the task onto the processor and required data, and actually to start the task. Termination of a task is not a simple chore, either: at the very least, results have to be combined or transferred, and operating system resources have to be freed before the processor can be used for other tasks.

Using parallel processing in modelling, a user can investigate the *granularity* of the work, i.e. a measure of the ratio of the amount of computation done in a parallel task to the amount of communication. Scale of granularity ranges from fine-grained (very little computation per communication-byte) to coarse-grained (extensive computation per communication-byte). The finer the granularity, the greater the limitation on speedup, due to the amount of synchronisation needed. Remember from Section 9.1 our discussion about considering how hard it would be to coordinate the activity of 30–40 students, who try to help sort minerals?

Some algorithms and codes are better suited to parallelism (i.e. more *scalable*) than others, and there are even economies of scale within parallel ones, i.e. algorithms can work quite well on a certain number of processors and work poorly at a higher number of processors, and vice versa. It should be emphasised here that great care should be taken to match the algorithm with the actual problem, and both of these with the actual size of the computer on which the problem will be solved by using that particular algorithm. Even if a user employs an algorithm well suited to specific purposes, the way the user implements it can determine how much parallelism is actually expressed in this application.

9.4 Shared and distributed memory

Memory access refers to the way in which the working storage is viewed by the user. The access method plays a very large role in determining the conceptualisation of the relationship of the program to its data.

9.4.1 Shared memory

Think of a single large blackboard, marked off so that all data elements have their own unique locations assigned, and all the members of a team are working together to test out a particular problem, all at the same time. This is an example of *shared memory* in action. The shared memory is accessible to multiple processors (Fig 9.2a). All processors associated with the same shared memory structure access the exact same storage, just as all the members of the team (in the above example) used the same unique data-element location on the blackboard to record any changes in those values.

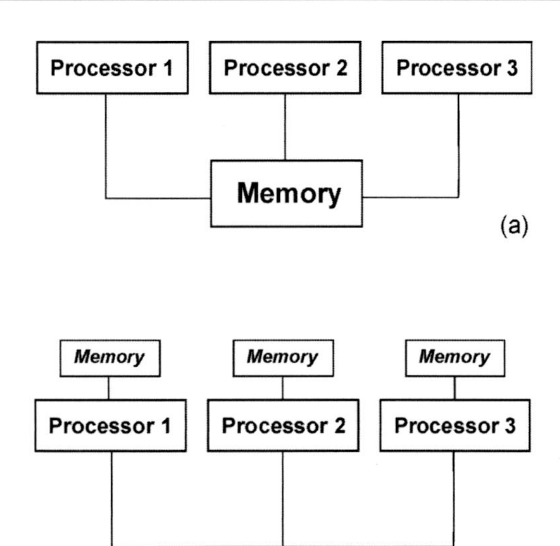

Fig. 9.2. Sketch of the shared (a) and distributed (b) memory.

Synchronisation is achieved by tasks' reading from and writing to the shared memory. In just the same way that the programmers would have to take turns writing into the blackboard locations, so the processors have to take turns accessing the shared memory cells. This makes it easy to implement *synchronisation* among all of the tasks, by simply coding them all to watch particular locations in the shared memory and do not make anything until certain values appear.

A shared memory location must not be changed by one task while another, concurrent task is accessing it. If one programmer is trying to use a value from the blackboard to calculate some other value, and sees another programmer begin to write over the one being copied, screams and shouts and thrown chalk and erasers can keep the needed value from being overwritten until it is no longer needed. Processors use more polite means of achieving the same ends, sometimes called *guards*: these are shared variables associated with the location in question, and a task can be programmed not to change the location before first gaining sole ownership of the *guard*. If all tasks have been programmed so that sole ownership of the *guard* is required before either reading or writing the associated location, this guarantees that no task will be attempting to read while another is busy changing that same value.

Data sharing among tasks provides a high speed of memory access. One of the most attractive features of shared memory, besides its conceptual simplicity, is that the time to communicate among the tasks is effectively a factor of a single fixed value, that being 'the time it takes a single task to read a single location'. There are, of course, limitations to this sharing. If you have more tasks than connections to memory, you have contention for access to the desired locations, and this amounts to increased latencies while all tasks obtain the required values. So the degree to which you can effectively scale a shared memory system is limited by the characteristics of the communication network coupling the processors to the memory units.

9.4.2 Distributed memory

The other major distinctive model of memory access is termed *distributed*. Memory is physically distributed among processors; each local memory is directly accessible only by its processor (Fig. 9.2b). Each component of a *distributed memory* parallel system is, in most cases, a self-contained environment, capable of acting independently of all other processors in the system. But in order to achieve the true benefits of this system, of course there must be a way for all of the processors to act in concert, which means 'control'. Synchronisation is achieved by moving data (even if it is just the message itself) between processors (communication). The only link between these distributed processors is the traffic along the communications network that couples them; therefore, any 'control' must take the form of data moving along that network to the processors. This is not all that different from the shared-memory case, in that you still have control information flowing back to processors, but now it is from other processors instead of from a central memory store.

A major concern is *data decomposition*, namely, how to divide arrays among local processors to minimise communication. Here is a major distinction between shared and distributed memory. In the case of shared memory, the processors do not need to worry about communicating with their peers, only with the central memory, while, in the case of distributed memory, the processors should properly communicate. A single large regular data structure, such as an array, can be left intact within shared memory, and each cooperating processor simply told which ranges of indices are its to deal with; for the distributed case, once the decision as to index ranges has been made, the data structure has to be *decomposed*, i.e. the data within a given set of ranges assigned to a particular processor must be physically *sent* to that processor in order for the processing to be done, and then any results must be *sent back* to whichever processor has responsibility for coordinating the final result.

Today supercomputers employ so-called *hybrid distributed–shared memory* (both shared and distributed memory architectures). This means that each component (node) contains multiple processors, each of which typically has multiple cores, all sharing the local memory and a network connection. Processors on a given multi-processor component of the supercomputer can address the component's memory as global (the shared memory part of the hybrid memory). This multi-processor component knows only about it own memory and not the memory on the neighbouring components, and therefore network communication is required to move data from one to another computer's components (the distributed memory part). The current trend seems to indicate that this type of memory architecture will continue to prevail and increase at the high end of computing for the foreseeable future (Barney, 2009).

9.5 Domain decomposition

When a spatial discretisation is used (finite difference, finite volume or finite element), the usual method of splitting the problem amongst N CPUs is simply to divide the physical domain into N sub-domains and assign one to each CPU, an approach that is called domain

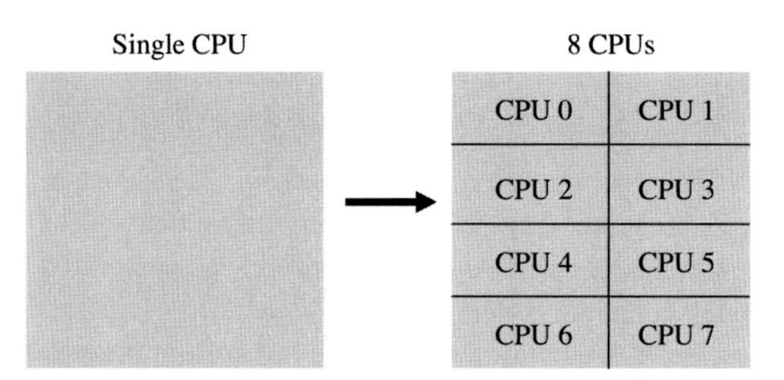

Fig. 9.3. Domain decomposition of the global domain into eight sub-domains, each assigned to a single CPU.

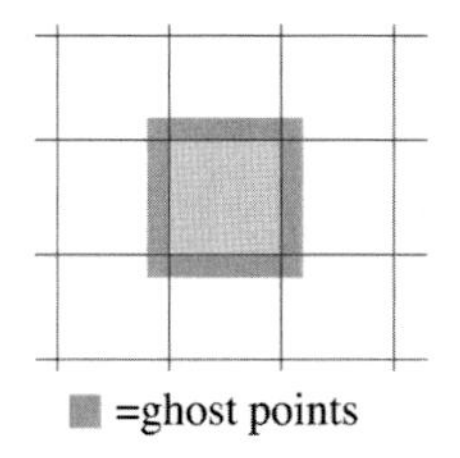

Fig. 9.4. Each CPU stores its own sub-domain (light grey) plus copies of the points at the edge of adjacent sub-domains, which act as the boundary condition to the local sub-domain.

decomposition (Figure 9.3). The CPUs work simultaneously on their local sub-domains. The boundary conditions to each local sub-domain are supplied by adjacent sub-domains. This approach works well for explicit time stepping and for iterative solvers, because performing one time step or iteration requires knowing only the values within the sub-domain and on the sides of the adjacent sub-domains. Therefore, on a distributed memory system each CPU holds in local memory its own sub-domain plus copies of the points on the boundaries of adjacent sub-domains: so-called *ghost points* (Figure 9.4). After each time step or iteration, which is performed simultaneously by each CPU, these ghost points must be updated, requiring communication between CPUs, which is typically performed by using the Message Passing Interface (MPI) (Section 9.7). Such a domain decomposition is straightforward to implement on a structured mesh but involves much more complexity for unstructured meshes or structured meshes with adaptive grid refinement, because it is important to balance the computational load of all CPUs, requiring a more complicated method of defining sub-domains.

If a structured mesh is divided in all three spatial directions, then to update the ghost points, which includes points in the corners and edges as well as on the faces of the sub-domain, communication with 26 other CPUs is apparently required (6 faces, 12 edges, 8 corners). This can, however, be accomplished by communicating with only six other CPUs using three sequential communication steps, one in each direction, as illustrated in Fig. 9.5. In the first step, the faces perpendicular to the x-direction are exchanged with the two adjacent nodes in the x-direction. In the second step, the y-faces are communicated, plus edge values that were just received by the x-communication. In the third step, involving communication

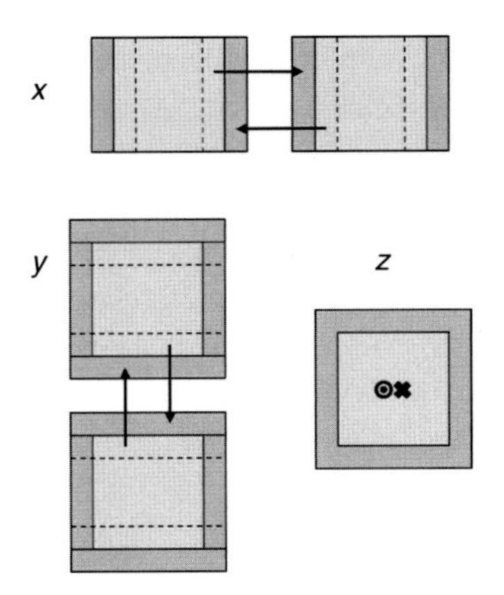

Fig. 9.5. **Three sequential communication steps are needed to update all ghost points including face, edge and corner, in a 3-D domain decomposition. Points within the sub-domain are in light grey, whereas ghost points are in dark grey.**

in the z-direction, entire faces including edges and corners are communicated. If tracers (markers) are used, these can be simply divided amongst the CPUs according to which sub-domain they are in. Tracers that cross sub-domain boundaries must be communicated to the relevant CPU.

The domain decomposition and the communication of ghost points must be explicitly programmed, which is typically done by using MPI calls (see Section 9.7). If this is not done, and a scalar program is run on N CPUs of a parallel computer, then it will duplicate the same task N times. A program that is parallelised using MPI must first call `MPI_Init` and then it should detect how many CPUs are being used using `MPI_Comm_size`, from which it can calculate how the domain is divided up. CPUs are given a logical number from 0 to the number of CPUs being used minus one, regardless of where they are physically located in the parallel computer. A CPU can find its logical number using `MPI_Comm_rank` and from this, calculate which sub-domain it is handling and the logical numbers of CPUs holding adjacent sub-domains. Ghost points can be updated by using message-passing calls such as `MPI_Send`, `MPI_Recv` and `MPI_Sendrecv`. Although this might sound complicated, in fact the iteration and time-stepping routines are barely changed from the equivalent scalar code: the main difference is that boundary conditions come from adjacent sub-domains rather than the external domain boundaries.

The efficiency of a parallel program is defined as

$$e = \frac{t_1}{N t_N}, \tag{9.1}$$

where t_1 is the execution time on a single CPU, N is the number of CPUs, and t_N is the execution time on N CPUs. If the parallel program scales perfectly, then $e = 1$ and the execution time is inversely proportional to N. In reality, the efficiency is normally significantly less than 1 because of time 'wasted' in communication, and much care must be taken to minimise this. The efficiency thus depends on the ratio of the time spent communicating to the time spent computing. To first order, the time spent computing is proportional to the number of grid points in each sub-domain (analogous to 'volume'), while the time spent communicating depends on the number of points on the boundary of each sub-domain (analogous to 'surface area'). Thus, the efficiency depends on the ratio of surface area to volume. If the domain is divided into cubic sub-domains each with n points in each direction, then the number of grid points per CPU is n^3 while the surface area is proportional to n^2. The ratio of communication to computation therefore scales as $1/n$. Clearly, the more points there are on each CPU, the more efficient the program will be.

Such logic can also guide decisions on whether to divide the domain along only one direction, or along two or three directions. As an example, consider a $256 \times 256 \times 256$ global grid distributed between 64 CPUs. Whatever the decomposition, there will be 262 144 points per CPU, but the number of boundary points depends on how they are split up. If split only in one direction ($256 \times 256 \times 4$ points/CPU), then each of the two internal sub-domain faces will contain 256×256 points, giving 131 072 ghost points to communicate per update. If split in two directions ($256 \times 32 \times 32$ points/CPU), then each of the four internal faces will contain 256×32 points, giving 32 768 ghost points. If split in three directions ($64 \times 64 \times 64$ points/CPU) then each of the six internal faces has 64×64 points, giving 24 576 ghost points. Clearly, a two- or three-dimensional domain decomposition is much better than a one-dimensional decomposition, but as the communication time is influenced not only by the number of points communicated but also the number of messages, additional analysis is needed.

It is useful to develop a model of how the code performance scales as a function of grid size and number of nodes. Here we assume a three-dimensional grid, decomposed in all three directions into approximately cubic sub-domains. The computation time t_{comp} can be written as:

$$t_{comp} = \frac{an^3}{f}, \tag{9.2}$$

where a is the number of floating point operations required per grid point and per iteration, n is the number of grid points in each direction on each CPU, and f is the number of floating point operations per second performed by the CPU.

The communication time depends both on the latency, which is the time taken to start sending a message, and the bandwidth, which is the number of bytes (or bits) that are sent per second along the communication network, once the message is started. The communication time can thus be written as:

$$t_{comm} = mL + \frac{b}{B}, \tag{9.3}$$

where L is the latency, B is the bandwidth, m is the number of messages and b is the number of bytes (or bits) communicated when updating the ghost points for one sub-domain. Note that b can be written as $b = psn^2$, where p is the precision (number of bytes or bits used to store each number: normally 4 or 8 bytes, equivalent to 32 or 64 bits), and s is the number of sides communicated, which is normally equal to m (6 in a 3-D decomposition). The total time needed per iteration is then:

$$t_{iter} = \frac{an^3}{f} + s\left(L + \frac{pn^2}{B}\right). \tag{9.4}$$

To give a specific example, consider an iterative solver for the scalar Poisson's equation, using a standard second-order finite-difference stencil. Each iteration takes 15 floating point operations per grid point. A CPU performance $f = 1$ Gflop/s in single precision ($p = 4$) is assumed. Using a 3-D domain composition, $m = s = 6$ on at least 64 or more CPUs and is progressively less than this on fewer than 64. A gigabit ethernet network gives approximately $L = 40$ μs and $B = 100$ MB/s, while a 'fast' network might give $L = 1$ μs and $B = 1$ GB/s (slightly faster than Quadrics QSII, but slower than the latest Infiniband networks). Figure 9.6 shows the calculated time per iteration and the parallel efficiency for up to 1024 CPUs, for different global grid sizes and on both gigabit and the 'fast' networks. On a gigabit network, the time per iteration has some minimum related to the latency time of the passed messages. This results in very low parallel efficiency for small grid sizes. On the coarsest grid of 16^3 points, it is actually fastest on a single CPU. Such coarse resolutions are encountered during multigrid cycles regardless of the number of fine grid points; the coarest levels are thus typically solved on a few or even one CPU. On 1024 CPUs, even a grid as large as 256^3 is predicted to be treated with only 40% efficiency. With a 'fast' network the situation is much improved, although global grids of 16^3 to 64^3 are still predicted to have low efficiency on 512 or more CPUs. This illustrates that parallel computers are not good for solving small problems faster, but rather are best for solving large problems at a similar speed to the speed that small problems can be solved on a single CPU.

This analysis assumes that each CPU has its own communication channel, which has often been the case. In recent years, however, nodes with multiple CPUs, each with several cores, have become the norm. A performance prediction then becomes more complicated because cores inside the same node compete for the external network connection, but can communicate with each other very quickly. Figure 9.6 (bottom row) shows the scaling of an actual 3-D spherical mantle convection code, StagYY (Tackley, 2008) run on up to 32 nodes (64 CPUs) of a Beowulf cluster in which each node contains two AMD Opteron CPUs and they are connected using a Quadrics QSII network. Perfect scaling would be a straight line with a slope of -1. For multigrid F-cycles the lines are not perfectly straight and communication takes a significant amount of time on 64 CPUs, but for advecting tracers efficiencies of $>90\%$ are achieved on all N.

9.6 Message passing

Distributed memory is synonymous with *message passing*, although the actual characteristics of the particular communication schemes used by different systems may hide that fact. The message-passing approach implies that tasks communicate by sending data packets to each other. *Messages* are discrete units of information, *discrete* meaning that they have a definite identity, and can be distinguished from all other messages. In practice, one of the

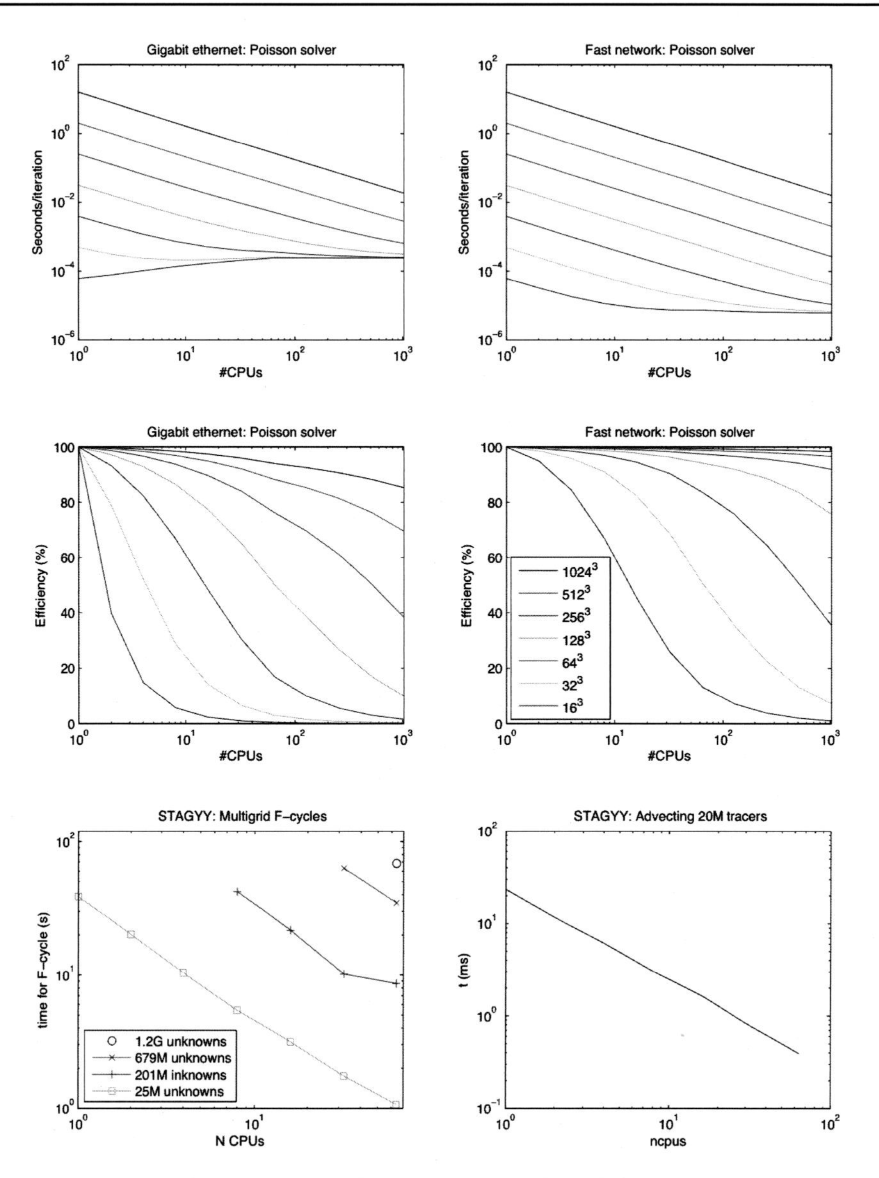

Fig. 9.6. **Top two rows: predicted performance scaling of an iterative finite difference 3-D Poisson solver on up to 1024 CPUs with up to $1024 \times 1024 \times 1024$ grid points, and on a gigabit ethernet network (left column) or a 'fast' network (right column). Each line is for a different grid resolution, from 1024^3 (top lines) to 32^3 (bottom lines), as indicated in the legend in the lower right plot. Bottom row: scaling of the 3-D spherical mantle convection code StagYY (Tackley, 2008) on up to 64 CPUs of a cluster.**

most common programming errors is to forget to actually make the messages distinctly different, by giving them unique identifiers or *tags*. Regardless, parallel tasks use these messages to send information and requests to their peers.

Overhead is proportional to the size and number of packets (more communication means greater costs; sending data is slower than accessing shared memory). Message passing is not cheap: every one of those messages has to be individually constructed, addressed, sent, delivered and read, all before the information it contains can be acted upon. Obviously, then, the more messages being sent, the more time and cycles spent in servicing message-oriented duties, and the less spent on the actual tasks that the messages are supposed to be subservient to. It is also clear that, in the general case, *message passing* will take more time and effort than *shared memory*. Having said that, it is often the case that shared memory scales less well than message passing, and, once past its maximum effective bandwidth utilisation, the latency associated with message passing may actually be lower than that encountered on an over-extended shared memory communications network.

There are ways to decrease the overhead associated with message passing, the most significant being to somehow arrange to do as much valuable computation as possible while communication is occurring. The most easily conceived method of doing this is to have two completely separate processors, each dedicated to either computation or communication, and coupled via a dual-ported DMA (*direct memory access*) in order to cooperate. This is something of the nature of *shared memory* being put in the service of *distributed memory*, and requires a multi-processor configuration for a single entity in the distributed system.

Other schemes involve time-slicing between the two tasks, or *active waiting* where a processor waiting for a communications event, such as receipt of an awaited message or acknowledgment of delivery of a sent message, arranges for a pre-emptive signal to be generated when the event occurs, and then goes off and does independent computation. These alternatives require considerably more sophistication in the control programs than simply sitting and twiddling one's thumbs until the communication process completes, but can be made to be very effective.

9.7 Basics of the Message Passing Interface

The Message Passing Interface (MPI) is a library of functions (and macros) that can be used in parallel programming. MPI is intended for employment in codes that exploit the existence of multiple processors by message passing (see Gropp *et al.*, 1994; Pacheco, 1997; Quinn, 2003; Petersen and Arbenz, 2004). Although several different message-passing libraries have existed, including the Parallel Virtual Machine (PVM), Intel's NX and Cray's SHMEM, MPI has now become the *de facto* standard, and can be used on everything from the largest supercomputers to a multi-core laptop. In MPI, each process involved in the execution of a parallel program is identified by a non-negative integer. If there are P processes executing a program, they have ranks $0, 1, 2, \ldots, P-2, P-1$. Although the details of what happens when the program is executed vary from computer to computer, the essential steps are the same for all computers, provided one process is run on one processor. Initially, the user issues a directive to the operating system to place a copy of the executable program on each processor. Then each processor begins execution of its copy of the executable file. Different processes can execute different statements by branching within the program based typically on process ranks. In this section, we describe briefly some of more important features of MPI for C programmers (MPI calls for Fortran, C++

and other languages are almost identical and can be found in several online guides, for example at https://computing.llnl.gov/tutorials/mpi/).

Every MPI program must contain the following pre-processor directive.

```
#include ''mpi.h''
```

This mpi.h file contains the definitions, macros and function prototypes required for compiling an MPI program. Before any other MPI functions can be called, the function `MPI_Init` should be called to allow using the MPI library. Once the library is not in use anymore, the function `MPI_Finalize` should be called to finish the MPI use. When an MPI program starts running, the first step is to determine the number of processes (typically equal to the number of CPUs or cores) being used. To get the number of processes in a communicator (the second argument), the following function should be called:

```
int MPI_Comm_size(MPI_Comm comm, int size).
```

The second step is to determine which process number (rank) this particular process has. MPI provides the function `MPI_Comm_rank` to return the rank of a process in its second argument, whereas the first argument is a communicator. The rank will be an integer from 0 to (`MPI_Comm_size` − 1). The function can be called

```
int MPI_Comm_rank(MPI_Comm comm, int rank).
```

A communicator is a collection of processes that can send messages to each other. The function `MPI_Comm_world` is predefined in MPI and consists of all the processes running when program execution begins.

The actual message passing in a parallel program is carried out by the functions such as `MPI_Send` and `MPI_Recv`. The first command sends a message to a designated process, and the second one receives a message from a process. These functions are the most basic message passing commands in MPI. To communicate a message, the system should append some information to the data that the application program wishes to transmit. This additional information contains (i) the rank of the receiver, (ii) the rank of the sender, (iii) a tag and (iv) a communicator. These items can be used by the receiver to distinguish among incoming messages. The tag is used to distinguish messages received from a single process. Their syntax is as follows:

```
int MPI_Send(void* message, int count, MPI_Datatype datatype,
int dest, int tag, MPI_Comm comm) ,
int MPI_Recv(void* message, int count, MPI_Datatype datatype,
int source, int tag, MPI_Comm comm, MPI_Status* status) .
```

The contents of the message are stored in a block of memory referenced by the argument message. The arguments `count` and `datatype` allow the system to identify the end of the message: it contains a sequence of `count` values, each having MPI type `datatype`. The arguments `dest` and `source` are the ranks of the receiving and the sending processes, respectively. The `source` argument is used to distinguish message received from different processes. The arguments `tag` and `comm` are the tag and communicator, respectively. The last argument status in `MPI_Recv` returns information on the data that was actually received.

A communication pattern that involves all the processes in a communicator is a collective communication. A broadcast is a collective communication in which a single process sends the same data to every process. In MPI the function for broadcasting data is referred to as `MPI_Bcast`.

Below we consider an example of solving the system of linear algebraic equations (SLAE) using parallel processors. Let us represent the SLAE in the following form:

$$\mathbf{Av} = \mathbf{LL}^T\mathbf{v} = \mathbf{b}, \tag{9.5}$$

where $\mathbf{A}$ is the matrix of SLAE; $\mathbf{L}$ is the lower triangular matrix, and the width of its band is equal to the band width of the original matrix; $\mathbf{L}^T$ is the conjugate matrix; $\mathbf{v}$ is a vector of unknowns; and $\mathbf{b}$ is a known vector. It should be noted that the representation (9.5) is unique for positive definite and symmetric matrices (Ortega, 1988; Golub and Van Loan, 1989). The algorithm for solution of Eq. (9.5) consists of three main parts: (i) factorisation of the original matrix $\mathbf{A}$, that is, the computation of elements of matrix $\mathbf{L}$; (ii) solver for a SLAE $\mathbf{Ly} = \mathbf{b}$; and (iii) solver for SLAE $\mathbf{L}^T\mathbf{v} = \mathbf{y}$.

We use a modification of the Cholesky factorisation for parallel computations with distributed memory for P processors. The matrix of the system is stored column by column. Because the matrix is symmetric, we store only the lower triangular matrix. For simplicity, we assume that $N = kP$. The columns of the original matrix with numbers $j, P+j, \ldots, (k-1)P+j$ are stored in a memory of processor j (Fig. 9.7). In this case, the data are distributed uniformly between the processors: the time for computations and the size of memory requested are reduced by a factor of P.

The algorithm of factorisation is presented in the following form

$$\mathbf{A} = \begin{pmatrix} A_{11} & A_{21}^T \\ A_{21} & A_{22} \end{pmatrix} = \begin{pmatrix} L_{11} & 0 \\ L_{21} & L_{22} \end{pmatrix} \begin{pmatrix} L_{11}^T & L_{21}^T \\ 0 & L_{22}^T \end{pmatrix} = \begin{pmatrix} L_{11}L_{11}^T & L_{11}L_{21}^T \\ L_{21}L_{11}^T & L_{21}L_{21}^T + L_{22}L_{22}^T \end{pmatrix}. \tag{9.6}$$

Three steps are required for the decomposition of matrix $\mathbf{A}$ in (9.6):

step 1. $A_{11} \rightarrow L_{11}L_{11}^T$, e.g. the Cholesky factorisation of matrix A_{11};

step 2. $A_{21}\left(L_{11}^T\right)^{-1} \rightarrow L_{21}$, e.g. the Cholesky factorisation of matrix A_{21}; and

step 3. $A_{22} - L_{21}L_{21}^T \rightarrow \hat{A}_{22}$, e.g. the modification of matrix A_{22}.

Figure 9.8 (algorithm ALG1) shows a code for matrix factorisation written in the C language using MPI. As one processor performs steps 1 and 2 all other processors are idle. The

Processor 0:	0,	P,	$2P$,	...	$(k-1)P$
Processor 1:	1,	$P+1$,	$2P+1$,	...	$(k-1)P+1$
...	...	...	...	...	...
Processor $P-1$:	$P-1$,	$2P-1$,	$3P-1$,	...	$kP-1$

Fig. 9.7. Data distribution and storage.

```
/*  l[k]- pointer to kth column of the matrix  */
(*l[0])[0]=sqrt((*l[0])[0]);
for (k=0; k<N-1; k++) {
/*  If k\%P==idtask  */
            for (s=k+1; s<MIN(k+Width,N); s++)
                (*l[k])[s-k]=(*l[k])[s-k]/(*l[k])[0];
for (proc=0; proc<P; proc++)
if (proc!=idtask)
MPI_Send(l[k], Width, MPI_DOUBLE, proc, 0, COMM); }
else
MPI_Recv(buffer, Width, MPI_DOUBLE, (k%P), 0, COMM, &status);
            for (j=k+1; j<MIN(k+Width,N); j++) {
                        for (i=j; i<MIN(k+Width,N); i++)
                            (*l[j])[i-k]-=(*l[k])[i-k]*(*l[k])[j-k];
(*l[k+1])[0]=sqrt((*l[k+1])[0]); }
```

Fig. 9.8. **Algorithm for matrix factorisation.**

```
/*  X - vector of unknowns  */
/*  B - vector of right-hand side  */
for (j=0; j<N-1; j++) {
/*  If k%P==idtask  */ {
          X[j]=B[j]/(*l[j])[0];
          for (i=j+1; i<MIN(j+Width,N); i++)
                    B[i]-=X[j]*(*l[j])[i-j];
          MPI_Send(B, N, MPI_DOUBLE, (j+1)%P, 0, COMM); }
          if ((j+1)%P==idtask)
          MPI_Recv(B, N, MPI_DOUBLE, j%P, 0, COMM, &status); }
if ((N-1)%P==idtask) X[N-1]=B[N-1]/(*l[N-1])[0];
```

Fig. 9.9. **Solution algorithm for a system with a lower triangular matrix.**

computed results are sent to all other processors, and each processor performs step 3. The entries of L are stored in the original matrix. Steps 1–3 are then applied to $\hat{A}_{22}$, modified matrix A_{22}. To reduce idle time, data transmission may be organised during the performance of steps 1 and 2. However, in this case the time for communications increases, and therefore an optimal regime of performance should be found to minimise the execution time.

The solution algorithm for SLAE with triangular matrices is controlled by the procedure for storage of matrix elements. Figure 9.9 (algorithm ALG2) shows a code to solve such a

```
(*l[0])[0]=sqrt((*l[0])[0]);
for (k=0; k<N-1; k++) {
/*  If k%P==idtask  */
          for (s=k+1; s<MIN(k+Width,N); s++) {
                    (*l[k])[s-k]=(*l[k])[s-k]/(*l[k])[0]; X[k]=B[k]/(*l[k])[0];
                    (*l[k])[Width]=X[k]; }
/*  Broadcast l[k] and X[k] to all processors  */
          for (j=k+1; j<MIN(k+Width,N); j++) {
/*  If k%P==idtask  */
          for (i=j; i<MIN(k+Width,N); i++)
              (*l[j])[j-k]-=(*l[k])[i-k]*(*l[k])[j-k];
          B[j]-=(*l[k])[j-k]*X[k]; }
          (*l[k+1])[0]=sqrt((*l[k+1])[0]); }
if ((N-1)%P==idtask) X[N-1]=B[N-1]/(*l[N-1])[0];
```

Fig. 9.10. **Algorithm for matrix factorisation and solution of a system with the lower triangular matrix.**

```
for (k=N-1; k>=0; k--) {
          if (k%P==idtask) {
            X[k]=B[k]/(*l[k])[0]; }
/*  Broadcast X[k] to all processors  */
          for (j=k-1; j>=MAX(0,(k-Width+1)); j--)
              if (j%P==idtask) B[j]-=(*l[j])[k-j]*X[k]; }
```

Fig. 9.11. **Solution algorithm for a system with an upper triangular matrix.**

system. Actually, the computations are performed successively by each processor, and in this case the efficiency of multi-processors is sufficiently low. However, the combination of matrix factorisation and solving the system with the lower triangular matrix optimises the overall execution time (Fig. 9.10, algorithm ALG3). While the entries of the lower triangular matrix are stored column by column, the entries of the upper triangular matrix are stored row by row. Figure 9.11 (algorithm ALG4) shows a code for solution of such a system. The implementation of algorithms ALG3 and ALG4 on multi-processors with an optimisation of time for communications is suitable for implementation on parallel computers, compared to the implementation of algorithms ALG1, ALG2 and ALG4.

9.8 Cost of parallel processing

Parallel processing has not only advantages with respect to serial processing, but also some essential disadvantages, which we discuss in this section.

Time to write a parallel code. The time is largely going to be spent on analysing a code for parallelism and recoding. The more significant parallelism you can find in the task that the code is intended to address, the more speedup you can expect to obtain for your efforts. Having discovered the places where you think parallelism will give results, you now have to put it in. This can be a very time-consuming process.

Complicated debugging. A serial code is designed in such a way as to execute only one instruction at a time. This simplifies significantly the work of a programmer in searching for bugs in the serial code. Debugging a parallel application is sometimes infuriating, because programmers have multiple instructions at the same time, and it is quite difficult to analyse the reasons causing the errors in computing.

Loss of code portability. Portability (or the possibility of running a code at different computers) is one of the important issues in programming. When a serial code is converted to parallel, there is no way to return back: the parallel code will never again run on a serial computer, unless MPI is installed on that computer, the code is 'tricked' into thinking MPI is present by linking with dummy MPI calls that simulate a single processor environment, or unless the make process includes a pre-processor that strips out MPI calls. That is not perhaps true, but it is better to keep two completely different packages, one for a serial environment, and another for *every* different kind of parallel environment. However, there are efforts underway to insure portability of parallel programs. For example, High Performance Fortran (HPF) runs on a variety of platforms, and the entire MPI (Message Passing Interface) effort was directed at being able to provide message-passing portability on top of a wide range of underlying transport environments. Thanks to MPI it is now possible to run the same parallel code on computers ranging from a dual-core laptop, to a cluster of PCs, to the largest supercomputers in the world.

Other users might wait longer for their work. The extra CPU time, disk space and memory that your parallel application requires will not be available to other users of the system while you are using them. The overall time to perform a job is larger than the time it takes for the job to execute, because of the time the job spends in the queue waiting for nodes to run on.

Total CPU time greater with parallel processing. A good job of parallelisation will end up reducing the time you spend waiting for your application to finish. However, the bill you get back from your service centre is not likely to be based on time as measured by your clock. The bill will indicate the CPU time you racked up over all of the processors used by a particular job. Even on a per-CPU basis, a parallel task runs up a higher bill than the equivalent serial one; as previously explained, this increase is the result of the additional instructions and time required to initialise and terminate tasks and to communicate among tasks.

Replication of code and data requires more memory. Your service centre bill may take into consideration things other than CPU cycles, such as how much disk and main memory you use. A serial task uses a fixed amount of memory. A skilfully written parallel one will distribute most of it across the processors, but there will always be some values that are replicated in all tasks and some buffers used for communication that will make the total memory used by a set of parallel tasks greater than what the serial task had used.

9.9 Concluding remarks

Parallel processing can significantly reduce the overall time for solving a problem. The whole reason for getting involved in parallelism is to reduce the time spent by waiting for results. The characteristics of algorithms virtually insure that there will be points in most applications where significant savings can be achieved by judicious use of parallelism. Meanwhile we should note that writing and debugging software is more complicated. Parallelism involves more complexity in your code compared to sequential processing and does not come for free. Not only should a user do more work, but so does the computing system. Parallelism itself involves additional efforts in terms of process control: starting, stopping, synchronising and killing.

Special acknowledgment. This chapter is partly based on the lecture course at the Center of Parallel Computers of the Royal Institute of Technology in Stockholm (Sweden), which one of the authors attended. Simplicity and clarity of these lectures assisted in writing this section. The detailed lecture course can be downloaded from the Center web-page: http://www.pdc.kth.se.

10 Modelling of geodynamic problems

10.1 Introduction and overview

In this chapter we review the numerical methods that have been, or are being, used for various geodynamic problems, pointing the reader to relevant sections in this book, giving additional implementation information, and referring to detailed information elsewhere. The applications are mostly related to deformation of the solid Earth (mantle and crust), with a short discussion in the last section on methods for the liquid outer core, where the geodynamo is generated. The intent here is to focus on the numerical methods not the scientific results, for which the reader is referred to comprehensive discussions elsewhere (Schubert *et al.*, 2001; and volumes 6–8 of Schubert, 2007). This is also not intended to be an exhaustive survey, but rather to highlight common modelling problems involving technical aspects that are not covered in earlier chapters.

Historically the mantle convection research community has been somewhat separate from the crust/lithosphere dynamics modelling community. This is partly a result of the different dynamical regimes, with mantle convection occurring by viscous creep over long time scales, while crust/lithosphere deformation involves more complex rheology (visco-elasto-plastic), widespread localisation of deformation, shorter time scales, and the possibility that a free surface and surface processes may be important. It is also because different sets of observations are being explained, with mantle dynamicists interested in the long-term, large-scale and often deep, while crust/lithosphere dynamicists' research is closely tied to geological observations and shallow structure. In recent years, however, there has been increasing interest in including the lithosphere in mantle convection models in order to self-consistently obtain plate tectonics, while crust/lithosphere modellers have become increasingly interested in the influence of the deep mantle on local and regional deformation, as evidenced by Topo-Europe program (Cloetingh *et al.*, 2007). This has been accompanied by something of a convergence of numerical codes, with some codes now available that can model either the lithosphere or whole mantle, although each field is still dominated by custom codes adapted to each purpose.

A great many studies have addressed mantle convection, both in Earth and in other terrestrial planets and even icy moons. Studies of global convection have variously focussed on basic dynamics and scaling, the influence of viscosity variations (temperature-dependent, depth-dependent, composition-dependent), phase transitions (particularly the olivine-system), thermo-chemical convection and mixing, and the effect of tectonic plates and continents. It has also been common to model individual convective features separately from the large-scale flow by using regional models, with particular attention paid to plumes and slabs. From the perspective of realistic modelling, mantle convection models have generally assumed that the sole deformation mechanism is viscous creep and have used a viscosity variation with temperature that is much reduced from realistic, owing to numerical limitations and the fact that a stagnant lid, which is unrealistic for Earth, is obtained unless other rheological complexities are introduced.

For modelling the lithosphere and crust, local (regional) models are always used, with great interest in the processes of convergence, continental collision and mountain building, extension and basin formation, etc. Such models typically have rheological parameters that are closer to laboratory measurements and may include elastic and plastic modes of deformation, as well as a free surface. Implementation of all these features is discussed in this chapter.

Understanding the geodynamo is another major problem in geodynamics. Numerical simulation of the geodynamo has been possible since the mid 1990s, and there have been many studies published since then. Compared to modelling the silicate Earth, a completely different dynamical regime is involved, with rotation and magnetic fields being important, but with a much simpler rheology. Nevertheless, the same basic numerical methods can be used. This is further discussed in Section 10.10.

10.2 Numerical methods used

10.2.1 Mantle convection

All discretisations covered in this book, i.e. finite difference, finite element, finite volume and spectral, have been applied to model mantle convection. A Eulerian grid is almost always used, although often with Lagrangian markers to track composition. Owing to their simplicity, and the speed of spectral methods, the finite difference method and spectral methods were the first to be used. Now, spectral methods are seldom used, with most actively used codes employing the finite element or finite volume/finite difference methods. This, and the great advance of computing technology, is illustrated in Figure 10.1, which shows a calculation of 3-D spherical compressible, phase transition modulated convection that required a supercomputer in 1993 (and used a spectral method) but can now be run on a laptop (using a finite volume multigrid method).

A number of benchmark papers have been published over the years, which give an overview of the methods used at the time and their relative accuracy, as well as providing test cases to validate new codes. Such comparisons have been published for thermal convection in 2-D (Blankenbach *et al.*, 1989; Travis *et al.*, 1990b), 3-D (Busse *et al.*, 1993), 3-D spherical geometry (Stemmer *et al.*, 2006), for thermo-chemical convection (van Keken *et al.*, 1997; Tackley and King, 2003) and for compressible convection (King *et al.*, 2009).

Finite difference method. Because of its simplicity, the finite difference method (Chapter 2) was the earliest method to be applied to geodynamic modelling, and was extensively used in the 1960s and 1970s to develop numerical algorithms and computer codes to model mantle convection (Minear and Toksöz, 1970; Torrance and Turcotte, 1971; Turcotte *et al.*, 1973;

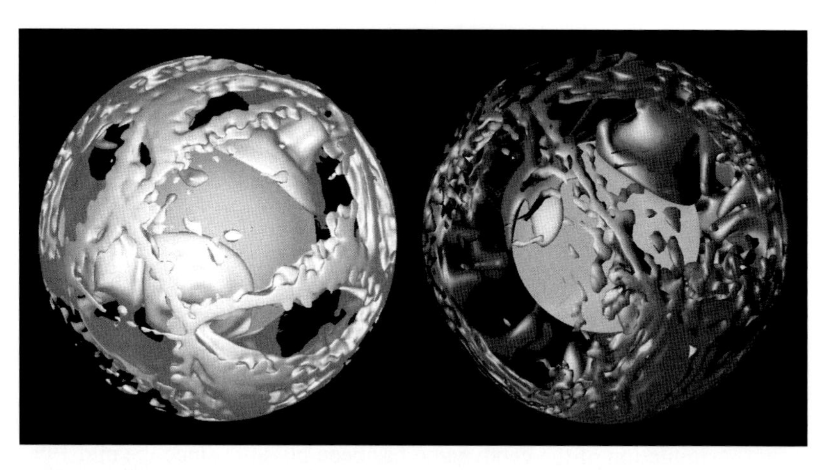

Fig. 10.1. **Compressible mantle convection simulations with an endothermic phase transition at 660 km depth, illustrating fifteen years of progress in numerical mantle convection modelling. Left: simulation from Tackley *et al.* (1993) run in 1992 using a spectral transform method on 128 CPUs of the Intel Touchstone Delta supercomputer, at one time the most powerful supercomputer in the world. Right: a similar simulation run in 2008 using a finite-volume discretisation and the Yin-Yang grid (after Tackley, 2008), which can be run on a modern laptop computer. Isosurfaces of residual temperature are plotted, showing where the temperature is significantly colder than the horizontal mean. (In colour as Plate 12. See colour plates section.)**

McKenzie *et al.*, 1974; Parmentier *et al.*, 1975; Jarvis and McKenzie, 1980; Schmeling and Jacoby, 1981). These typically used second-order finite differences with the stream function or stream function–vorticity formulation (for example, six codes in the benchmark paper by Blankenbach *et al.*, 1989). The stream function formulation is a way of eliminating pressure and reducing two velocity components to a single scalar, in two-dimensional geometry, and is discussed in Section 1.3.7.

A more accurate, high-order finite difference technique was implemented by Larsen *et al.* (1997) based on Fornberg (1990; 1995). Subsequently, the finite difference method was used in 3-D modelling of mantle convection, generally by solving for the poloidal velocity potential (see Section 1.3.8 and Travis *et al.*, 1990a; Parmentier *et al.*, 1994). More recent finite difference codes have typically used a staggered velocity–pressure grid, which is equivalent to the finite volume approximation, as discussed below.

Finite volume method. On an orthogonal Cartesian grid, application of the finite volume method to the momentum and pressure equations results in a straightforward finite difference approximation in which the velocity components and pressure are defined at staggered locations (Section 3.5). This is sometimes referred to as 'conservative finite differences' because stress is conserved. The first code to use such a discretisation was the 3-D Cartesian code of Ogawa *et al.* (1991), which used an iterative solver (Section 3.5.2). Subsequently, a multigrid solver for velocity–pressure on this grid (Section 3.5.3) was implemented in the 3-D Cartesian code Stag3D by Tackley (1993; 1994), adapted to model compressible convection (Section 3.6.2 and Tackley, 1996a) and used in many scientific studies. Stag3D uses the MPDATA advection method, one of many conservative finite volume methods

(Section 3.4.2). Several subsequent codes (Trompert and Hansen, 1996; Auth and Harder, 1999; Albers, 2000; Choblet and Sotin, 2000; Kameyama *et al.*, 2005) used the same basic finite volume/difference multigrid method but with various improvements, as discussed in Section 3.5.3. The 2-D code of Gerya and Yuen (2003; 2007) uses the same staggered grid and includes many additional physical complexities such as elasticity, a free surface and tracking of many compositional species, so is discussed later with lithosphere/crust codes in Section 10.2.2.

In 3-D spherical geometry, the finite volume method has been implemented on a (longitude, latitude, radius) mesh and non-multigrid iterative solver by Zebib *et al.* (1980) and Iwase and Honda (1997). Multigrid versions have also been implemented using the 'cubed sphere' grid (Hernlund and Tackley, 2003; Choblet, 2005), 'Yin-Yang' grid (Yoshida and Kageyama, 2004; Kameyama *et al.*, 2008; Tackley, 2008), and spiral grid (Huettig and Stemmer, 2008b) as discussed further in Appendix B; and these are active production codes at the time of writing.

Spectral methods. Spectral methods have been popular for 3-D modelling because of the relatively fast solution time afforded by decomposing the 3-D problem into a number of 1-D problems, the high accuracy (convergence rate), and the ease of modelling a sphere using spherical harmonics. The method and its application to geodynamic modelling are reviewed in detail in Chapter 5. To give a quick summary of applications, spectral methods have been used to model 3-D Cartesian geometry (Frick *et al.*, 1983; Cserepes *et al.*, 1988; Houseman, 1988; Christensen and Harder, 1991; Gable *et al.*, 1991), 3-D spherical geometry (Young, 1974; Machetel *et al.*, 1986; Glatzmaier, 1988), compressible convection (Glatzmaier, 1988; Bercovici *et al.*, 1989, 1992; Balachandar *et al.*, 1993; Tackley *et al.*, 1993), variable-viscosity convection (Balachandar *et al.*, 1993; Zhang and Christensen, 1993), viscoelasticity (Harder, 1991), mantle flow with rigid surface plates (Hager and O'Connell, 1981; Ricard and Vigny, 1989; Forte *et al.*, 1991; Gable *et al.*, 1991; Monnereau and Quere, 2001), and lithospheric dynamics (Schmalholz *et al.*, 2001). Owing to the difficulty in handling large lateral variations in viscosity and the growing use of multigrid solvers in 3-D, spectral methods for modelling solid Earth dynamics have largely been superseded by codes using a multigrid solver in conjunction with a spatial discretisation such as finite element, finite difference or finite volume. For geodynamo modelling, however, spectral methods are still widespread (see Section 10.10).

Finite element method. In the early 1980s, several 2-D Eulerian finite element codes were developed. Some of these used the stream function formulation to eliminate the continuity equation and reduce the momentum equation to a scalar equation. Codes using splines (Section 4.8) were developed by Kopitzke (1979) and Christensen (1981; 1983; 1984), who expanded the stream function in bicubic splines, temperature in biquadratic splines, and an upwind weighted residual method for the energy equation. Naimark and Malevsky (1984; 1988) and Malevsky and Yuen (1991) used a similar approach but expanded the stream function and temperature in bicubic splines to get a higher accuracy, and implemented a characteristics-based advection scheme (Section 7.7). A code developed by Hansen and Ebel (1984) used quadratic elements for the stream function and linear elements for temperature.

Many codes solve for velocity and pressure rather than a stream function. Schubert and Anderson (1985) used a variable mesh consisting of triangular elements with six-node quadratic shape functions for velocity and temperature and linear shape functions for pressure. The finite element code ConMan (King *et al.*, 1990), widely used in the geodynamic community in the 1990s, solves 2-D thermal convection of incompressible fluid and uses the penalty-function formulation to eliminate pressure from the momentum equation, replacing pressure by the divergence of velocity multiplied by a large number (the penalty parameter) (Section 4.7). The Galerkin formulation with bilinear functions as basis function is used. The system of linear algebraic equations is solved by Cholesky factorisation (Section 6.2.3). The heat equation is solved by the Streamline-Upwind Petrov–Galerkin (SUPG) method (Section 4.6).

In three dimensions, the finite element code TERRA (Baumgardner, 1985, 1988) was pioneering both by being one of the first codes of any type to model mantle convection in 3-D spherical geometry, and also by using a multigrid solver, which did not enter widespread use in the mantle convection community until the mid 1990s. TERRA uses an isocahedral discretisation in the azumithal direction (Appendix B), giving nearly uniform resolution over the sphere using triangular elements. This code has been used in a number of studies since the 1990s to the time of writing (Bunge *et al.*, 1996; Reese *et al.*, 1999; Davies *et al.*, 2007). A 2-D Cartesian version also exists (benchmarked in Travis *et al.*, 1990b) and following the implementation of matrix dependent transfers and the Galerkin coarse-grid approximation (Section 6.4.2) into this 2-D version, the multigrid solver showed remarkable robustness to large viscosity variations (Yang and Baumgardner, 2000).

Several 3-D Cartesian Eulerian finite element solvers for mantle convection have also been implemented. Malevsky (1996) developed a 3-D code and used a poloidal mass flux potential to represent the velocity, expanding it in tricubic splines and treating Rayleigh numbers of up to 10^8 (Malevsky and Yuen, 1993). In the code of Ismail-Zadeh *et al.* (1998; 2001a), the velocity vector is replaced by the vector potential eliminating pressure from the Stokes equations. The vector potential is then represented by a two-component vector and the resulting equations are solved by a Eulerian finite element method using tricubic splines as the basis functions (Section 4.10.2). A set of linear algebraic equations is then solved by an iterative solver (Section 6.3). The heat equation is discretised by finite differences and solved by a semi-Lagrangian method (Section 7.7).

The code CitCom, which solves convection with large viscosity variations using a multigrid method, is now widely used, partly because it is freely available through the Computational Infrastructure for Geodynamics, CIG (http://www.geodynamics.org). Developed in the mid 1990s for use in 2-D (Moresi and Solomatov, 1995; Moresi *et al.*, 1996) and 3-D (Moresi and Gurnis, 1996) geometries, CitCom employs the Eulerian finite element method, an iterative multigrid method (Section 6.4) and the Uzawa algorithm for pressure (Section 6.5.3) to solve the Stokes equations, and the Petrov–Galerkin method (Section 4.6) to solve the heat equation. A spherical version, called CitComS, was developed by Zhong *et al.* (2000). CitComS is parallelised and uses the Full Multigrid algorithm (Section 6.4.5), is also freely available through CIG, and has been used in several studies (see, for example, McNamara and Zhong, 2005). CitCom has also evolved into the lithospheric modelling code Ellipsis and the Underworld framework, which forms the basis for Gale (discussed later). Recently an improved preconditioning scheme for this type of solver has been found (May and Moresi, 2008).

Also in 3-D spherical geometry, Suzuki *et al.* (1999) and Tabata and Suzuki (2000) divided the sphere using a tetrahedral mesh, subdivided to the required refinement level and refined in the outer boundary layer. In order to reduce memory needs, they approximated velocity, pressure and temperature by piecewise linear elements. A Galerkin least squares method was used to stabilise this discretisation for the Stokes problem. The SUPG method was used for advection (Section 4.6). Iterative solvers were used to solve the resulting linear equations: conjugate gradient (Section 6.3.3) for the Stokes equation and the generalised conjugate gradient method for the convection–diffusion equation.

So far the mantle dynamics community has nearly always used meshes that are spatially uniform and do not change with time. Grid refinement, particularly adaptive grid refinement, offers several advantages to this, as demonstrated in 2-D and 3-D finite element codes by Davies (2008) and Davies *et al.* (2007), and it is a promising technology to pursue in the future.

10.2.2 Crust and lithosphere dynamics

Modelling the crust and lithosphere places different demands on a numerical solver than modelling mantle convection, because typically one has to deal with more complex rheology, including elasticity and plasticity in addition to viscous creep, and a free surface. On the other hand, the modelled time duration is much smaller: millions rather than billions of years. For a survey of currently used methods benchmark papers (Buiter *et al.*, 2006; Schmeling *et al.*, 2008) are useful, as well as a recent special volume on Computational Geodynamics (*Physics of the Earth and Planetary Interiors*, vol. 171, December 2008).

Although many different codes exist using a range of methods, almost all currently used lithosphere dynamics codes employ (i) Lagrangian markers to advect material properties (i.e. particle-in-cell method; Harlow and Welch (1965)) and (ii) a finite element method to solve for velocity/pressure, usually using a deforming mesh that is either fully Lagrangian or Arbitrary Lagrangian Eulerian (ALE; Section 4.2). Notable exceptions to (ii) are the Eulerian finite difference/volume codes of Gerya and Yuen (2003; 2007), the Eulerian finite element codes DOUAR (Braun *et al.*, 2008) and ELLIPSIS3D (O'Neill *et al.*, 2006). Sometimes commercial codes are used, e.g. three codes in the benchmark comparison of Buiter *et al.* (2006), but here we focus on codes that have been developed in the community.

FLAC. One popular approach, referred to as FLAC, Fast Lagrangian Analysis of Continua (Cundall, 1989), is to solve the momentum and continuity equations explicitly, i.e. retaining the $\rho\partial\mathbf{v}/\partial t$ and $\partial\rho/\partial t$ terms. This has the advantages that time steps are computationally very fast and little memory is required because no matrix problem is solved, and that it is straightforward to handle complex non-linear rheology (elasto-visco-plastic with free surface) without iterations, but the disadvantage that very small time steps must be taken. In practice, damping is applied by multiplying the velocity time derivative by a density that is lower than the physical density. An implementation of this in 2-D is PARAVOZ by Poliakov *et al.* (1993a;b,c), which has been used in many studies by different authors (see

Lyakhovsky *et al.*, 1993; Lavier *et al.*, 2000; Burov *et al.*, 2001; Lavier and Buck, 2002). PARAVOZ uses a Lagrangian mesh that extends to the free surface, with remeshing when the mesh gets distorted, and the discretised equations obtained by volume integrals over the cells, which although finite element-like is often referred to as finite difference. This code was modified and enhanced to produce the code Lapex-2D (Babeyko *et al.*, 2002), which was then extended to model 3-D problems as Lapex-2.5D (Sobolev *et al.*, 2005; Petrunin and Sobolev, 2006; 2008) by modelling a series of 2-D slices connected by a finite difference approximation in the perpendicular direction, a so-called 2.5D approximation. A fully 3-D implementation of FLAC called SNAC (Choi *et al.*, 2008) has been made freely available by CIG.

Implicit Lagrangian codes have also been widely used. Some examples are the 2-D viscous code of Poliakov and Podladchikov (1992), which uses a finite element, penalty function velocity-based discretisation plus remeshing, markers to track material properties (which avoids numerical diffusion of physical properties when remeshing), and a treatment of surface processes (Poliakov *et al.*, 1993a). A 3-D elasto-plastic finite element solver using quadratic parallelepiped elements with iterative solution using a conjugate gradient element-by-element method that avoids storing the whole matrix was implemented by Braun (1993; 1994) and Braun and Beaumont (1995). In their algorithm, remeshing occurs at the end of each time step. A Dynamic Lagrangian Remeshing (DLR) method, in which the nodes move and Delauney triangulation is used to join them together with an optimal mesh, was implemented by Braun and Sambridge (1994). Unusually, the 2-D code FEMS-2D of Frehner and Schmalholz (2006) and Schmalholz (2006) is written in MATLAB, and has recently been adapted to use the matrix assembly and solver MILAMIN (Dabrowski *et al.*, 2008), which is also written in MATLAB. The code SloMo (Kaus, 2005) is a 2-D plane-strain finite element code for non-linear visco-elasto-plastic rheologies.

Arbitrary Lagrangian Eulerian (ALE) method. A technique that combines advantages of both Lagrangian and Eulerian methods is the Arbitrary Lagrangian Eulerian (ALE) method, first introduced by Hirt *et al.* (1974) (see also Section 4.2). The basic idea is to solve the flow on an Eulerian grid and advect material properties on Lagrangian markers, but the grid can also deform in order to track interfaces, most importantly the free surface. Fullsack (1995) used this to model in 2-D a visco-plastic domain including a crustal layer that is an elastic beam and surface processes, and termed the method ALE-R (because it includes regridding). This code has been used in many subsequent papers. This code is referred to as microFEM in Buiter *et al.* (2006), while a successor to it, with various enhancements, is referred to as Sopale. Another 2-D ALE implementation is Plasti (Fuller *et al.*, 2006), which is freely available from CIG. At the time of writing, 3-D ALE implementations are beginning to appear, particularly SLIM3D (Popov and Sobolev, 2008), which uses hexahedral finite elements with linear shape functions, and Gale, which is based on the method described in Moresi *et al.* (2003), and the Underworld framework, and is a joint effort between CIG, Victoria Partnership for Advanced Computing (VPAC) and Monash University, and is freely available from CIG. Other codes in this category include the 3-D LaMEM by B. Kaus (outlined in Schmeling *et al.*, 2008), which uses with PETSc package for solution, giving much flexibility in choice of solver, and can also be used in fully Lagrangian or Eulerian modes.

Eulerian meshes. Some codes use a completely Eulerian mesh, again with Lagrangian particles to track material properties. A free surface, if included, can be implemented by introducing a layer of weak, low density material at the top ('sticky air'; see Section 10.8). Examples are Weinberg and Schmeling (1992), who used a finite difference code with particle-in-cell to track composition to model crustal diapirs with a Newtonian viscous rheology, Willett *et al.* (1993), who used an Eulerian velocity-based finite element technique to study a mechanical model for the tectonics of doubly vergent compressional orogens. Christensen (1992) demonstrated how such a method can be applied to lithospheric extension. Naimark *et al.* (1998) and Ismail-Zadeh *et al.* (2001b; 2004b) used a Eulerian finite element method to study sedimentary basin evolution and viscous deformations due to salt diapirism.

One such code is Ellipsis3D (O'Neill *et al.*, 2006), a finite element code that is descended from CitCom (Moresi and Solomatov, 1995), and is also available from CIG. The code DOUAR (Braun *et al.*, 2008) models visco-plastic deformation in 3-D using a Eulerian mesh consisting of cubic finite elements. The elements can be locally refined by dividing them in two in each direction (an octree). Interfaces are tracked using particles on the interface, while clouds of particles track material properties that depend on deformation history. An efficient direct solver is used.

The code of Gerya and Yuen (2003, 2007) is unusual in using a conservative finite difference (finite volume) discretisation rather than finite elements. The staggered grid is the same as that used by the finite volume mantle convection codes discussed in Section 10.1, but a direct solver is used rather than a multigrid solver, giving greater robustness to large viscosity variations at the expense of higher CPU and memory requirements. While markers (tracers) are used in many codes to track composition (see van Keken *et al.*, 1997; Tackley and King, 2003), this code has taken their use further than before, by (i) advecting temperature on markers rather than using a grid-based approach (diffusion is calculated on grid points then interpolated to marker positions, as discussed in Section 7.10), (ii) including viscoelasticity (see Section 10.6.2) with stress advected on markers, and (iii) calculating and advecting viscosity on markers then averaging to the grid points, rather than calculating viscosity directly on grid points. As discussed in Section 10.5.2, the use of markers for advecting quantities minimises numerical diffusion which is particularly important for elastic stress and allows tracking of fine-scale structures. When used for viscosity it helps to maintain sharper plastic shear zones then if viscosity is calculated directly at grid points.

10.3 Compressible flow

10.3.1 Importance, extended Bounssinesq and anelastic approximations

Here, the approximations and equations used to model compressible flow in the mantle are discussed, together with general implementation issues. Additional implementation issues related to finite volume and spectral discretisations are discussed in Sections 3.6.2 and 5.4.3, respectively. We do not discuss scientific findings regarding the effects of compressibility: for these the reader is referred to Jarvis and McKenzie (1980), Bercovici *et al.* (1992) and other relevant literature (see Peltier, 1972; Parmentier and Turcotte, 1974; Yuen *et al.*, 1987; Machetel and Yuen, 1989; Schmeling, 1989; Glatzmaier *et al.*, 1990; Solheim and Peltier, 1990; Balachandar *et al.*, 1993; Tackley, 1996a; Zhang and Yuen, 1996a,b).

The increasing pressure with depth leads to significant compression of rocks – to about 60% higher density by the core–mantle boundary (including phase changes), which can lead to important geometrical effects in whole mantle flow. Furthermore, the inclusion of compressibility implies the inclusion of viscous dissipation (shear heating) and adiabatic heating/cooling terms in the energy equation, which can have an important influence on the temperature distribution. Compressibility is usually incorporated by using either the *extended Boussinesq approximation*, in which the density is still assumed constant in the continuity equation but the extra terms are included in the energy equation, or the *anelastic approximation* (and its variants anelastic-liquid and truncated anelastic), in which the density is assumed to vary with position (usually depth) but not time. Compressibility also implies the inclusion of other depth-dependent physical properties such as thermal expansivity and thermal conductivity.

10.3.2 Continuity equation

The full compressible continuity equation expressed in the Eulerian frame is:

$$\frac{\partial \rho}{\partial t} = -\nabla \cdot (\rho \mathbf{v}) \tag{10.1}$$

and that in the Lagrangian frame is

$$\frac{D\rho}{Dt} = -\rho \nabla \cdot \mathbf{v}, \tag{10.2}$$

but these are generally approximated in some way. Both the extended Boussinesq and anelastic approximations assume, in the continuity equation, that density variations associated with thermal expansion and dynamic pressure are small (e.g. no more than a few per cent of the total density), which is the case for thermal convection and for thermo-chemical convection when compositional density variations of no more than a few per cent are considered. The extended Boussinesq approximation assumes the incompressible continuity equation.

In the anelastic approximation, a depth-dependent reference state density, rather than the full density, is used in the continuity equation. In this case, the time-derivative is zero and the Eulerian continuity equation simplifies to:

$$\nabla \cdot (\bar{\rho} \mathbf{v}) = 0, \tag{10.3}$$

where $\bar{\rho}$ is a reference density that depends on depth. This approximation is also generally applicable for situations where the density at a fixed location does not change with time, or changes very slowly with time, such as when material is moving through a phase transition.

The usual justification of the anelastic approximation is that the Mach number (i.e. ratio of velocity to sound velocity) multipled by the Prandtl number (i.e. ratio of heat flux to conductive heat flux) is much less than unity (Bercovici *et al.*, 1992). It is important to note, however, that it is valid only if the density variations associated with the flow are much smaller than the density, which for thermal convection can be written as $\alpha\Delta T \ll 1$. This condition is clearly violated when the advection of large (~100%) compositional density differences is considered, such as metal–silicate separation during core formation or deflection of the free surface. An end-member case is when all density variations move with the material, for example owing to compositional variations, and none is the result of hydrostatic compression. In this case, the time derivative in the Lagrangian continuity equation is zero and the incompressible continuity equation is recovered:

$$\nabla \cdot \mathbf{v} = 0. \tag{10.4}$$

So, this is appropriate either when the density variations are negligible or when they move with the flow. If both pressure dependence and a large compositional dependence (e.g. owing to a free surface) are required, then it is necessary to use the full compressible continuity equation, either in the Eulerian frame with the time derivative calculated by considering compositional variations moving past the fixed point, or in the Lagrangian frame with the time derivative calculated from the rate of change in pressure, and possibly temperature, in which case care must be taken to avoid the possibility of acoustic waves in the solution.

10.3.3 Momentum equation

In the momentum equation, the main influence of compressibility is that the normal stresses contain an additional $\nabla \cdot \mathbf{v}$ term, which needs special care if an iterative solver is used, as discussed in Section 3.6.2., e.g.

$$\tau_{xx} = 2\eta \left(\frac{\partial u}{\partial x} - \frac{1}{3} \nabla \cdot \mathbf{v} \right), \tag{10.5}$$

where τ_{xx} is the component of the deviatoric stress tensor, and η is viscosity. Density variations due to variations in temperature and composition are included in the buoyancy term. In the anelastic approximation the influence of non-hydrostatic (i.e. dynamic) pressure variations on density is also included here, but this is neglected in the truncated anelastic approximation and the extended Boussinesq approximation.

10.3.4 Energy equation

In the compressible energy equation, the main change is the addition of terms describing viscous dissipation and adiabatic heating/cooling. The compressible energy equation is given by:

$$\rho C_p \frac{DT}{Dt} = \alpha T \frac{DP}{Dt} + \nabla \cdot (k \nabla T) + \rho H + \tau : \dot{\varepsilon}, \tag{10.6}$$

where the terms on the right-hand side are adiabatic heating, conduction, radiogenic heating and viscous dissipation, respectively. Here T is the absolute temperature (in Kelvin if dimensional units are used), ρ is density, C_p is the coefficient of specific heat capacity, α is the coefficient of thermal expansivity, P is pressure, k is the coefficient of thermal conductivity, H is radiogenic heating per unit mass, τ is the deviatoric stress tensor, and $\dot{\varepsilon}$ is the strain rate tensor. The treatment of the adiabatic heating term depends on the approximation being used. While the anelastic approximation uses the full pressure, in the anelastic-liquid, truncated anelastic and extended Boussinesq approximations, the hydrostatic pressure is used here rather than the total pressure, in which case the term becomes proportional to vertical (radial) velocity v_3:

$$\alpha T \frac{DP}{Dt} \approx -\alpha T \rho g v_3, \tag{10.7}$$

where g is the acceleration due to gravity. The viscous dissipation term may equivalently be written in terms of velocity $\mathbf{v} = (v_1, v_2, v_3)$:

$$\tau : \dot{\varepsilon} = \tau_{ij} \frac{\partial v_i}{\partial x_j}. \tag{10.8}$$

This term is often referred to as 'shear heating', because only shear deformation contributes to it and because it is associated with not only viscous deformation but also plastic deformation (but not elastic).

10.3.5 Non-dimensionalisation and dissipation number

In summary, while both extended Boussinesq and anelastic approximations assume a time-invariant density in the continuity equation (ignoring the effect of pressure and temperature variations from the reference state), they differ in their treatments of the momentum and energy equations. The anelastic approximation includes the effect of dynamic pressure on the buoyancy term density and on adiabatic heating, while the anelastic-liquid approximation uses $\alpha T \ll 1$ to justify ignoring the influence of dynamic pressure in the energy equation (Jarvis and McKenzie, 1980) and the truncated anelastic and extended Boussinesq approximations ignore it in both places.

The equations are often non-dimensionalised to the length scale of the mantle depth and related thermal diffusion time scales and velocity scales, as in Section 1.3.2. When this is done, a new non-dimensional parameter, the dissipation number, appears in the energy equation and governs the magnitude of non-dimensional viscous dissipation, adiabatic heating and (for phase changes) latent heat terms (Christensen and Yuen, 1985). Physically, the dissipation number is the ratio of mantle depth to the temperature scale height (the height over which temperature increases adiabatically by a factor e) (Jarvis and McKenzie, 1980):

$$Di = \frac{h}{H_T} = \frac{\alpha g h}{C_P}. \tag{10.9}$$

Here, Di is the dissipation number, h is the mantle depth, H_T is the temperature scale height, and the other symbols were defined above. For the Earth's mantle, the depth-averaged value of Di is in the range 0.5–0.7. The dissipation number is also the upper bound on the ratio of volume-integrated viscous dissipation to convective heat flux (Hewitt *et al.*, 1975), which means that for mantle convection in the Earth, a substantial amount of viscous dissipation takes place. For a convecting system in secular equilibrium, the (volume-integrated) viscous dissipation and adiabatic heating/cooling are equal and opposite, so cancel out. Therefore, it is important to include both terms to avoid a large error in energy conservation. This is not always done in lithosphere/crust models, but as such models treat transient evolution often driven by external forces, rather than convection in equilibrium, the error in energy conservation is probably small.

10.4 Phase transitions

10.4.1 Background

Solid–solid phase transitions have been the topic of many modelling studies because they have an important effect on mantle convection by modulating the flow of material across themselves. Phase transitions introduce two main effects to be included in numerical models: variations in density, and latent heat release or absorption. Where these appear in the governing equations depends on the approximation being made. For the Boussinesq approximation, phase-related density variations are included in the buoyancy term of the momentum equation but not elsewhere, and latent heat is not included (Christensen and Yuen, 1985). For the extended Boussinesq and anelastic approximations, latent heat is added to the energy equation. Strictly speaking, the extended Boussinesq and anelastic approximations use a time-invariant density in the continuity equation, although sometimes codes have included time-varying phase transition related density variations here.

Most studies of phase transitions to date have parametrised the effect of individual phase transitions, commonly the ones in the olivine system that cause the seismic discontinuities at 410 km and 660 km depth. Phase diagrams for mantle materials can, however, be very complicated and dependent on composition, particularly in the transition zone and in the (temperature, pressure) conditions of subduction zones so, recently, some studies have adopted a more comprehensive approach in which phase assemblages are calculated by phase equilibria as a function of temperature, pressure and composition, so that all complexities of phase diagrams are automatically included without having to parametrise each transition.

10.4.2 Effective expansivity and heat capacity

In either case, a common approach to adding the necessary terms to the governing equations, particularly the energy equation, is to use an *effective thermal expansivity* and *effective specific heat capacity*. The extended Boussinesq, non-dimensional version of the energy equation that uses these is (assuming constant thermal diffusivity):

$$C'_P \frac{DT}{Dt} = \alpha' T \, Di \, u_2 + \nabla^2 T + \frac{Di}{Ra} \tau_{ij} \frac{\partial u_i}{\partial x_j}, \tag{10.10}$$

where the dismensionless quantities are effective heat capacity C'_P, effective thermal expansivity α', absolute temperature T, time t, dissipation number Di, Rayleigh number Ra, velocity (u_1, u_2), and stress tensor τ_{ij} (Christensen and Yuen, 1985). This equation is also valid in 3-D with u_2 replaced by u_3.

The dimensionless forms of α' and C'_P are

$$\alpha' = \alpha + P_{ph} \frac{d\Gamma}{d\pi}, \qquad C'_P = C_p \left[1 + \bar{\gamma} P_{ph} Di \, T \frac{d\Gamma}{d\pi} \right], \tag{10.11}$$

where $\Gamma(\pi)$ is the phase function discussed below, $\bar{\gamma}$ is the dimensionless Clapeyron slope and P_{ph} is the phase buoyancy parameter:

$$\bar{\gamma} = \gamma \frac{\Delta T}{\rho_0 g D}, \qquad P_{ph} = \frac{\gamma \Delta \rho}{\rho_0^2 g \alpha D}. \tag{10.12}$$

Here ΔT is the temperature scale, ρ_0 is density, g is the acceleration due to gravity, D is the length scale, γ is the Clapeyron slope, and $\Delta\rho$ is the phase change density jump. The effective expansivity and heat capacity vary from the standard expansivity and heat capacity only when a phase transition is taking place. A conceptual interpretation of α' and C'_p is that deviations in the effective heat capacity from the actual heat capacity account for latent heat absorption or release as heat is added or subtracted at constant pressure, while the effective thermal expansivity takes into account latent heat release or absorption as the pressure is changed at constant temperature. The above equations assume that the density and heat capacity are constant; if they are not, then appropriate modifications must be made to the equations for α' and C'_p. If dimensional units are being used, the effective quantities are given by Schubert *et al.* (1975):

$$\alpha' = \alpha + \frac{\Delta\rho}{\rho} \left(\frac{\partial \Gamma}{\partial T} \right)_p ; \qquad C'_p = C'_p + \left(\frac{\partial \Gamma}{\partial T} \right)_p T \Delta s, \tag{10.13}$$

where Δs is the entropy change of the phase transition.

The function $\Gamma(\pi)$ above indicates the fraction of the denser phase and changes smoothly from 0 to 1 as a function of temperature and pressure, which are here combined into the 'excess pressure' π, defined dimensionally as

$$\pi = p - p_0 - \gamma(T - T_0), \tag{10.14}$$

where (p_0, T_0) is a point in (p, T) space lying in the middle of the phase transition region, i.e. with equal fractions of both phases. The equivalent dimensionless form is

$$\pi = z_0 - z - \bar{\gamma}(T - T_0). \tag{10.15}$$

The phase function is typically chosen to be a hyperbolic tangent (Richter, 1973), i.e.

$$\Gamma(\pi) = \frac{1}{2}\left(1 + \tanh\frac{\pi}{d}\right), \qquad \frac{d\Gamma}{d\pi} = \frac{2}{d}(\Gamma - \Gamma^2), \tag{10.16}$$

where d is the width of the phase transition in pressure or depth space. This must be chosen such that the phase transition is spread out over several grid spacings, otherwise phase transition effects will not be correctly resolved, which is unfortunate because phase transitions have more effect when they are sharper (Tackley, 1995), and the two major phase transitions are observed seismologically to be spread over less than 10 km, less than the typical grid spacing in a global simulation.

10.4.3 Fixed depth approximations

The above parametrisation correctly allows phase transitions to deflect vertically depending on the local temperature and is widely used (see Peltier and Solheim, 1992; Kameyama and Yuen, 2006; Nakagawa and Tackley, 2006; Monnereau and Yuen, 2007). For numerical convenience, however, several researchers have used parametrisations of phase transitions in which they are fixed in grid space. One way in which this can be accomplished is by setting $\bar{\gamma} = 0$ in the above expressions but using the correct value of P_{ph}. The buoyancy due to phase change deflection and the latent heat are then both represented by an effective thermal expansivity, but the heat capacity is not modified. Christensen and Yuen (1985) showed that this gives very similar results to the full approximation, and it has been used in many subsequent research papers (see, for example, Steinbach and Yuen, 1992; Honda *et al.*, 1993; Weinstein, 1993).

Reducing the width of a phase transition as much as possible while keeping it fixed in grid space leads to the *sheet mass anomaly* approximation for the buoyancy caused by its deflection. In this, the anomalous mass is placed at a fixed depth, either between grids (Tackley *et al.*, 1993; 1994) or in one level of cells or elements (Tackley, 1996b; Bunge *et al.*, 1997). Depending on the numerical method, it may be necessary to spread out latent heat over a finite depth range (Tackley *et al.*, 1994). One approach to treating latent heat discontinuously is to advect potential temperature rather than total temperature (Xie and Tackley, 2004b), although this does not account correctly for the effect of heat addition or subtraction at constant pressure.

10.4.4 Phase equilibria approach

To be able to include complex phase diagrams, the *phase equilibria approach* is becoming increasingly popular. In this, phase assemblages are calculated by minimising Gibbs free energy as a function of temperature, pressure and composition, with composition expressed as the ratio of the five or six most abundant oxides. Physical properties are then calculated by suitable averaging over the properties of the individual minerals present.

For crustal dynamics, physical properties calculated by phase equilibria have been applied to sedimentary basin subsidence and uplift by Petrini *et al.* (2001) and Kaus *et al.* (2005) using the petrological approach described in Connolly (1990) and Connolly and Petrini (2002), to the gravitational instability of hot continental crust by Gerya *et al.* (2004) and to continental collision by Gerya *et al.* (2007) and Faccenda *et al.* (2009). Moving to the upper mantle, this has been applied to slab dehydration (Ruepke *et al.*, 2004), subduction channel and wedge dynamics (Gerya *et al.*, 2006; Gorczyk *et al.*, 2006; Castro and Gerya, 2007; Gorcyzk *et al.*, 2007a, b, c; Gerya *et al.*, 2008; Mishin *et al.*, 2008; Nikolaeva *et al.*, 2008), and to plume–lithosphere interaction (Ueda *et al.*, 2008). For calculating the phase equilibria, a widely used tool is Perple_X (Connolly, 2005).

At the global scale, several groups have used phase equilibria to calculate physical properties up to deep mantle pressures and temperatures (Ricard *et al.*, 2005; Stixrude and Lithgow-Bertelloni, 2005a, b; 2007; Piazzoni *et al.*, 2007; Xu *et al.*, 2008), although dynamical simulations using the resulting physical properties have only recently been performed (Hebert *et al.*, 2009; Nakagawa *et al.*, 2009; Schuberth *et al.*, 2009; Tirone *et al.*, 2009).

Physical properties calculated by phase equilibria, for example using Perple_X (Connolly, 2005), can be straightforwardly included in a thermo-mechanical code. Density is used in the buoyancy term and in other places depending on the approximation being used. In the energy equation, an effective heat capacity and thermal expansivity ensures that latent heat is included. The dimensional energy equation can be written (Gerya *et al.*, 2004) as:

$$\rho C_p' \left(\frac{DT}{Dt}\right) = \nabla \cdot (k\nabla T) + \alpha' T \left(\frac{DP}{Dt}\right) + Q_{sh} + Q_{ra}, \tag{10.17}$$

where Q_{sh} and Q_{ra} are is shear and radiogenic heating, respectively. The effective heat capacity and effective thermal expansivity are given by

$$C_p' = \left(\frac{\partial H}{\partial T}\right)_p, \qquad \alpha' = \frac{1}{T}\left[1 - \rho\left(\frac{\partial H}{\partial P}\right)_T\right], \tag{10.18}$$

where H is enthalpy, calculated by Perple_X, for example. The transport properties thermal conductivity and viscosity are not calculated during this process and must be separately specified. Because of the significant amount of computer time needed to compute stable mineral assemblages, it is computationally most efficient to precompute tables of the relevant physical properties and use them as look-up tables during the dynamical calculation. The spacing of the tabulated values in (P, T) space should be large enough that it effectively smooths sharp transitions over a depth range that can be resolved on the numerical grid. Calculating thermodynamic properties during the actual simulation takes much longer, but has the advantages of being able to treat equilibration between solid and (possibly mobile) fluid phases, as well as a continuous range of composition instead of a fixed number of predefined compositions, as illustrated by Tirone *et al.* (2009) and Hebert *et al.* (2009).

10.5 Compositional variations

Accurate modelling of flow and deformation with compositional variations requires additional equations to trace different chemical components and additional numerical methods to solve these equations coupled to the set of the mantle convection equations. Compositional variations may be active, i.e. contributing to the density anomalies that drive the flow and possibly changing the physical properties, or passive. The major numerical challenge is that compositional variations have extremely low diffusivity compared to temperature (essentially zero for the purpose of modelling large-scale mantle convection). This can lead to sharp interfaces and very narrow lamella, both of which are very difficult to treat numerically. The main approaches used to represent compositional heterogeneities are tracers (markers), marker chain and field. Various benchmark tests of these have been performed in the mantle convection community (van Keken *et al.*, 1997; Tackley and King, 2003).

Another aspect of composition is tracking the strain (stretching) of tracers, which can be done by integrating the velocity gradient tensor. This is not discussed here, rather the reader is referred to relevant references (McKenzie, 1979; Christensen, 1989; Kellogg and Turcotte, 1990; Tackley, 2007).

10.5.1 Field-based approaches

Here the standard advection–diffusion equation for composition C is solved by using similar methods to those used to treat the temperature field, with values of the compositional field held on nodes. The appropriate non-dimensional equation is:

$$\frac{\partial C}{\partial t} = \frac{1}{Le}\nabla^2 C - \mathbf{v} \cdot \nabla C. \tag{10.19}$$

where Le is the Lewis number, which is the ratio of thermal diffusivity to compositional diffusivity. The Lewis number should be as large as possible, and is sometimes set to infinity so that the only diffusion is numerical. Field-based approaches have the advantage of computational speed and little additional computational complexity, but the disadvantages of numerical diffusion, which smears out sharp features, and numerical dispersion, which causes artifacts like overshoots and ripples. Conservative finite volume advection methods are often used for advection; schemes designed to minimise such numerical artefacts are discussed in Section 3.4.2. Many other accurate, low-diffusion advection methods also exist (see Yabe *et al.*, 2002; Katz *et al.*, 2007). For advecting sharp interfaces, the interface-sharpening scheme of Lenardic and Kaula (1993) can be applied as a post-processing step to any standard advection method, and yields a dramatic improvement, as illustrated for example in the comparisons (see Tackley and King, 2003), some of which are shown in the right-hand columns of Figure 10.2.

10.5.2 Tracer (marker)-based approaches

In the *tracer method*, tracers (often called *markers*) carrying compositional information are simply advected with the flow with almost zero diffusion (a small amount of numerical

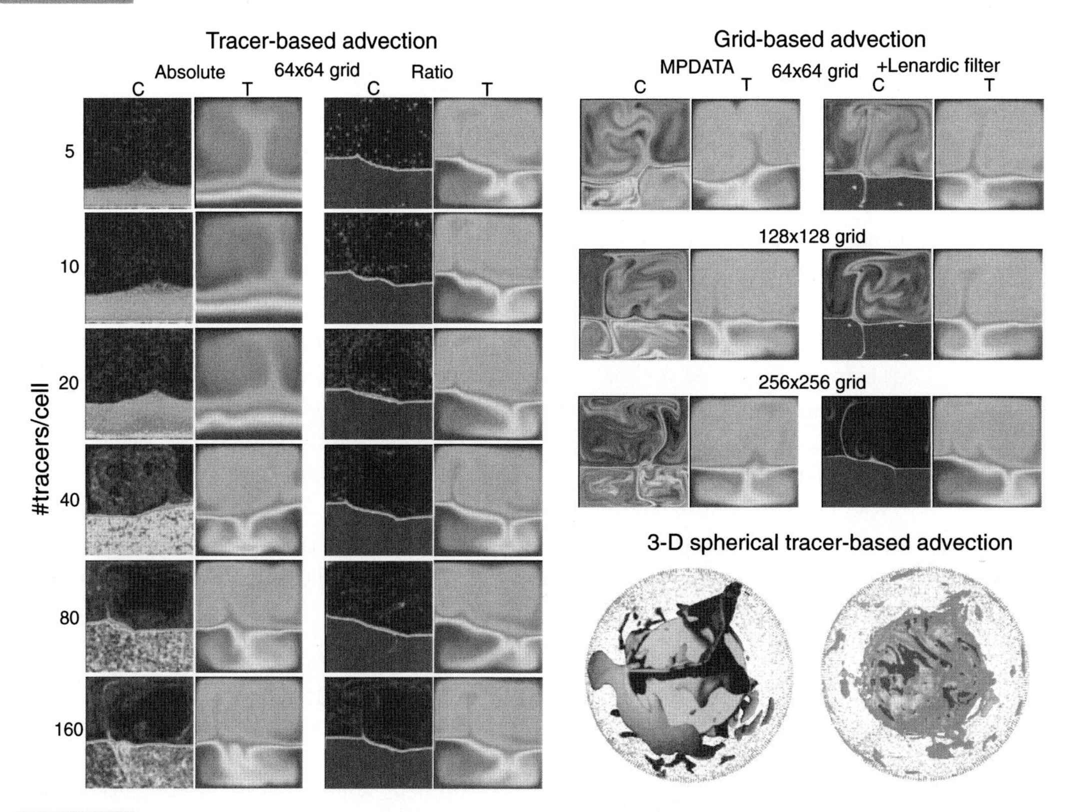

Fig. 10.2. The effect of the numerical treatment of compositional variations on thermo-chemical convection, comparing a tracer (marker) based treatment (left half and 3-D spherical case) with grid-based method (right Cartesian cases). The 2-D Cartesian cases are from Tackley and King (2003). Using tracers and the ratio method to calculate composition from the tracers, a very similar solution is obtained regardless of the number of tracers per cell, from 5 to 160 tracers/cell. If the absolute method is used, however, spurious settling of tracers to the base and compositions of larger than 1.0 are observed for less than 40 tracers/cell, and even then the compositional field is very noisy. When using the finite volume MPDATA (Smolarkiewicz, 1984) advection method for composition, substantially more entrainment is observed. If the filter of Lenardic and Kaula (1993) is added, this entrainment is greatly reduced, but four times as many grid points in each direction are needed to obtain the same result as when using tracers. The 3-D spherical case from Nakagawa and Tackley (2008) illustrates a practical application of the tracer ratio method to investigate the influence of dense material on lateral variations in heat flux at the core–mantle boundary: plotted are composition (green) and isosurfaces of anomalously hot (red) and cold (blue) material. (In colour as Plate13. See colour plates section.)

diffusion does exist because of error in updating each tracer's position). This is the major advantage of the method, and gives the apparent ability to represent discontinuities and sub-grid scale features. A disadvantage is the computational cost of storing and advecting the tracers, as several are needed in each grid cell: 5–50 depending on the exact method being used (Christensen and Hofmann, 1994; Tackley and King, 2003). Advection of the tracers is typically done using Runge–Kutta or predictor–corrector methods, of which versions with different mathematical orders are available (Section 7.3). This requires interpolating the velocity field from the grid points to the tracer positions, which can also be performed with different levels of accuracy. Care must be taken to interpolate velocities in such a way that the divergence of velocity still satisfies the continuity equation. The mathematical order chosen for the velocity interpolation and advection makes a very large difference to the computational time, so a careful compromise between accuracy and speed must be chosen.

The tracer method involves the conversion of tracer distributions to values on grid points, in order to calculate density and other physical properties, or for analysis or visualisation. Two main methods for this exist (Tackley and King, 2003).

(i) In the *absolute* method, tracers are placed where material of a particular composition exists but not elsewhere, and the concentration of tracers (i.e. number per unit volume) indicates the concentration of that compositional component, which means that each tracer carries a particular mass. Unfortunately, because the number of tracers per unit volume has a natural statistical fluctuation, the resulting compositional field is noisy, even in regions filled by a single component. Concentrations spuriously greater than 1 can exist, and there is a tendency for spurious settling of dense tracers to the base of the domain. These problems are clearly visible in the left-hand column of Figure 10.2. The number of tracers per cell needs to be quite high, up to 50, to reduce this noise to an acceptable level (Christensen and Hofmann, 1994; Tackley and King, 2003). On the other hand, this method has the advantage that mass is conserved because each tracer carries a particular mass.

(ii) In the *ratio* method, tracers are placed everywhere in the domain, and if the local composition is assumed to be one of two or more discrete compositions, each tracer is given a number to represent the composition that exists at that point. The concentration of each component in a particular region (such as an element or volume cell) is given by the fraction of tracers of that component in the region rather than the absolute number. This avoids concentrations greater than 1 and noise in regions of single component, so that far fewer tracers per cell or element are required, such as 5, regardless of the underlying numerical scheme (Tackley and King, 2003). These advantages are visible in the second column of Figure 10.2. Placing tracers everywhere also has the advantage that continuously varying fields, such as temperature or the concentration of a trace elements or the accumulated strain, can also be handled (see Gerya and Yuen, 2007). To generalise this, the value C_i of a field C (nominally composition) at a grid point $\mathbf{x}_i$ can be regarded as the average value carried by the tracers C_j^{tr} with positions $\mathbf{x}_j^{tr}$ in the vicinity of that grid point, with a shape function $S_i(\mathbf{x})$ defining the form of local averaging

$$C_i = \frac{1}{N_i} \int_{domain} S_i(\mathbf{x}) \sum_{j=1}^{Ntr} C_j^{tr} \delta(\mathbf{x} - \mathbf{x}_j^{tr}) dV = \frac{1}{N_i} \sum_{j=1}^{Ntr} C_j^{tr} S_i(\mathbf{x}_t^{tr}). \tag{10.20}$$

Here, N_i is the number of tracers in the local region, δ is the Dirac delta function, and Ntr is the total number of tracers in the domain.

$$N_i = \int_{domain} S_i(\mathbf{x}) \sum_{j=1}^{Ntr} \delta(\mathbf{x} - \mathbf{x}_j^{tr}) dV = \sum_{j=1}^{Ntr} S_i(\mathbf{x}_j^{tr}). \tag{10.21}$$

The simplest forms of the shape function are boxcar, for which $S_i = 1$ inside the cell or element and 0 outside, or linear, which on a Cartesian mesh can be written as

$$S_i(x,y,z) = \max\left(0, 1 - \frac{|x - x_i|}{\Delta x}\right) \max\left(0, 1 - \frac{|y - y_i|}{\Delta y}\right) \max\left(0, 1 - \frac{|z - z_i|}{\Delta z}\right). \tag{10.22}$$

where (x_i, y_i, z_i) are the coordinates of node i and Δx, Δy and Δz are the grid spacing in the three directions.

Additionally, tracers may carry several independent compositional components, such as the concentrations of several trace elements (Christensen and Hofmann, 1994; Xie and Tackley, 2004a,b). This ratio method is used in the code Stag3D (Tackley and King, 2003), which subsequently became StagYY (Tackley, 2008), and in CitComS (McNamara and Zhong, 2004; Zhong, 2006).

There are various other considerations to be made when using a tracer method. One is how to initialise their positions. Placing them on a regular grid is one possibility, but this has the disadvantage that severe bunching of tracers can occur under certain types of flow. Completely random initial positions is another possibility, but then large statistical variations in the number of tracers per grid cell occur. A compromise method is to initialise them on regular grid but with their actual positions perturbed randomly in each direction by ±half a grid spacing. When density varies with depth, it is most appropriate to keep a constant number of tracers per unit mass rather than per unit volume. Another consideration relates to the number of tracers per grid cell or element. Ideally, this should be approximately constant, but it will not be if the cells/elements are different sizes but the number of tracers per unit volume is constant. Possible solutions are simply to use enough tracers that there are enough even in the smallest cells/elements, to add or delete tracers during a simulation in order to keep a roughly constant number of tracers per cell/element, or to use a scheme for going back and forth between tracers and nodes that works well for variable number of tracers per cell/element.

The *marker-chain method* is designed to track sharp interfaces between layers or regions of different composition. The interface is discretised using a set of markers connected together in a 1-D chain (in a 2-D domain) or a 2-D surface (in a 3-D domain), and the markers are advected with the flow (Christensen and Yuen, 1984; Naimark and Ismail-Zadeh, 1995; Naimark *et al.*, 1998; Schmalzl and Loddoch, 2003; Lin and Van Keken, 2006). This is far more efficient than placing tracers everywhere and also has the advantage of zero diffusion. In all of the papers referenced here the compositional differences are active, i.e. they influence the buoyancy, although this could also be used to track passive interfaces. If mixing of the two materials occurs, for example by entrainment, then the interface becomes more complex with time and additional markers must be used to track it, resulting in computational needs that increase as much as exponentially with time (Lin and Van Keken, 2006). To counteract this, algorithms exist to simplify the interface, as applied in three dimensions by Schmalzl and Loddoch, (2003).

Benchmark comparisons of these various methods for cases with an active compositional layer (van Keken *et al.*, 1997; Tackley and King, 2003), indicate a sensitivity of the results to the details of the method being used, including a strong sensitivity of the entrainment rate of a stable compositional layer to resolution (both grid spacing and number of tracers) (Tackley and King, 2003). Thus quantitative conclusions from thermo-chemical convection simulations, such as entrainment rates, must be interpreted with great care, particularly as more calculations are performed in 3-D. Several tests from Tackley and King (2003) are shown in Figure 10.2, together with an example application from Nakagawa and Tackley (2008).

10.6 Complex rheologies

10.6.1 Non-Newtonian creep and plasticity

Non-Newtonian deformation involves a non-linear relationship between stress and strain rate. Dislocation creep and plastic failure are the commonly modelled non-linear deformation mechanisms. Dislocation creep is characterised by a power-law relationship between stress and strain rate, while plasticity is modelled either using the Peierls creep law (Equation (1.31)), in which strain rate increases rapidly with stress as it approaches the Peierls stress, or a yield criterion with a yield stress that is typically dependent on depth or pressure. As the solvers discussed in this book are for linear problems, they cannot directly solve problems with non-linear rheology, but rather an iterative method must be used.

The usual approach is to define an *effective viscosity* that depends on the stress or strain rate. For a given effective viscosity field a velocity solution is obtained, then this velocity field is used to recalculate stresses (and/or strain rates) and update the effective viscosity field. This procedure is iterated until an appropriate convergence criterion is reached, such as the velocity or viscosity field changing by less than some amount, or the residue falling below a certain threshold. These are *Picard iterations*, which are commonly used in the geoscience community. They often, however, lead to very slow convergence, with many iterations being necessary and possible oscillations. In the engineering community more robust and efficient iterative methods are used, such as Newton–Raphson iterations (e.g. used in geoscience by Muhlhaus and Regenauer-Lieb, 2005; Popov and Sobolev, 2008); for a full discussion see de Souza Neto *et al.*, (2008).

For non-Newtonian deformation with a power-law relationship between the stress and strain rate, which is relevant to dislocation creep, the rheology can be written as:

$$\dot{\varepsilon}_{ij} = B(T,P)\tau^{n-1}\tau_{ij}, \tag{10.23}$$

which, using $\tau_{ij} = 2\eta_{eff}\dot{\varepsilon}_{ij}$, leads to an effective viscosity:

$$\eta_{eff} = \frac{1}{2}B^{-1}\tau^{1-n}, \tag{10.24}$$

or equivalently, in terms of the strain rate:

$$\eta_{eff} = \frac{1}{2}B^{-\frac{1}{n}}\dot{\varepsilon}^{\frac{1}{n}-1}. \tag{10.25}$$

where $\dot{\varepsilon}$ is the second invariant of the strain rate tensor, τ is the second invariant of the deviatoric stress tensor τ_{ij}, n is the power-law exponent, $\dot{\varepsilon}_{ij}$ is the strain rate tensor, and $B(T, P)$ is the function dependent on temperature and pressure (and is the inverse of C in Eq. (1.30)). If both diffusion creep (with $n = 1$) and dislocation creep (with $n \approx 3.5$) operate simultaneously, then it is usually assumed that both mechanisms experience the same stress and their strain rates are added, leading to a total effective viscosity η_{eff} that is the harmonic sum of the viscosities of the two individual mechanisms:

$$\eta_{eff} = \frac{1}{(1/\eta_{eff,dis} + 1/\eta_{eff,dif})}. \tag{10.26}$$

where $\eta_{eff,dis}$ and $\eta_{eff,dif}$ are the diffusion and dislocation creep viscosities, respectively.

Plastic yielding is equivalent to power-law rheology with infinite exponent. Plasticity is normally implemented in geodynamic codes by assuming that the second invariant of the stress tensor cannot exceed a yield stress that either increases linearly with depth or pressure (Drucker–Prager yield criterion, Section 1.3.4) in order to mimic brittle failure (e.g. the Mohr–Coulomb criterion, Eqs. (1.36)–(1.37)), or is constant (von Mises yield criterion) in order to mimic semi-brittle, semi-ductile deformation in the mid-lithosphere (Kohlstedt *et al.*, 1995). This is implemented using an effective viscosity. The 'plastic' viscosity η_{pla} is given by

$$\eta_{pla} = \frac{\sigma_{yield}}{2\dot{\varepsilon}}. \tag{10.27}$$

and is combined with the creep viscosity η_{vis} using either

$$\eta_{eff} = \min[\eta_{vis}, \eta_{pla}]. \tag{10.28}$$

which gives a sharp transition between viscous and plastic deformation, or

$$\eta_{eff} = \frac{1}{(1/\eta_{vis} + 1/\eta_{pla})}, \tag{10.29}$$

which gives a smooth transition that may be easier for numerical schemes to handle. There are many examples of applications, both for whole mantle and for lithospheric deformation. For lithospheric deformation, strain weakening is often added to plasticity treatments. Mohr–Coulomb failure is typically treated using a Drucker–Prager yield criterion; however, Moresi and Muhlhaus (2006) implemented a more complex treatment in which brittle failure using the Mohr–Coulomb criterion is treated using an anisotropic viscosity.

10.6.2 Viscoelasticity

Elasticity is important in cold regions of the crust and lithosphere for some geodynamic problems. A method of adding elasticity to an existing viscous flow solver in order to model large-strain viscoelastic deformation is outlined here, following the approach introduced by Moresi *et al.* (2003; 2007), and also implemented by Gerya and Yuen (2007) and Kaus

(2010). Several Eulerian codes use this approach, usually with markers to advect stress (and sometimes other quantities).

The total strain rate is the sum of elastic $\dot{e}_{ij}^{e}$ and viscous $\dot{e}_{ij}^{v}$ parts:

$$\dot{e}_{ij} = \dot{e}_{ij}^{e} + \dot{e}_{ij}^{v} = \frac{1}{2\mu}\frac{D\tau_{ij}}{Dt} + \frac{\tau_{ij}}{2\eta}, \tag{10.30}$$

where μ is shear modulus, η is viscosity, and τ_{ij} is the deviatoric stress tensor. The material derivative of the deviatoric stress is commonly written using the Jaumann derivative, which includes advection and rotation of the stress tensor:

$$\frac{D\tau_{ij}}{Dt} = \frac{\partial \tau_{ij}}{\partial t} + v_k \frac{\partial \tau_{ij}}{\partial x_k} + \tau_{ik} W_{kj} - W_{ik}\tau_{kj}, \tag{10.31}$$

where $\mathbf{v} = \{v_i\}$ is velocity, and $\mathbf{W} = \{W_{ij}\}$ is the rotation tensor given by

$$W_{ij} = \frac{1}{2}\left[\frac{\partial v_i}{\partial x_j} - \frac{\partial v_j}{\partial x_i}\right]. \tag{10.32}$$

The Eulerian time derivative is now approximated using first-order finite differences, with explicit treatment of the advection and rotation terms:

$$\frac{D\tau_{ij}}{Dt} \approx \frac{\tau_{ij} - \tau_{ij}^{old}}{\Delta t^{e}} + v_k \frac{\partial \tau_{ij}^{old}}{\partial x_k} + \tau_{ik}^{old} W_{kj} - W_{ik}\tau_{kj}^{old}, \tag{10.33}$$

where Δt^{e} is the elastic time step, τ_{ij} is the present deviatoric stress tensor and τ_{ij}^{old} is the deviatoric stress tensor at a time Δt^{e} before the present time. Combining Eq. (10.30) and (10.33), the current stress tensor can be expressed in terms of the strain rate tensor and the 'old' advected and rotated stress tensor:

$$\tau_{ij} = 2\frac{\eta\mu\Delta t^{e}}{\eta + \mu\Delta t^{e}}\left[\dot{\varepsilon}_{ij} + \frac{1}{2\mu}\left(\frac{\tau_{ij}^{old}}{\Delta t^{e}} + v_k \frac{\partial \tau_{ij}^{old}}{\partial x_k} - \tau_{ik}^{old} W_{kj} + W_{ik}\tau_{kj}^{old}\right)\right]. \tag{10.34}$$

This can be rewritten in much simpler form by defining an effective viscosity and an effective strain rate:

$$\tau_{ij} = 2\eta_{eff}\dot{\varepsilon}_{ij}^{eff}, \tag{10.35}$$

where

$$\eta_{eff} = \frac{\eta\mu\Delta t^{e}}{\eta + \mu\Delta t^{e}} = \eta\left[\frac{1}{\Delta t^{Mxl}/\Delta t^{e} + 1}\right] \tag{10.36}$$

and $\dot{\varepsilon}_{ij}^{eff}$ is the term in square brackets in (10.34). Note that Δt^{Mxl} is the Maxwell relaxation time, given by η/μ. If the Maxwell time is very small compared to the elastic time step then the effective viscosity reverts to the actual viscosity. Conversely, if the Maxwell time is much longer than the elastic time step then the effective viscosity becomes much lower than the actual viscosity and tends to $\mu\Delta t^{e}$.

The momentum equation then becomes:

$$\nabla \cdot (2\eta_{eff}\dot{\varepsilon}_{ij}) - \nabla P = \rho g \mathbf{e} - \nabla \cdot \left(\frac{\eta_{eff}}{\mu} \left[\frac{\tau_{ij}^{old}}{\Delta t^e} - v_k \frac{\partial \tau_{ij}^{old}}{\partial x_k} - \tau_{ik}^{old} W_{kj} + W_{ik} \tau_{kj}^{old} \right] \right). \tag{10.37}$$

This is the same as the usual Stokes equation, except with an effective viscosity instead of the actual viscosity, and with additional terms on the right-hand side representing elastic stress. Therefore, by using this method viscoelasticity can be straightforwardly added to an existing viscous flow solver, to model either the lithosphere or the entire mantle convection–lithosphere system. When advecting stress, it is vitally important to minimise numerical diffusion. One way to accomplish this is to advect stress in a Lagrangian manner, either with a Lagrangian mesh or on Lagrangian particles moving through an Eulerian mesh (Moresi *et al.*, 2003; Gerya and Yuen, 2007). In this case the $v_k \frac{\partial \tau_{ij}^{old}}{\partial x_k}$ term in the above equations does not need to be explicitly evaluated. Some authors have used a Eulerian field advection technique that has low numerical diffusion (see Harder, 1991; Furuichi *et al.*, 2008). The latter approach is based on a completely Eulerian frame of reference and employs a multigrid method (Section 6.4) and the conservative semi-Lagrangian advection scheme (Section 7.8) called Constrained Interpolation Profile method with rational functions (CIP-CSLR; see Xiao and Yabe, 2001). The CIP-CSLP scheme is used to numerically solve advection problems while keeping the sharpness of the profile of the transported quantity without oscillations.

Often plasticity is also included, making the rheology visco-elasto-plastic. Plasticity can be added to the above treatment in one of two ways: either by adding the plastic strain rate to Eq. (10.30) and iterating on it until the stress does not exceed the yield stress, or by reducing the viscous creep viscosity to keep the stress on the yield envelope. These two methods give identical results. Examples of visco-elasto-plastic calculations that also include a free surface are shown in Figure 10.3, which show the widespread strain localisation that occurs as a result.

10.7 Continents and lithospheric plates in mantle convection models

10.7.1 Overview

In 1912, A. Wegener proposed the hypothesis that continents move, and were once joined together into a single supercontinent called Pangea, which split up. Half a century later (Wilson, 1966) proposed the consumption of the proto-Atlantic ocean along a system of island arcs off the east coast of North America, and the regeneration of the Atlantic ocean by flow at the mid-Atlantic ridge (this cycle of closing and re-opening of a palaeo-ocean basin is now referred to as the Wilson cycle).

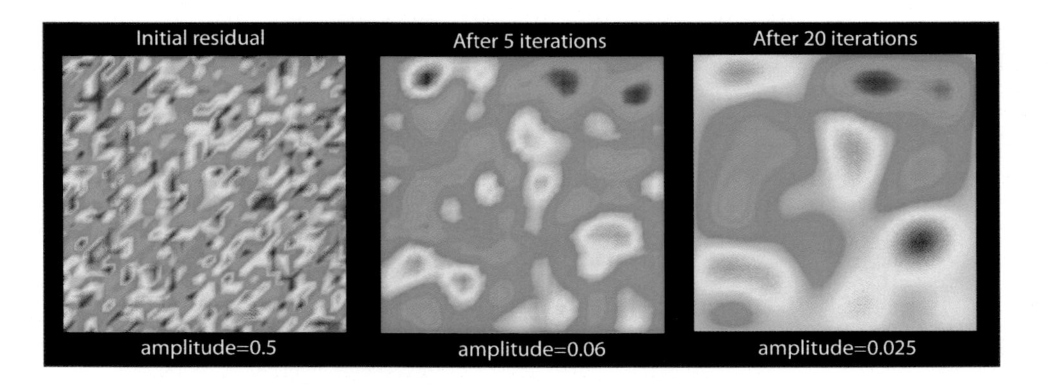

Plate 1. Spatial distribution of initial residual and residual after 5 and 20 Gauss–Seidel iterations on a 32×32 grid, for a Poisson equation with random source term. The iterations smooth the residual. (See Chapter 6.)

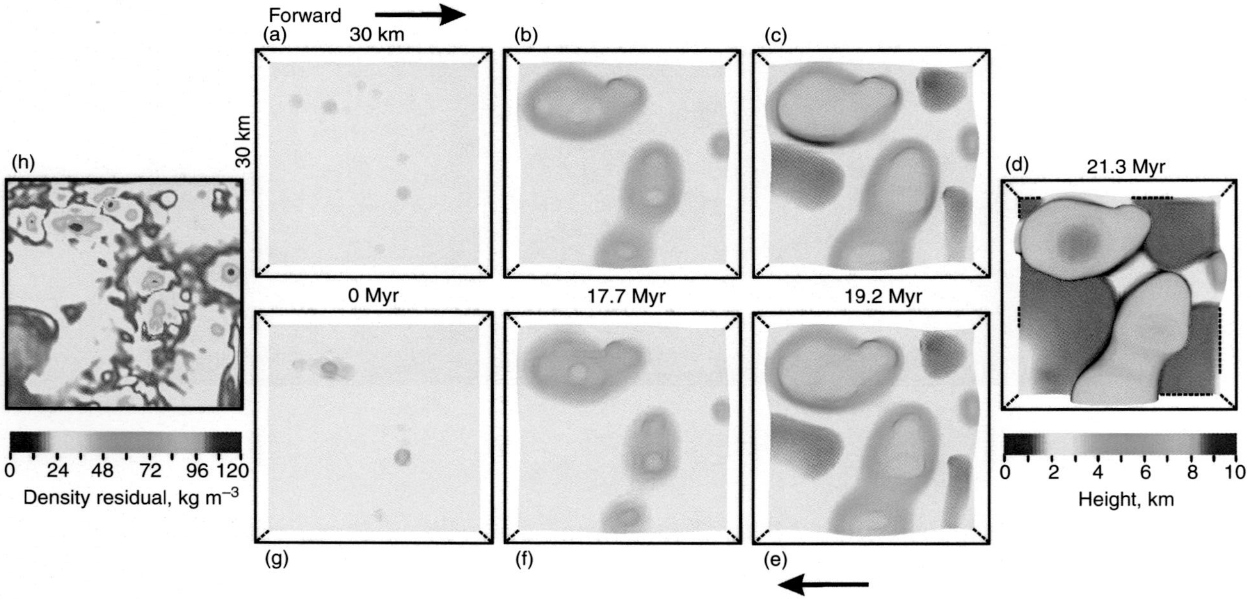

Plate 2. Evolution (top view) of salt diapirs toward increasing maturity (a)–(d) and restoration of the evolution (d)–(g) at the same times as in Fig. 8.2. (h) Restoration errors. After Ismail-Zadeh *et al.* (2004b). (See Chapter 8.)

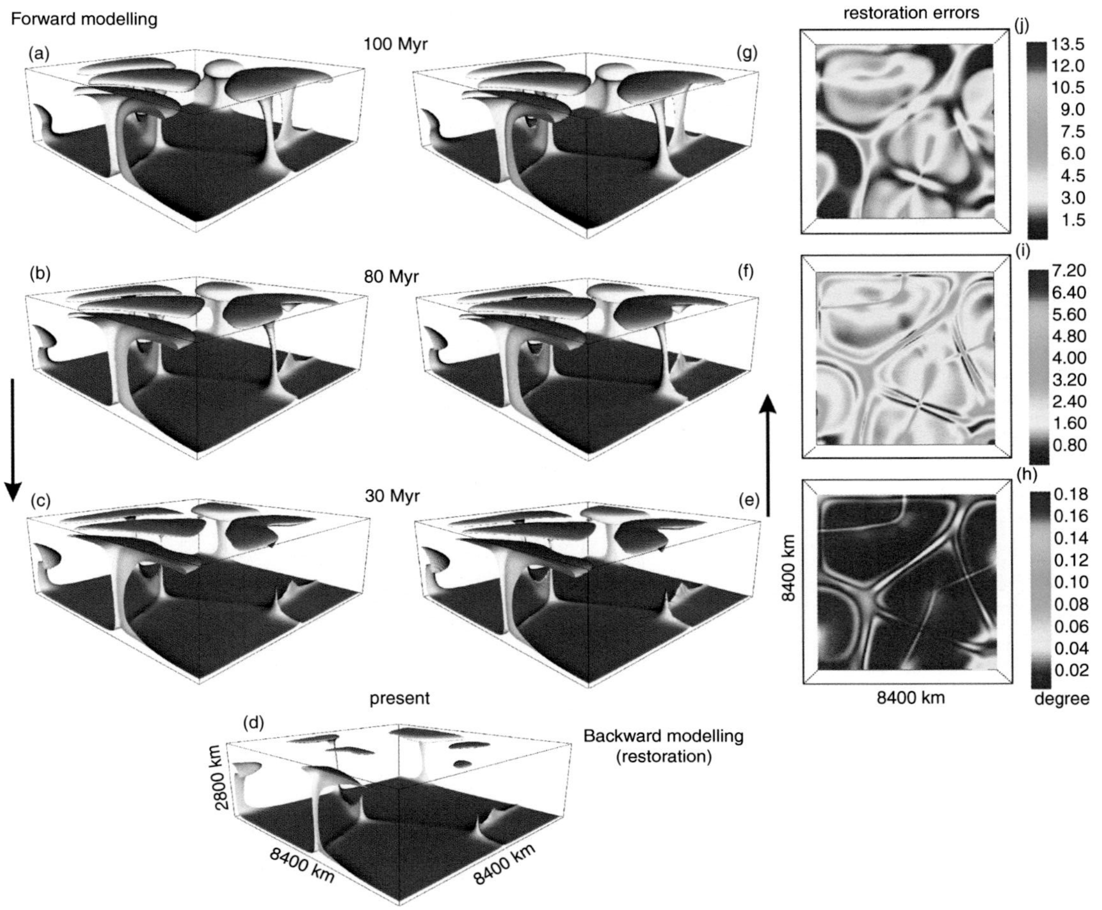

Plate 3. Mantle plume diffusion ($r = 20$ and $Ra = 9.5 \times 10^5$) in the forward modelling at successive diffusion times: from 100 Myr ago to the 'present' state of the plumes (left panel, a–d). Restored mantle plumes in the backward modelling (central panel, e–g) and restoration errors (right panel, h–j). After Ismail-Zadeh *et al.* (2006). (See Chapter 8.)

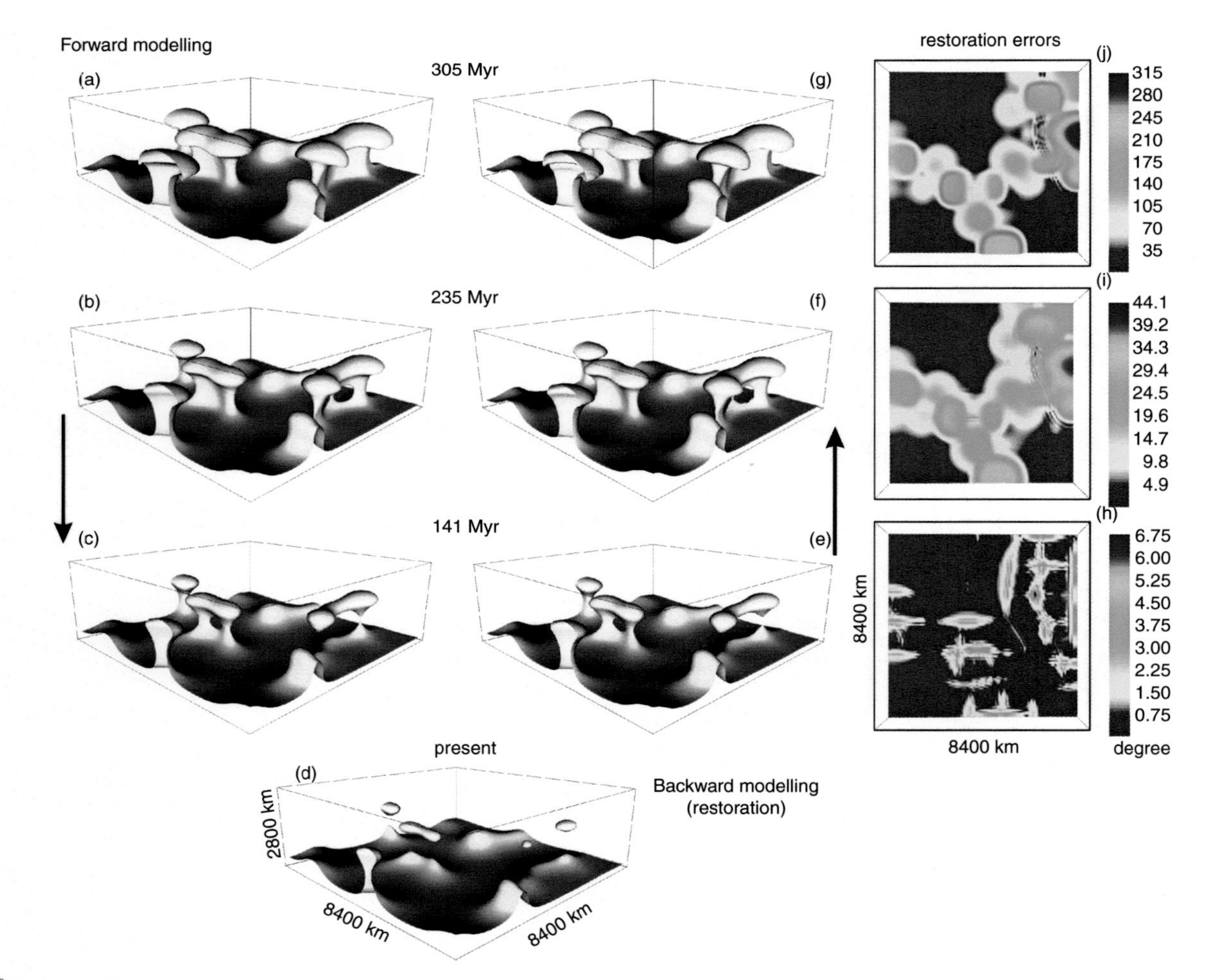

Plate 4. Mantle plume diffusion ($r = 200$ and $Ra = 9.5 \times 10^3$) in the forward modelling at successive diffusion times: from 305 Myr ago to the

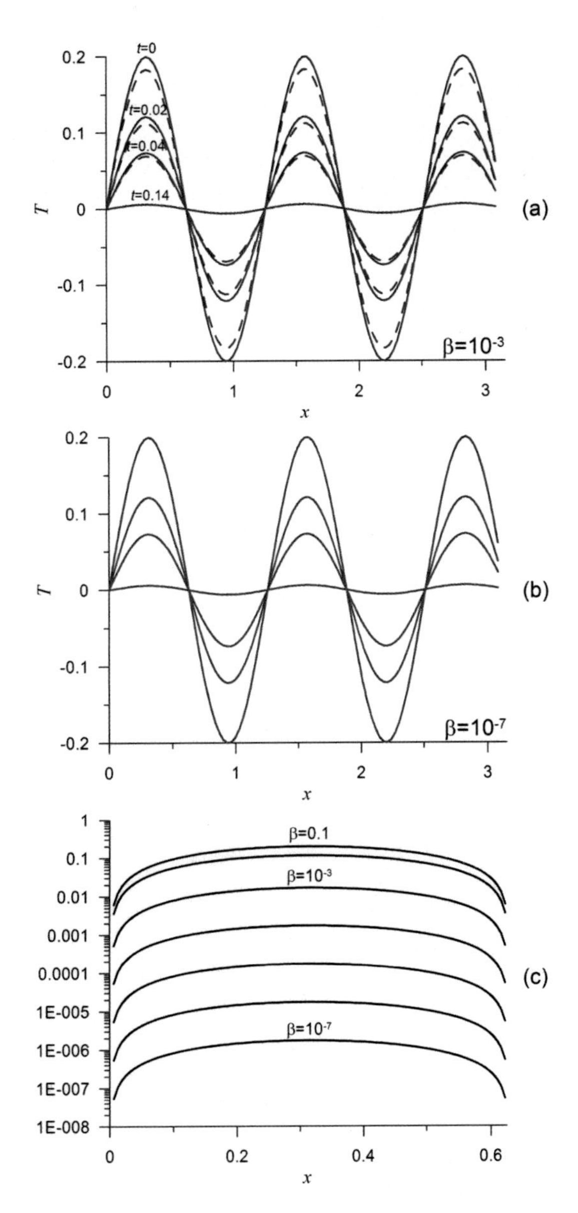

Plate 5. Comparison of the exact solutions to the heat conduction problem (red solid curves; a and b) and to the regularised backward heat conduction problem (a: $\beta = 10^{-3}$ and b: $\beta = 10^{-7}$; blue dashed curves). The temperature residual between two solutions is presented in panel c at various values of the regularisation parameter β. After Ismail-Zadeh *et al.* (2007). (See Chapter 8.)

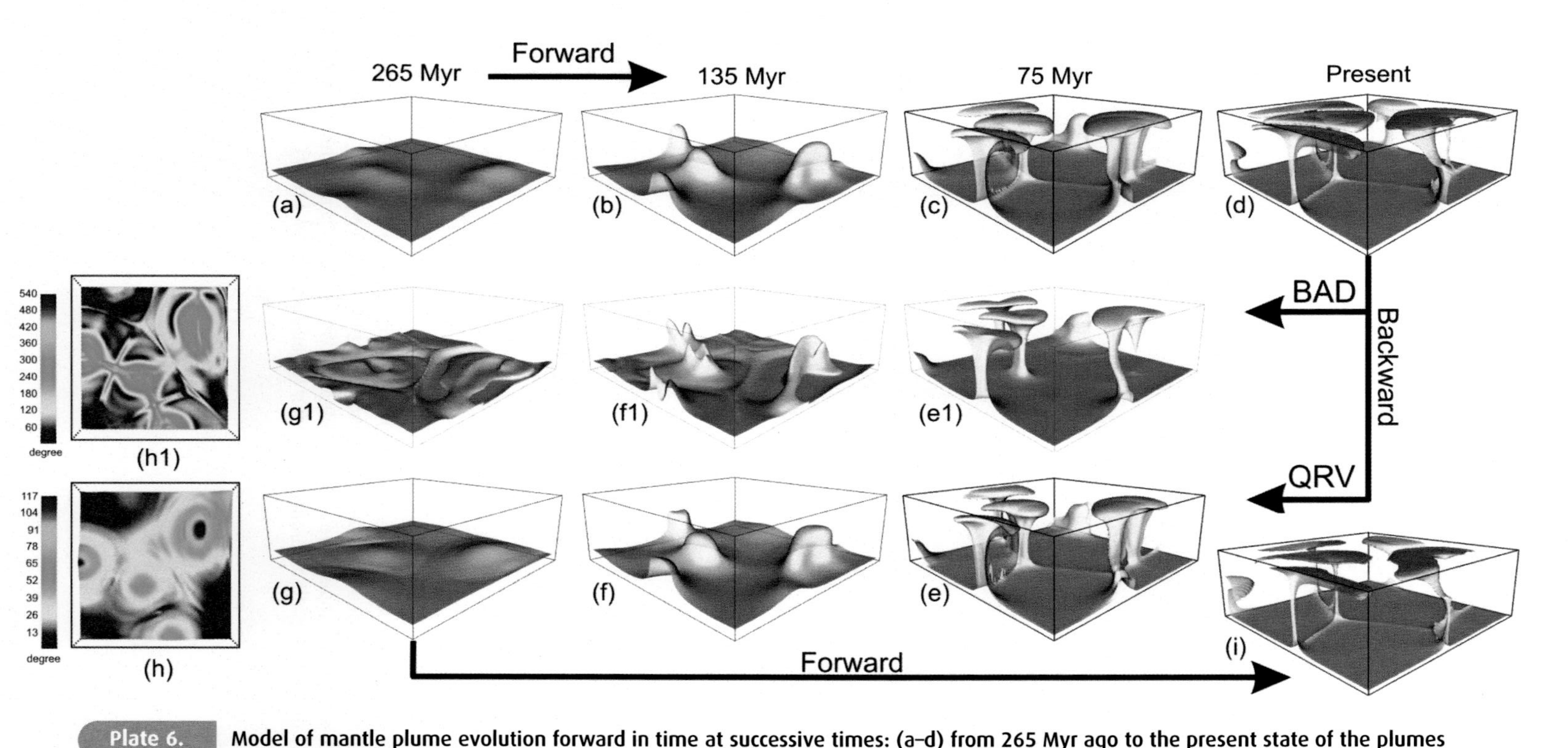

Plate 6. Model of mantle plume evolution forward in time at successive times: (a–d) from 265 Myr ago to the present state of the plumes ($r = 20$). Assimilation of the mantle temperature and flow from the present state back to the geological past using the QRV (d–g; $\beta = 10^{-7}$) and BAD (d, e1–g1) methods. Verification of the QRV assimilation accuracy: Forward model of the plume evolution starting from the initial (restored) state of the plumes (g) to their present state (i). Temperature residuals between the initial temperature for the forward model and the temperature assimilated to the same age using the QRV and BAD methods are presented in panels (h) and (h1), respectively. After Ismail-Zadeh *et al.* (2007). (See Chapter 8.)

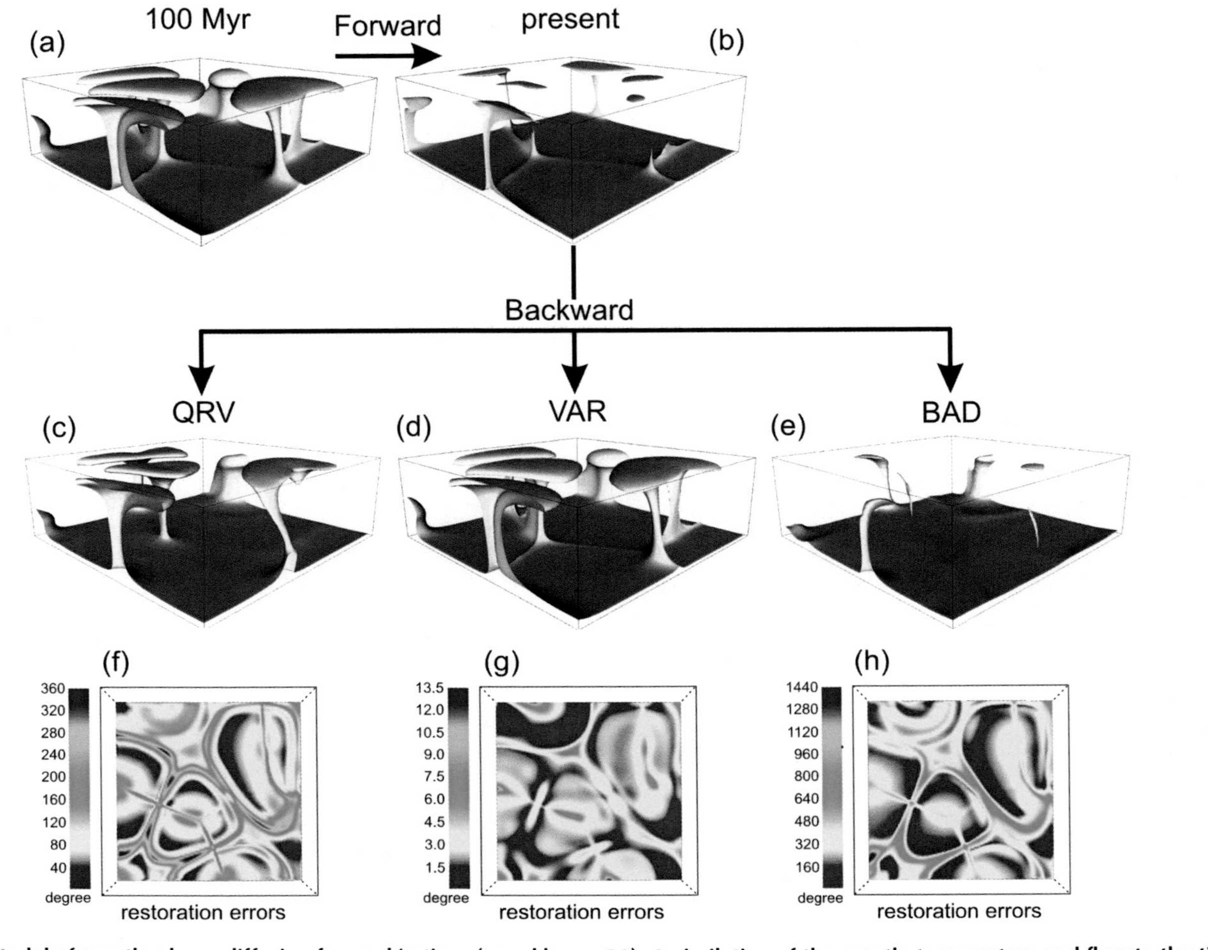

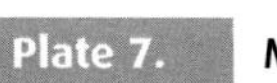

Plate 7. Model of mantle plume diffusion forward in time (a and b; $r = 20$). Assimilation of the mantle temperature and flow to the time of 100 Myrs ago and temperature residuals between the present temperature model (b) and the temperature assimilated to the same age, using the QRV (c and f; $\beta = 10^{-7}$), VAR (d and g), and BAD (e and h) methods, respectively. After Ismail-Zadeh *et al.* (2007). (See Chapter 8.)

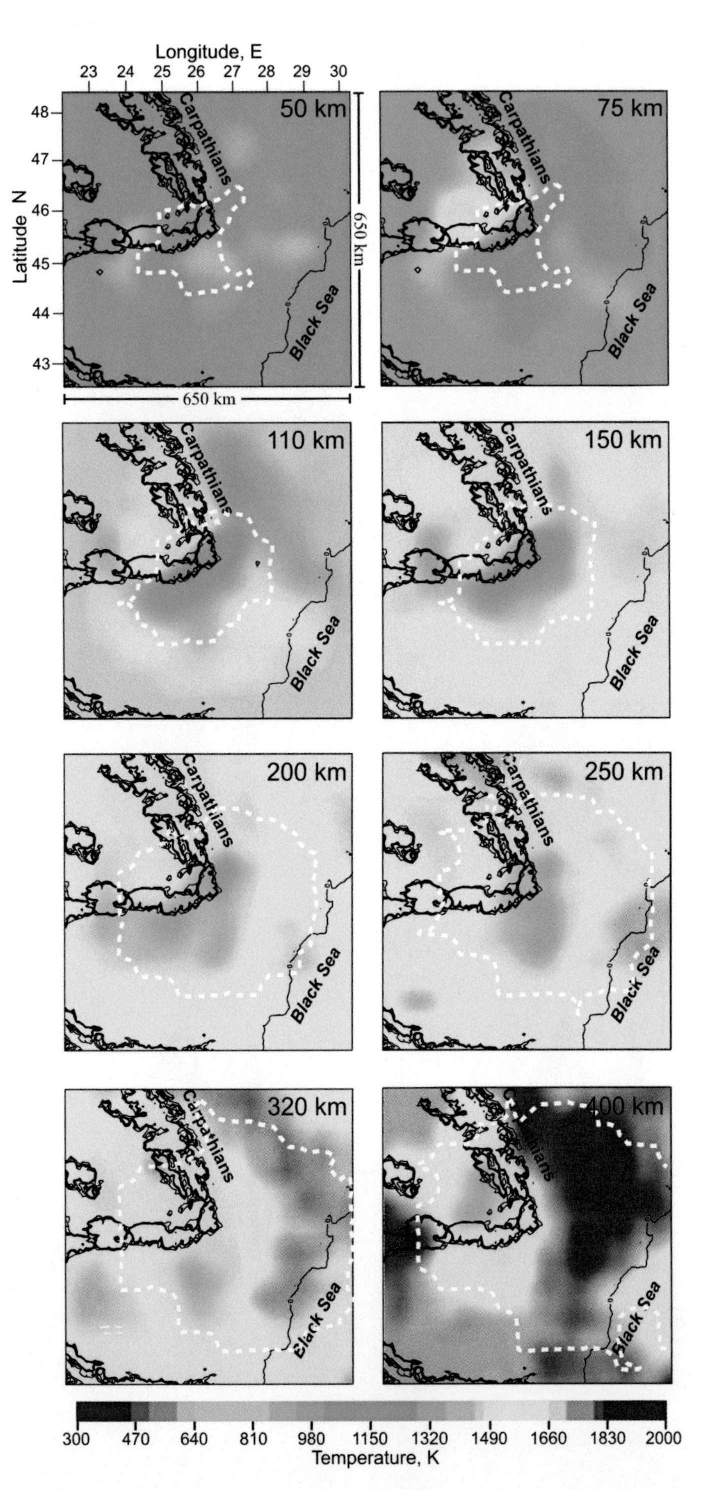

Plate 8. Present temperature model as the result of the inversion of the *P*-wave velocity model. Theoretically well-resolved regions are bounded by dashed line (see text and Martin *et al.*, 2006). Each slice presents a part of the horizontal section of the model domain Ω corresponding to [x_1 = 177.5 km, x_1 = 825.5 km] × [x_2 = 177.5 km, x_2 = 825.5 km], and the isolines present the surface topography (also in Figs. 8.15 and 8.17). After Ismail-Zadeh *et al.* (2008). (See Chapter 8.)

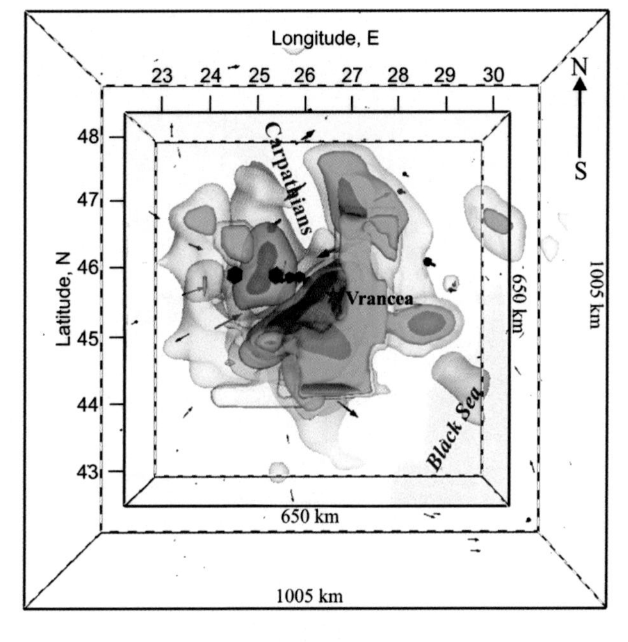

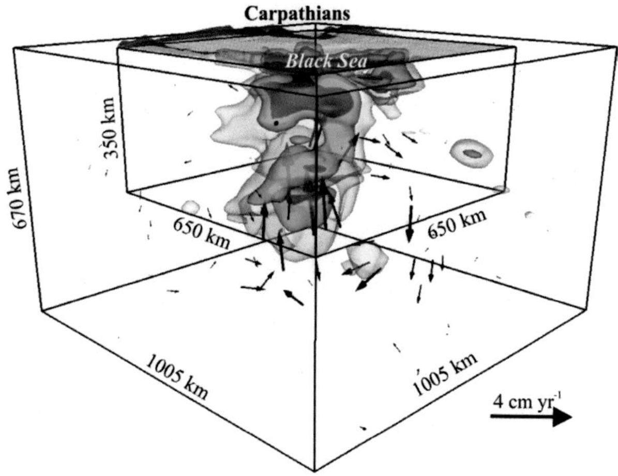

Plate 9. A 3-D thermal shape of the Vrancea slab and contemporary flow induced by the descending slab beneath the SE-Carpathians. Upper panel: top view. Lower panel: side view from the SE toward NW. Arrows illustrate the direction and magnitude of the flow. The marked sub-domain of the model domain presents the region around the Vrancea shown in Fig. 8.17 (in horizontal slices) and in Fig. 8.18. The surfaces marked by blue, dark cyan and light cyan illustrate the surfaces of 0.07, 0.14 and 0.21 temperature anomaly δT, respectively, where $\delta T = (T_{hav} - T)/T_{hav}$ and T_{hav} is the horizontally averaged temperature. The top surface presents the topography, and the red star marks the location of the intermediate-depth earthquakes. After Ismail-Zadeh *et al.* (2008). (See Chapter 8.)

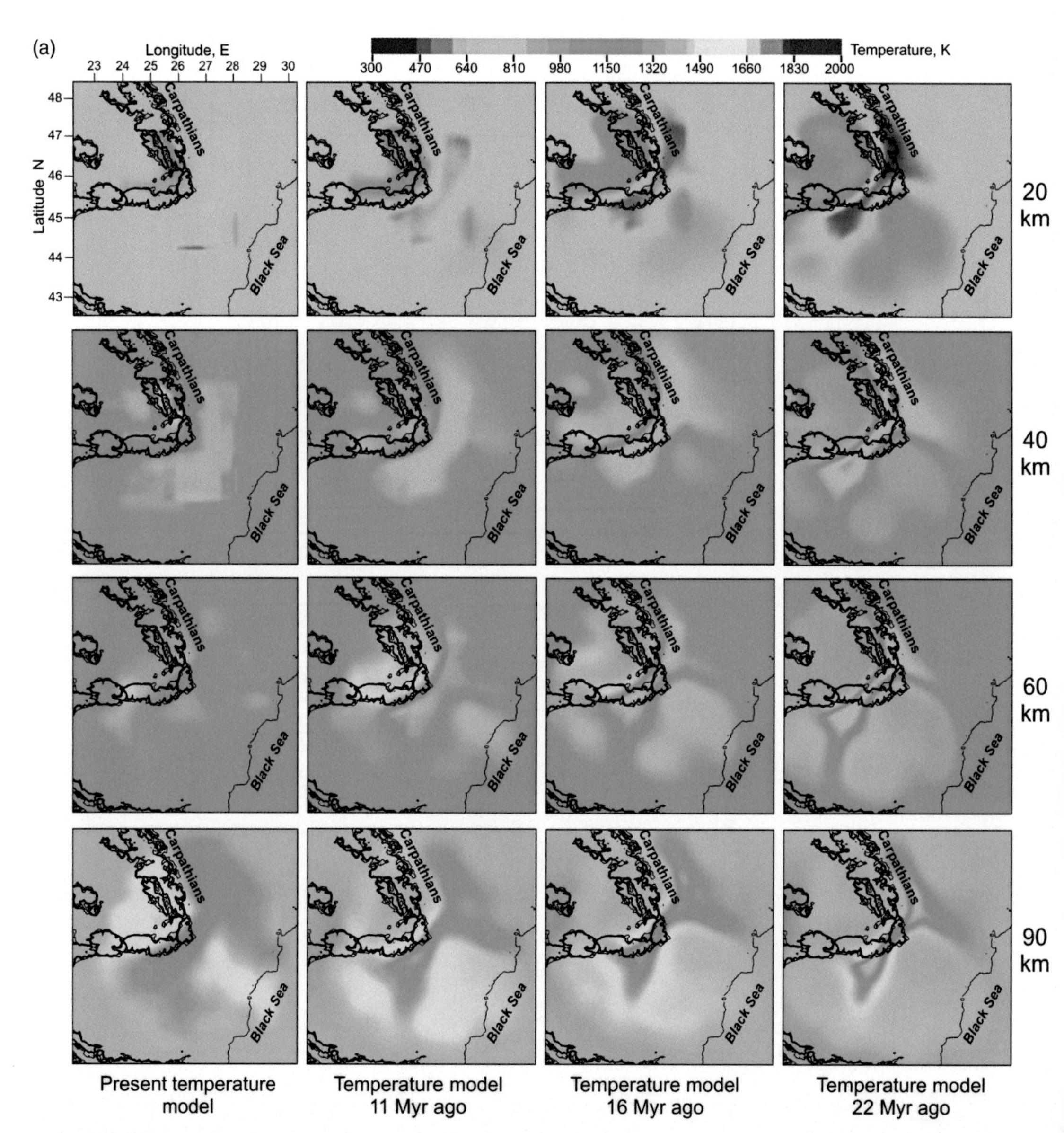

Plate 10. Thermal evolution of the crust and mantle beneath the SE-Carpathians. Horizontal sections of temperature obtained by the assimilation of the present temperature to the Miocene times. After Ismail-Zadeh *et al.* (2008). (See Chapter 8.)

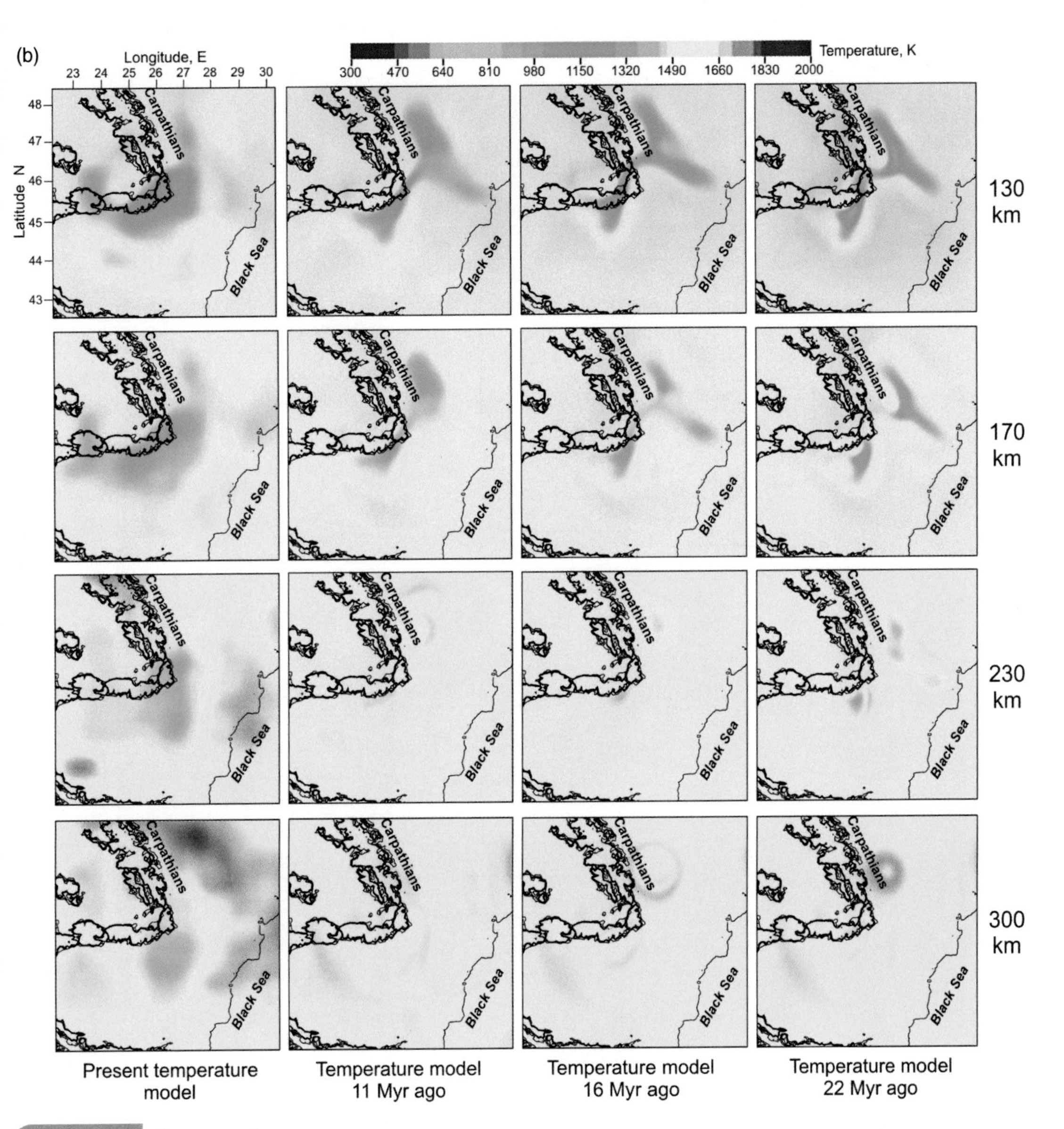

Plate 10. (*Continued*)

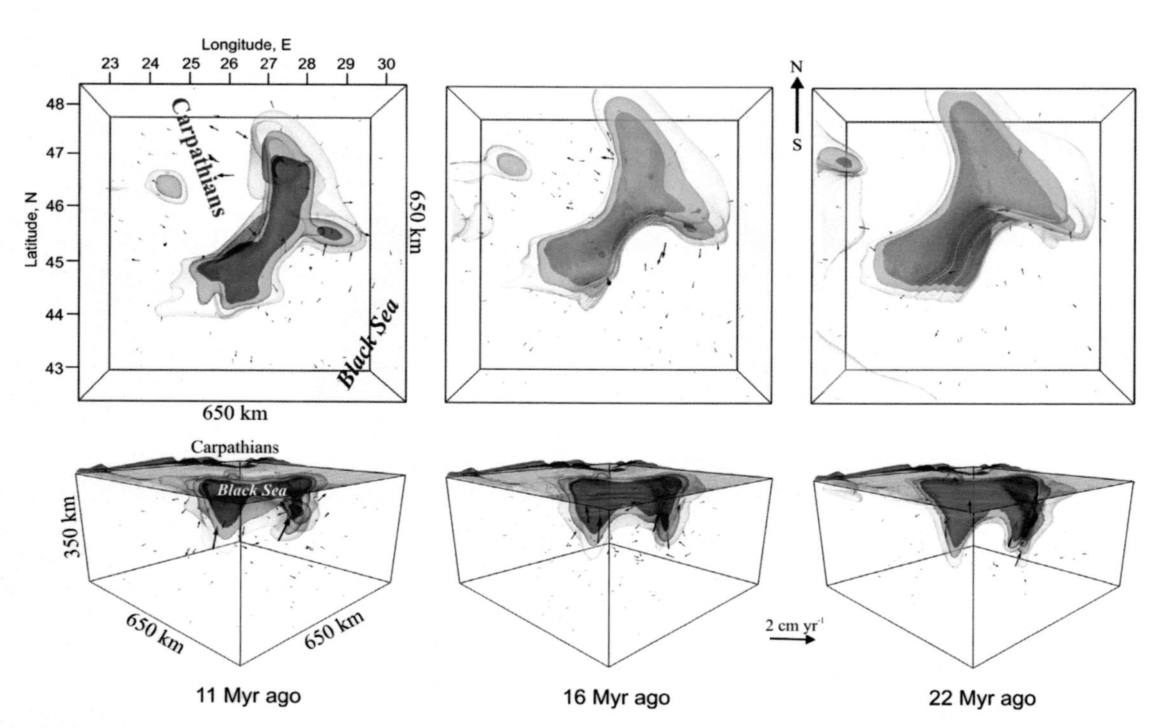

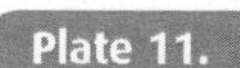

Plate 11. Snapshots of the 3-D thermal shape of the Vrancea slab and pattern of mantle flow beneath the SE-Carpathians in the Miocene times. See Fig. 8.16 for other notations. After Ismail-Zadeh *et al.* (2008). (See Chapter 8.)

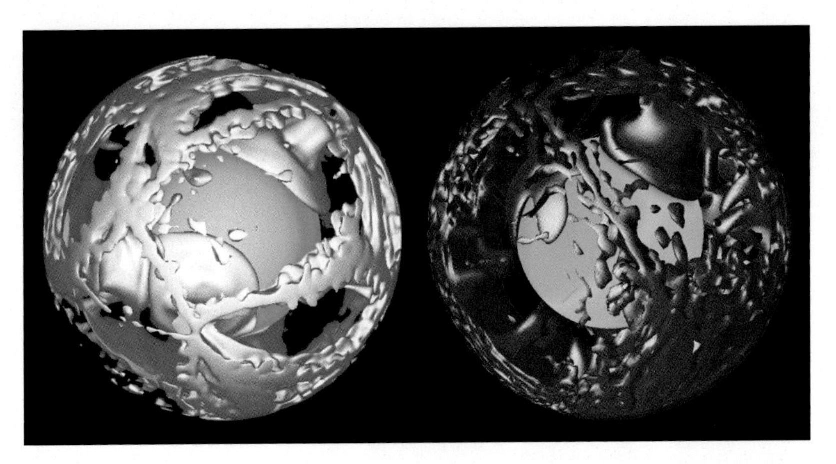

Plate 12. Compressible mantle convection simulations with an endothermic phase transition at 660 km depth, illustrating fifteen years of progress in numerical mantle convection modelling. Left: simulation from Tackley *et al.* (1993) run in 1992 using a spectral transform method on 128 CPUs of the Intel Touchstone Delta supercomputer, at one time the most powerful supercomputer in the world. Right: a similar simulation run in 2008 using a finite-volume discretisation and the Yin-Yang grid (after Tackley, 2008), which can be run on a modern laptop computer. Isosurfaces of residual temperature are plotted, showing where the temperature is significantly colder than the horizontal mean. (See Chapter 10.)

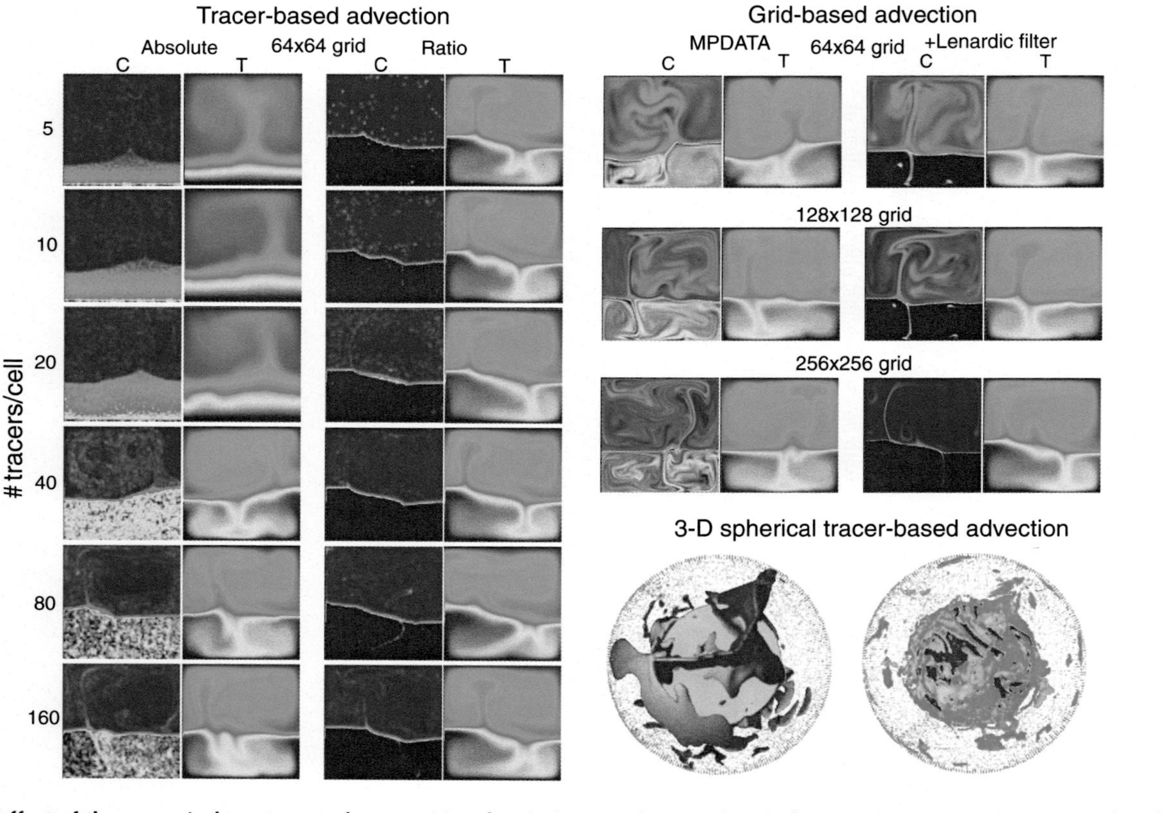

Plate 13. The effect of the numerical treatment of compositional variations on thermo-chemical convection, comparing a tracer (marker) based treatment (left half and 3-D spherical case) with grid-based method (right Cartesian cases). The 2-D Cartesian cases are from Tackley and King (2003). Using tracers and the ratio method to calculate composition from the tracers, a very similar solution is obtained regardless of the number of tracers per cell, from 5 to 160 tracers/cell. If the absolute method is used, however, spurious settling of tracers to the base and compositions of larger than 1.0 are observed for less than 40 tracers/cell, and even then the compositional field is very noisy. When using the finite volume MPDATA (Smolarkiewicz, 1984) advection method for composition, substantially more entrainment is observed. If the filter of Lenardic and Kaula (1993) is added, this entrainment is greatly reduced, but four times as many grid points in each direction are needed to obtain the same result as when using tracers. The 3-D spherical case from Nakagawa and Tackley (2008) illustrates a practical application of the tracer ratio method to investigate the influence of dense material on lateral

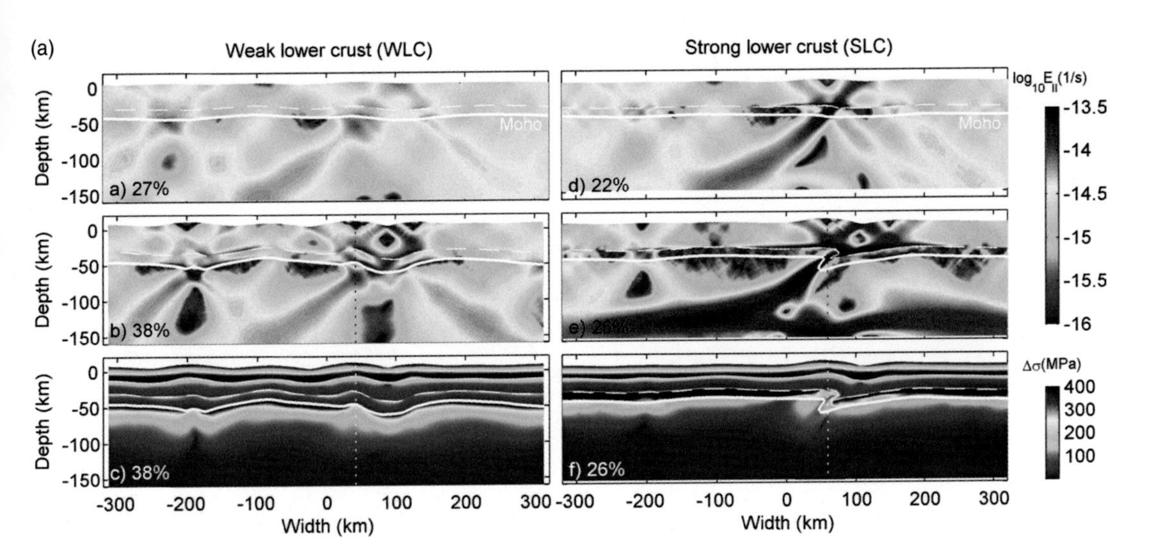

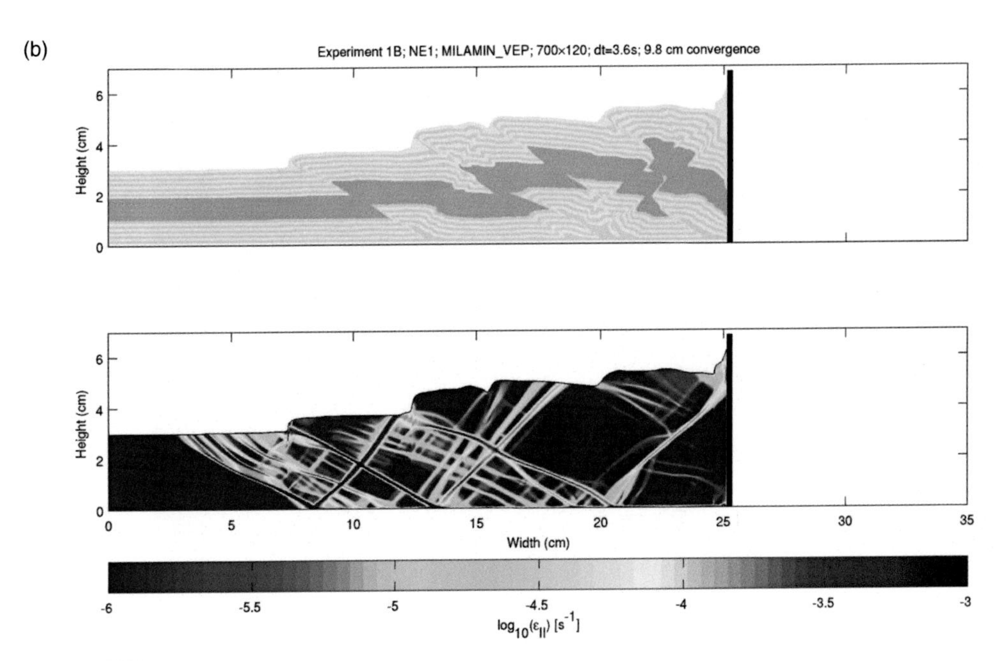

Plate 14. Visco-elasto-plastic models with a free surface and widespread strain localisation, relevant to understanding lithosphere and crust dynamics. (a) The influence of lower crustal strength of strain rate and stress distributions during continental collision (after Schmalholz *et al.*, 2009). (b) The 'numerical sandbox' benchmark test (Buiter *et al.*, 2006) performed by the code MILAMIN_VEP (a visco-elasto-plastic code developed from MILAMIN (Dabrowski *et al.*, 2008) by B. Kaus), showing the formation of shear bands (image courtesy of B. Kaus). (See Chapter 10.)

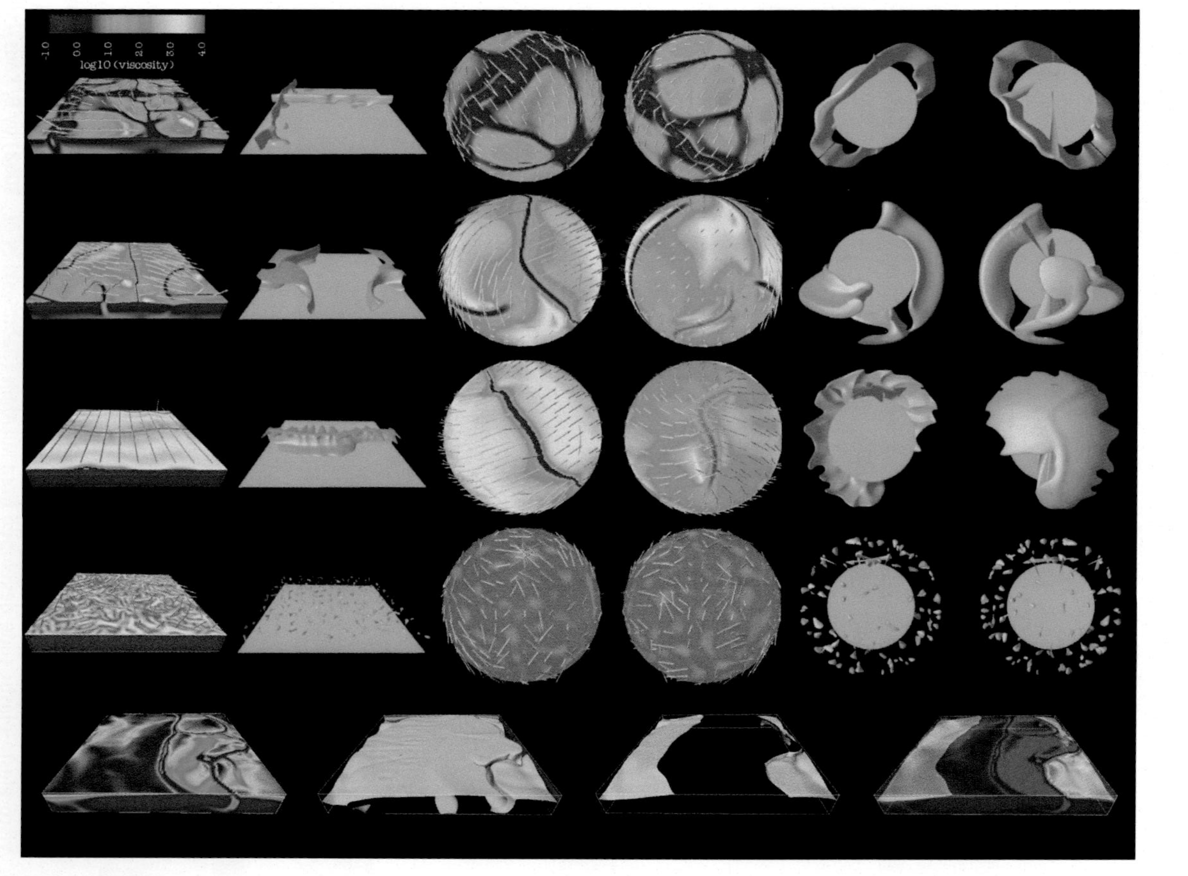

Plate 15. Internally heated convection with a strongly temperature-dependent visco-plastic rheology, in Cartesian and spherical geometry using the finite volume code StagYY (Tackley, 2008). The top four rows from van Heck and Tackley (2008) illustrate the effect of yield stress. Plotted are effective viscosity (columns 1, 3 and 4) and a residual temperature isosurface (columns 2, 5 and 6) showing cold downwellings. As the yield stress is increased the lithospheric behaviour goes from weak, distributed deformation (top row), to mobile lid with reasonably strong plates and weak plate boundaries (rows 2 and 3) to stagnant lid (row 4). The bottom row shows the effect of

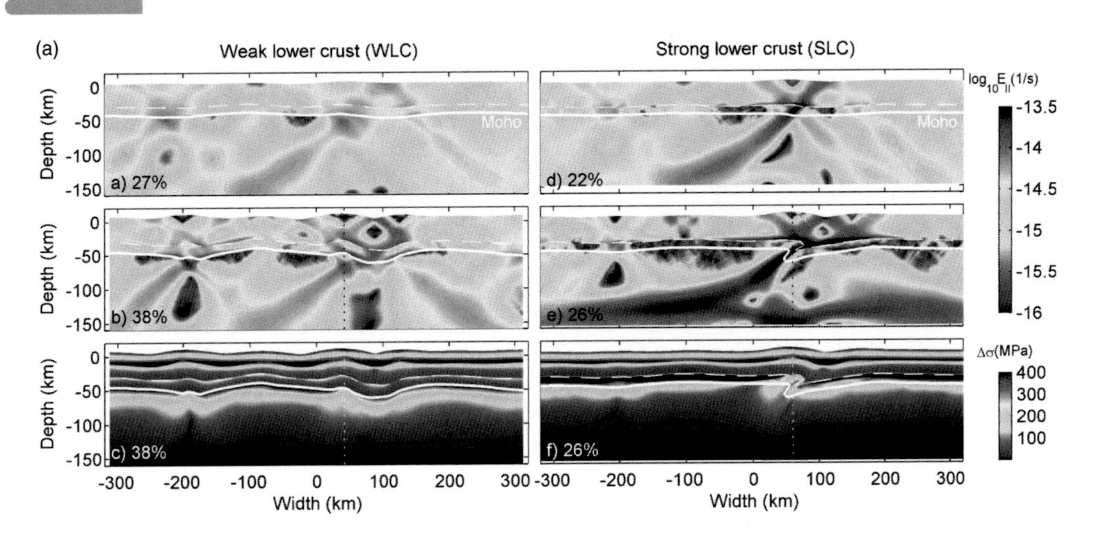

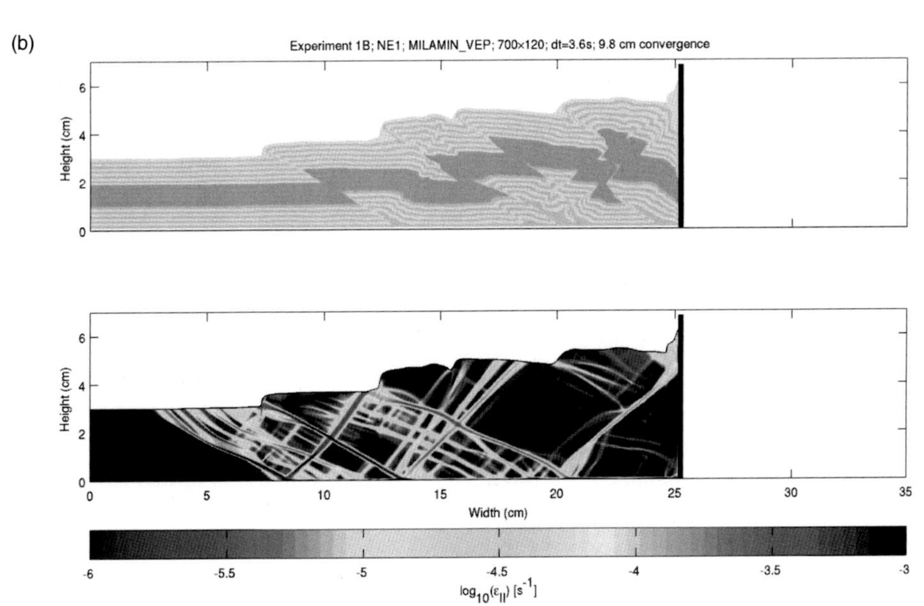

Fig. 10.3. Visco-elasto-plastic models with a free surface and widespread strain localisation, relevant to understanding lithosphere and crust dynamics. (a) The influence of lower crustal strength of strain rate and stress distributions during continental collision (after Schmalholz *et al.*, 2009). (b) The 'numerical sandbox' benchmark test (Buiter *et al.*, 2006) performed by the code MILAMIN_VEP (a visco-elasto-plastic code developed from MILAMIN (Dabrowski *et al.*, 2008) by B. Kaus), showing the formation of shear bands (image courtesy of B. Kaus). (In colour as Plate 14. See colour plates section.)

One of the first models of mantle convection accounting for the effect of the lithospheric plates was an analytical fluid dynamic model proposed by Busse (1978). In Busse's model, the lithosphere was introduced as a less dense layer of higher viscosity, lower thermal conductivity and higher concentration of radiogenic heat sources compared with the underlying mantle. Since then, plates have been approximated in various different ways in numerical models of global mantle convection. In some models only the continental parts of plates are included, while in others plates cover the entire surface. In some studes plates are represented as a surface boundary condition, while in others they extend to some depth into the computational domain. Perhaps the most important distinction is whether plates are assumed to be completely rigid (the *rigid block approach*), or whether they obey the same physical equations as regular mantle and deform less simply because their viscosity is higher (the *rheological approach*). Since the late 1990s, models in which tectonic plates form self-consistently from mantle convection by a combination of temperature-dependent viscosity and plastic yielding have appeared, which is surely the way forwards.

In this section, these various approaches are briefly reviewed. A useful comparison of some of the methods discussed below is given in King *et al.* (1992). Some additional details on the implementation of plates using spectral methods is given in Section 5.4.5.

10.7.2 Rigid block approach

In this approach plates are assumed to be completely rigid and their velocity is either imposed or calculated to be consistent with buoyancy-driven flow beneath, usually using the criterion that the net shear stress on the base of the plate is zero. The rigid regions cover either only part of the surface to mimic continents, or the entire surface to mimic the whole plates. In some models they are applied only at the surface, whereas in other models they extend into the domain. Sometimes rigid blocks that are intended to mimic continents also have different thermal properties, i.e. different thermal conductivity or a different surface thermal boundary condition.

Rigid continents. Numerically, it is straightforward to implement a single rigid continent in a periodic domain, because it can be fixed in grid space, i.e. it does not move relative to the numerical grid. This is because in a periodic domain with free-slip horizontal boundaries the mean horizontal flow is unconstrained. Adding a rigid block that is fixed in grid space removes the non-uniqueness of the solution. Thus, the usual conceptual picture of a continent moving around over a fixed mantle reference frame is in this case replaced by the mantle reference frame moving relative to the fixed continent (the mean flow of the mantle is minus the velocity of the continent). For the purposes of visualisation it can be plotted with the continent moving and the mantle fixed. This approach was used by Gurnis and Zhong (1991) and Zhong and Gurnis (1993) to study mantle convection with a single continent-like raft in 2-D cylindrical geometry. Also along these lines, Yoshida *et al.* (1999) included a high viscosity 3-D circular region mimicking a supercontinent into a 3-D spherical model. In a somewhat different approach, Grigne *et al.* (2007a, b) added a continent by employing a separate conductive grid above the main grid used to calculate convection, with a free slip mechanical boundary condition between the continent and underlying mantle, a configuration that was designed to replicate a laboratory study.

The treatment of two or more rigid continents is more complicated, as at least one of them has to be advected relative to the numerical grid, collisions must be considered and, in 3-D, continents can rotate as well as advect. The velocity of a moving rigid continent is typically calculated by using the criterion that there is no net shear force on its base, which may require an iterative procedure. In three dimensions, the combined rotation and translation of a continent is best treated by calculating its Euler pole, i.e. the pole about which it rotates. Care must be taken when rigid continents collide to avoid them overlapping; a common method is to add a repulsive force between the continents when they collide.

Gurnis (1988) was perhaps the first to introduce two rigid continental blocks into a 2-D numerical model of mantle convection, demonstrating that mantle flow rearranges on the global scale and that continents aggregate and disperse. Three-dimensional models using this approach were presented by Trubitsyn and Rykov (1995), who developed finite difference models of mantle convection with floating rigid surface continents in both 2-D and 3-D Cartesian geometry, with a viscous crust and root underlying the rigid surface treatment. The Newton and Euler solid body equations for 2-D and 3-D models were employed to describe the motion of each continental plate, and plate velocities were determined by applying force balance conditions at the plate bottom. Their numerical approach (with weakly temperature-dependent viscosity in addition to rigid continents) is described in some detail by Rykov and Trubitsyn (2006) and Trubitsyn *et al.* (2006). This model was subsequently extended to spherical geometry using a (longitude, latitude, radius) finite difference grid (Trubitsyn *et al.*, 2008), with continents treated as thin rigid spherical caps. Following a very similar methodology, i.e. calculating Euler poles through force balance and applying repulsive forces to prevent overlap, a single rigid continent (Phillips and Bunge, 2005) and subsequently three to six circular, insulating, rigid continents (Phillips and Bunge, 2007) were implemented into the spherical finite element multigrid code TERRA.

Rigid plates. Many modelling studies have applied one to several plates covering the whole surface of the domain as a surface boundary condition. In some of these plate velocities are imposed, but as this is trivial to implement we do not discuss it. It is more satisfactory from a physical standpoint for plates to move with velocities that are determined by buoyancy forces in the system, particularly in global calculations. This is commonly done by calculating the velocity of each plate based on the criterion that the net shear stress on the base of the plate is zero. This was first done by Jarvis and Peltier (1981) with a single plate in a 2-D box. Several authors subsequently performed 3-D Cartesian (Gable *et al.*, 1991; Lowman *et al.*, 2004) or spherical (Ricard and Vigny, 1989; Monnereau and Quere, 2001) models using the criterion that the net torque on the plate is zero, and some also accounted for plate boundary forces (Forte and Peltier, 1991; 1994) (Section 5.4.5). This approach was also used for plates that contain a mixture of oceanic and continental parts with different physical properties for each, as in the 2-D finite difference code of Lowman and Jarvis (1993; 1995).

10.7.3 Rheological approach

In this approach, plates are modelled as obeying the same physical equations as the rest of the domain, behaving in a plate-like manner because they have a higher viscosity. This approach has been applied to just continents, to the entire lithosphere with weak plate boundaries inserted by hand, or to the entire lithosphere with weak plate boundaries arising self-consistently from the rheology. It has the advantages of representing physical reality and avoiding the need for additional programming.

Deformable continents. This approach of modelling continents as deformable, compositionally distinct entities with a different (higher) viscosity than normal mantle was pioneered in a series of 2-D papers (Lenardic and Kaula, 1995a,b, 1996; Lenardic *et al.*, 1995) and subsequently in 3-D (Moresi and Lenardic, 1997; Tackley, 2000c). Advantages of this over the rigid block approach are that no special numerical treatment is required and that it better represents physical reality: continents do undergo substantial deformation, and although they are stronger than the underlying mantle, they are weaker than oceanic lithosphere. In these models, regions outside continents are relatively weak with at best, weakly temperature-dependent viscosity. In reality, oceanic lithosphere is stronger than continental lithosphere. Models in which rheological oceanic plates are combined with rheological continents are discussed below.

High-viscosity plates with imposed weak boundaries. When a realistically high temperature-dependence of viscosity is included, a stagnant lid forms over the entire planet, with no mobile plates. In order to create mobile plates, a common method is to specify weak zones or weak faults at plate boundaries. This has been done in a 2-D Cartesian box by using finite differences (Jacoby and Schmeling, 1982), finite elements (Gurnis, 1989; King and Hager, 1990), and finite volumes (Honda, 1997). If such weak zones are used, then their width must be similar to the thickness of the plate. Rather than using such broad weak zones, it is perhaps more satisfactory to impose narrow weak faults, which can be done by using a finite element method, either with inclined weak zones representing the subduction channel (Zhong and Gurnis, 1992, 1995), or as vertical faults representing transform faults in oceanic lithosphere (Zhong and Gurnis, 1996). A weak channel separating the subducting and overlying plates may also arise self-consistently through the advection of weak sediments (Lenardic and Kaula, 1994; Gerya *et al.*, 2008).

'Self-consistent' plates. The ultimate goal is for plates to arise self-consistently from the material properties (rheology) rather than being inserted by hand, as happens on the Earth. This is a topic of active research by several groups. Although it is far from solved how Earth-like plate tectonics arises, it is clear that necessary ingredients are an effective rheology that is non-Newtonian and, most likely, history- and composition-dependent. The numerical implementation of non-Newtonian rheology including plasticity is discussed in Section 10.6.1.

Early experiments with a 1-D non-Newtonian layer inserted on top of a 2-D Newtonian domain (Weinstein and Olson, 1992; Weinstein, 1996) showed that power-law rheology with a high exponent localises deformation at plate boundaries. Extending this further, plastic yielding is similar to power-law rheology with an infinite exponent. Mantle convection with a visco-plastic temperature-dependent rheology was investigated systematically in 2-D by Moresi and Solomatov (1998) and Richards *et al.* (2001), who found that plastic yielding produces a 'mobile lid' behaviour for a certain range of yield stress. Three-dimensional investigations followed, in Cartesian (Trompert and Hansen, 1998a,b; Tackley, 2000a,b) and spherical (van Heck and Tackley, 2008; Foley and Becker, 2009) geometry. All of these use finite element or finite volume discretisations. Examples of plate tectonics-like behaviour self-consistently generated from mantle convection are shown in Figure 10.4.

Rheologies in which the viscosity becomes weaker with increasing strain (strain weakening) or strain rate (strain rate weakening) have also been applied in global simulations.

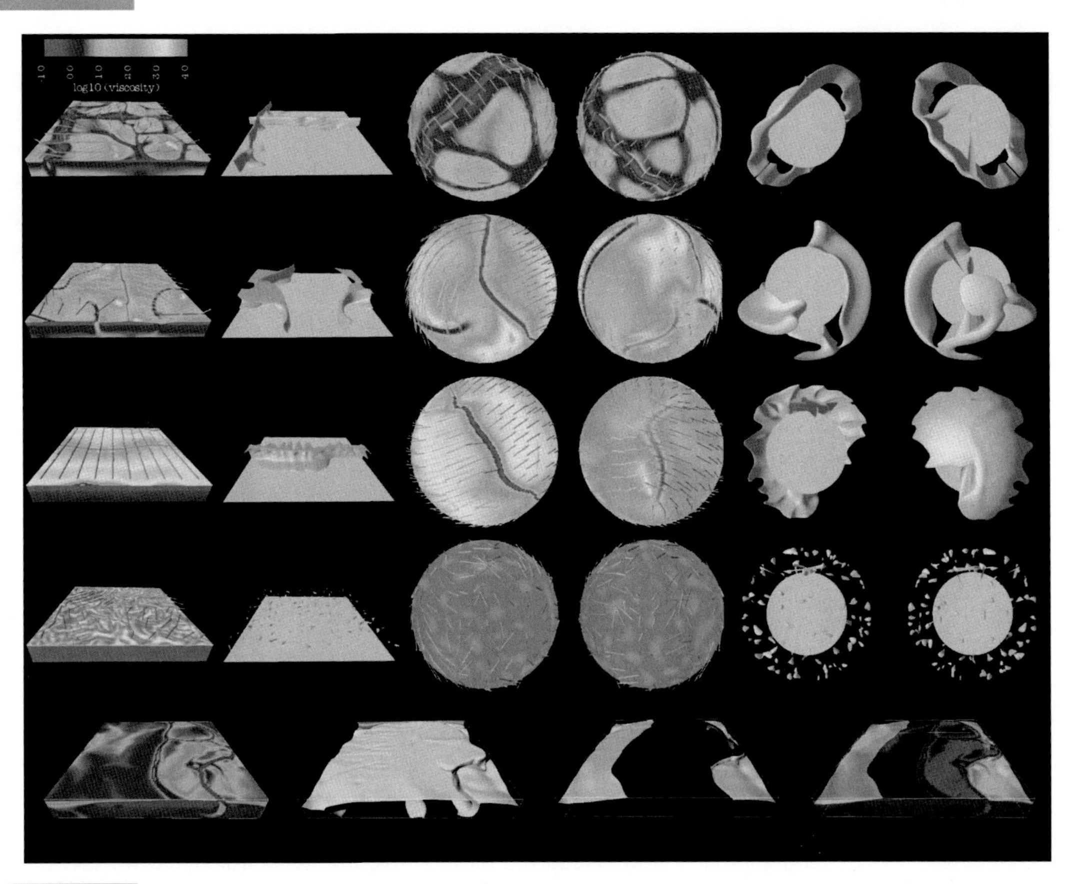

Fig. 10.4. **Internally heated convection with a strongly temperature-dependent visco-plastic rheology, in Cartesian and spherical geometry using the finite volume code StagYY (Tackley, 2008). The top four rows from van Heck and Tackley (2008) illustrate the effect of yield stress. Plotted are effective viscosity (columns 1, 3 and 4) and a residual temperature isosurface (columns 2, 5 and 6) showing cold downwellings. As the yield stress is increased the lithospheric behaviour goes from weak, distributed deformation (top row), to mobile lid with reasonably strong plates and weak plate boundaries (rows 2 and 3) to stagnant lid (row 4). The bottom row shows the effect of a buoyant, compositionally distinct continent (yellow isosurface) on inducing single-sided subduction at its edge. (In colour as Plate 15. See colour plates section.)**

Narrow plate boundaries formed in the models of Tackley (1998, 2000c) consisting of a horizontal 2-D sheet with a rheology based on the 'stick-slip' rheology of Bercovici (1993, 1995), inserted on top of a 3-D Newtonian box. Other studies have considered the evolution of 'damage' in time-dependent mantle convection simulations in 2-D (Auth *et al.*, 2003; Ogawa, 2003; Landuyt *et al.*, 2008) and 3-D (Yoshida and Ogawa, 2004), again using finite element or finite difference formulations.

Models where both rheological oceanic plates (owing to plastic yielding) and rheological continents exist simultaneously were pioneered in 2-D by Lenardic and Moresi (1999) and continued in Lenardic *et al.* (2003); Cooper *et al.* (2004); O'Neill *et al.* (2008). Recent models have integrated rheological treatment of both oceanic and continental plates with physical properties including rheology that depend on temperature, depth, composition and stress.

10.8 Treatment of a free surface and surface processes

For many problems it is desirable to model correctly the free outer surface of the Earth, i.e. with zero shear and normal stress and vertical deflection, rather than assuming a free-slip boundary condition, in which the surface is forced to be flat. This can be problematic because the density contrast at the outer surface is about two orders of magnitude larger than thermal density contrasts driving flow, so small deflections of the free surface cause very large normal stresses. This makes it desirable that vertical motion of the free surface be treated in an implicit manner, because if an explicit time integration is used then the time step must be small, to avoid overshoots and oscillations. Several methods of implementing a free surface have been used. A useful summary and comparison of seven codes using different approaches is given in Schmeling *et al.* (2008). The methods can be divided into two categories: using an Eulerian mesh with a free surface that lies within the mesh, or using a Lagrangian mesh in which the boundary of the mesh follows the free surface.

There are two approaches using an Eulerian mesh.

(i) If the wavelength of the surface topography is long compared to its amplitude, then the *pseudo-free surface* approach (Zhong *et al.*, 1996) can be used. In this, the topography of the free surface is applied as a normal stress to the top of the Eulerian mesh, and vertical flow through the top boundary of the mesh causes changes in the surface topography. Typically, however, on Earth features like subduction zones involve rapid variations in topography so the application of this method is limited.

(ii) To treat arbitrary deflections of a free surface using an Eulerian mesh, an increasingly common approach is to extend the mesh above the free surface and fill this region with material representing air or water, i.e. very weak and with low or zero mass. This method dates back to at least 1965 (Harlow and Welch, 1965), and has recently been referred to as '*sticky air*' or '*sticky water*' (Naimark *et al.*, 1998; Gerya and Yuen, 2007). Clearly, the viscosity of this material cannot be set to a realistically low value, but experiments indicate that provided its viscosity is low compared to that of the crust and lithosphere, the exact value does not matter. If the 'sticky air' viscosity is set too low, then spurious and undesirable high velocities can occur in this material. Tests in Schmeling *et al.* (2008) show that this can give similar results to a 'real' free surface.

Using a Lagrangian mesh allows the mesh to follow the free surface.

(i) Some codes use a fully Lagrangian mesh, in which the nodes move with the material, and the mesh tracks the free surface, typically implemented using a finite element discretisation (Poliakov *et al.*, 1993c; Babeyko *et al.*, 2002). This requires additional computational effort to deal with the constantly changing mesh, and re-meshing may be required when elements get too distorted.

(ii) In the Arbitrary Lagrangian-Eulerian (ALE) approach (Hughes *et al.*, 1981; Fullsack, 1995), some nodes move vertically to follow the free surface, while others remain fixed. Lagrangian markers may be used to advect material properies. This combines some advantages of Eulerian and Lagrangian approaches.

It is often desirable to take into account redistribution of surface mass by erosion and sedimentation. Unfortunately the correct quantitative description of these processes is not known. Various parametrisations exist, taking into account hillslope diffusion and fluvial incision (see Willett, 1999; Beaumont *et al.*, 2000; Godard *et al.*, 2004; Simpson, 2004; Godard *et al.*, 2006; Simpson and Castelltort, 2006; and references therein).

Two approaches have been used to model orogeny. One approach assumes that rocks behave like a viscoplastic material on geological time scales and requires the solution of the Stokes equation adapted to take into consideration the non-linear rheology (Fullsack, 1995; Moresi and Solomatov, 1998). The other approach assumes that rock behaves intrinsically as an elastic solid, which displays a brittle response to large stresses and creep at elevated temperature, and requires the solution of the basic equations of structural mechanics (Chery *et al.*, 1991; Braun, 1993; Batt and Braun, 1997). The main advantage of the viscous approach is that large deformations are easily handled, but the tracking of material interfaces is difficult and requires the introduction of methods such as the Arbitrary Langrangian Eulerian (ALE) method. In the elastic approach, finite deformation requires the introduction of non-linear strains and the tracking of element rotation (Braun, 1994). Most elastic-solid methods are based on a Lagrangian representation of the deformation. The advection of nodes with the flow of material facilitates greatly the tracking of material particles, but large deformations lead to mesh deformation and the need frequently to redefine the mesh, as is done in the Dynamic Lagrangian Remeshing by Braun and Sambridge (1994). The main improvement of both approaches resides in their ability to predict the path that a rock particle follows during its transit through an orogen. By coupling the equation of force balance to the heat advection–diffusion equation (Jamieson *et al.*, 1996; Batt and Braun, 1997), one can make accurate predictions on the pressure and temperature conditions experienced by a given rock particle through the development of the orogen.

10.9 Porous flow through a deformable matrix

The flow of liquid through solid is of great interest in several geodynamical settings, e.g. partial melt flowing through solid rock in the asthenosphere, molten iron flowing through solid rock during core formation and groundwater flow (e.g. in sedimentary basins or hydrothermal flow under volcanoes or mid-ocean ridges). For relatively low-pressure applications such as the flow of groundwater or hydrocarbons, the matrix is typically assumed to be rigid, in which case solution is straightforward for a single fluid phase, although it presents challenges in the case of multiple fluid phases (Geiger *et al.*, 2006; Coumou *et al.*, 2009) and/or dissolution or precipitation processes between fluid and matrix (Aharonov *et al.*, 1997). This section focuses, however, on the case where the solid matrix is *deformable*, in which case there is a stress associated with its compaction or dilation, which couples together the solid equation of motion and the liquid equation of motion and makes the resulting set of equations numerically challenging to solve. We also focus on the case of a single fluid phase, such as molten rock or metal.

The most commonly used version of the appropriate two-phase flow equations (given below) was presented by McKenzie (1984) while another set, based on somewhat different physical reasoning although sharing a similar structure, was proposed by Bercovici *et al.* (2001); for a comparison see Appendix A of Ricard *et al.* (2001). For simplicity, the version below assumes density of liquid ρ_{liq} and of solid ρ_{sol} to be constant (and $\rho_{liq} = \rho_{sol}$) except in the buoyancy term, and transfer of mass between liquid and solid by melting or freezing is not included. In this case the continuity equations for liquid and solid, Darcy equation for segregation of liquid and solid, and solid momentum (Stokes) equation become:

$$\frac{\partial \phi}{\partial t} = -\nabla \cdot (\phi \mathbf{v}) = \nabla((1-\phi)\mathbf{V}), \tag{10.38}$$

$$\phi(\mathbf{v} - \mathbf{V}) = -\frac{K}{\eta_{liq}}(\nabla P - \rho_{liq} g \mathbf{e}), \tag{10.39}$$

$$-\nabla P + \nabla[(\varsigma - \frac{2}{3}\eta)\nabla \cdot \mathbf{V}] + \nabla \cdot [\eta(V_{i,j} + V_{j,i})] = g\rho\mathbf{e}, \tag{10.40}$$

where $\mathbf{v}$ and $\mathbf{V}$ are the liquid and solid velocities, respectively; ϕ is porosity; P is the pressure in the liquid; η_{liq} is the liquid viscosity; g is the acceleration due to gravity, ρ is the total density, ς is the bulk viscosity, K is permeability, and $\mathbf{e}$ is the unit vector. The energy equation is not given because is straightforward to solve. The permeability is normally taken to be a power-law function of porosity, such as

$$K = k_\phi \phi^2, \tag{10.41}$$

where k_ϕ is a constant. The bulk viscosity parametrises the resistance of the matrix to compaction or dilation, and is often assumed to be inversely proportional to porosity. The momentum term involving bulk viscosity and the divergence of velocity gives the stress associated with compaction or dilation, and couples the solid and fluid flow fields, making this set of equations much more difficult to solve than those of simple Stokes flow. At each time instant the porosity is known, and the equations must be solved for $\mathbf{V}$, $\mathbf{v}$ and P. Various approaches have been taken to simplify and solve these equations.

Some researchers neglect the compaction term and use the bulk velocity (i.e. a weighted average of the solid and liquid velocities), rather than solid velocity, in the Stokes-like equation, which then becomes a bulk momentum equation (Scott and Stevenson, 1989; Tackley and Stevenson, 1993; Hernlund *et al.*, 2008). With these assumptions, the bulk Stokes and continuity equations are completely decoupled from the Darcy equation and can be solved first, with liquid–solid segregation calculated as a separate step. This is justified when the length scales of the numerical calculation are much larger than the compaction length, which is the characteristic length scale for compaction.

A more accurate approximation is the *Compaction Boussinesq Approximation* (Schmeling, 2000). In this, the compaction term in the solid momentum equation (10.40) is neglected. Compaction stress is, however, included in the segregation equation (10.39) by eliminating the pressure gradient using the full momentum equation, i.e. including the compaction stress. The required divergence of solid velocity is obtained from the continuity equations

(10.38). This approximation is valid for melt fractions of up to 10–20% (Schmeling, 2000).

To solve the full equations for the case of magma migration, Spiegelman (1993) rewrote them by decomposing the solid pressure into hydrostatic, compaction and dynamic components. By using this approach, fluid velocity can be eliminated from the equations, which are then expressed in terms of (and solved for) solid velocity, dynamic pressure and compaction pressure. This method also works in three dimensions (Wiggins and Spiegelman, 1995).

Another approach to solving the full equations, used for the purpose of liquid metal–solid silicate segregation, was taken by Sramek (2007) and Sramek *et al.* (2007). In two dimensions, velocity was decomposed into incompressible and irrotational parts, with each part expressed in terms of a scalar function (stream function and velocity potential, respectively). Solving the coupled equations then required iterating between the solution of velocity scalars and the Darcy equation, using a direct solver at each step.

The complexity of the solid matrix being viscoelastic was considered by Connolly and Podladchikov (1998). They simplified the equations by neglecting shear force balance (i.e. the solid momentum equation), solving only for bulk viscoelasticity. A similar approach was taken by Connolly and Podladchikov (2007) to test the consequences of a bulk viscosity that is different in compaction than in dilation, for a viscous matrix. Starting from the usual equations they introduced the 'effective pressure', which is identical to the 'compaction pressure' defined by Spiegelman (1993). By using this, solid velocity was eliminated from the continuity and liquid momentum equations, which were then straightforwardly solved for porosity and effective pressure. This required neglecting porosity advection in the continuity equation in addition to shear stresses (the solid momentum equation). These two assumptions were justified by a scaling analysis for their physical setup, in which the solid is at rest except for melt-migration related motion.

10.10 Geodynamo modelling

Flow in the liquid outer core, which generates the geodynamo, is in a quite different fluid dynamical regime from the lithosphere and mantle and presents different modelling challenges, although the same basic methods can be used. Here we give a brief summary of this field; for a detailed review of the modelling assumptions, numerical approaches and results the reader is referred to Christensen and Wicht (2007).

Major complexities in core dynamics that are not present in mantle dynamics are that rotation is very important, requiring the Coriolis term to be included in the momentum (Navier–Stokes) equation, and that the magnetic field is both generated by and modulates the flow, requiring an additional vector equation to describe the evolution of the magnetic field $\mathbf{B}$ (which is also subject to $\nabla \cdot \mathbf{B} = 0$) plus the inclusion of the Lorentz force in the momentum equation. Additionally, the Prandtl number is low (and less than unity) rather than effectively infinite, which nominally requires the inclusion of inertial terms in the momentum equation, although they are often neglected because other terms are considered to be much larger. In any case, because the viscosity of liquid iron is very low, the parameter space, for example expressed in terms of Rayleigh number and Ekman number, is many orders of magnitude beyond what can be reached by numerical simulations in the foreseeable future.

In other ways, however, core flow is easier to model than mantle flow, firstly because material properties such as viscosity do not vary by much, whereas large viscosity variations are the bane of the mantle convection modeller. Secondly, the low Prandtl number means that the momentum equation can be time stepped explicitly like the energy equation, rather than needing to obtain an 'implicit' global solution at each time step as is necessary for mantle modelling.

Spectral methods have generally been the method of choice for geodynamo modelling, as discussed in Chapter 5. Reasons for this are the straightforward handling of the boundary conditions for the magnetic field, which are similar to those of the gravitational field given in Equation (5.46), the high convergence rate (accuracy) of the spectral method, and the natural treatment of spherical geometry using spherical harmonics. A typical approach for a 3-D spherical shell is to expand variables in spherical harmonics azimuthally and Chebyshev polynomials radially, use a toroidal-poloidal decomposition (Eq. (5.26)) to represent velocity and magnetic field, and a spectral transform method (Section 5.4) to calculate non-linear terms in the momentum and energy equations. Time integration is typically performed by using a mixed scheme with non-linear terms treated explicitly and linear terms treated implicitly, and with at least second order accuracy in time. A key decision is whether to treat the Coriolis term ($2\Omega\times\mathbf{v}$, where $\mathbf{v}$ is velocity and Ω is the rotation (angular velocity) vector) implicitly or explicitly, because although this term is linear it couples together harmonics with different degrees but the same order and also toroidal and poloidal components (Christensen and Wicht, 2007), so that individual modes do not completely decouple and a larger matrix equation must be solved. This is generally treated explicitly, with the penalty that a smaller time step must be taken.

The first two studies to achieve self-sustaining 'strong' dynamos in fully 3-D geometry were Glatzmaier and Roberts (1995b) and Kageyama and Sato (1995), and these have been followed by numerous subsequent studies by these and various other researchers, as reviewed in Christensen and Wicht (2007). The code of Glatzmaier and Roberts (1995b) follows the above numerical scheme with an implicit treatment of the Coriolis term and inertia neglected (Glatzmaier and Roberts, 1995a), while that of Kageyama and Sato (1995) uses finite differences on a (longitude, latitude, radius) mesh, as discussed below.

Here we mention variations in the numerical approaches that have been used. Most studies have used spectral methods similar to that summarised above. The code used by Christensen *et al.* (1999) and others is similar to that of Glatzmaier and Roberts (1995a) but with explicit treatment of the Coriolis term. Different discretisations have been used in the radial direction, such as finite differences (Dormy *et al.*, 1998; Katayama *et al.*, 1999; Kuang and Bloxham, 1999), trigonometric series (Wicht and Busse, 1997) or non-linearly-mapped Chebyshev polynomials (Tilgner, 1999). The code used by (Wicht and Busse, 1997) calculates non-linear terms in spectral space and uses implicit time integration. A comparison of the accuracy of six spectral codes is given in the benchmark paper of Christensen *et al.* (2001).

Some codes have used a *Cartesian* rather than spherical geometry. The code used by Jones and Roberts (2000) and Rotvig and Jones (2002) employs a spectral method, with a Fourier expansion in the horizontal direction and Chebyshev collocation in the vertical direction. A finite-volume, multigrid method with fully implicit time stepping (even for advection terms) was implemented by Schmalzl and Hansen (2000) and used in Stellmach and Hansen (2004). Finite elements were used by Matsui and Buffett (2005) to investigate parametrisations of subgrid-scale processes.

There is increasing interest in developing grid-based geodynamo codes in spherical geometry, i.e. using local basis functions, because they scale better with problem size than the spectral transform method owing to the optimal scaling of the multigrid method (Section 6.3), compared to the $(\text{degree})^3$ scaling of the Legendre transform, which dominates computation time for the spherical spectral transform method at large problem size. A problem with grid-based methods is treating the boundary condition for magnetic field, which is usually done by either using a different boundary condition such as zero azimuthal magnetic field, or extending the grid into the external region and setting $\mathbf{B} = 0$ at a large distance from the surface.

A finite difference approximation on a (longitude, latitude, radius) mesh was used in the code of Kageyama and Sato (1995, 1997) and Li *et al.* (2002). Because of the converging grid lines at the poles, it was necessary to apply smoothing in these regions. This code was completely explicit and assumed that the fluid obeys the ideal gas equation. The magnetic field was taken to be radial only at the outer boundary, something that is referred to as 'quasi-vacuum conditions'. To avoid the pole problem, this method was subsequently implemented on the Yin-Yang grid (Appendix B), and run on the Earth Simulator supercomputer (Kageyama and Yoshida, 2005) with as many as 800 million grid points (Kageyama *et al.*, 2008), allowing the viscosity (i.e. Ekman number) to be reduced by an order of magnitude compared to previous models.

The finite volume method has been used in two spherical codes. Hejda and Reshatnyak (2003) used a (longitude, latitude, radius) grid, whereas Harder and Hansen (2005) used the 'cubed sphere' grid (Appendix B) to avoid the pole problem and obtain reasonably even resolution. In contrast to the usual staggered arrangement of velocity and pressure components (Section 3.5.1), Harder and Hansen (2005) used a collocated grid and a pressure weighted interpolation scheme (Rhie and Chow, 1983) to avoid checkerboard pressure oscillations. They also assumed that the magnetic field at the outer surface is radial only.

Two spherical dynamo codes use the finite element method. An isocahedral mesh was used by Chan *et al.* (2001; 2007), with the magnetic boundary condition treated by imposing $\mathbf{B} = 0$ on a spherical surface with a radius three times the radius of the core. A cubed sphere mesh was used by Matsui and Okuda (2004), with an external region extending to six times the outer core radius used to treat the magnetic field, while the grid extended to the centre of the Earth using a cube in the centre.

A spectral element method, in which a spectral expansion is used inside each element, was implemented by Fournier *et al.* (2005) to model rotating convection, though not yet the geodynamo. They used the spectral element approach in the radial and latitude directions, with a Fourier expansion in the longitude direction.

A

Appendix A
Definitions and relations from vector and matrix algebra

A1 Introduction

Computational methods for geodynamics use many basic elements of linear algebra. Mathematical descriptions of geodynamical problems can often be reduced to the problem of solving systems of linear algebraic equations. Among the problems that can be so treated are the solution of partial differential equations by finite difference methods. The use of matrix notation is not only convenient, but also extremely powerful, in bringing out fundamental relationships. In this section we introduce some basic notation, some of the important relations between vectors and matrices, and some simple properties of vector and matrix analysis.

A system of m linear algebraic equations with n unknowns has the general form

$$
\begin{aligned}
a_{11}x_1 + a_{12}x_2 + \cdots + a_{1n}x_n &= b_1, \\
a_{21}x_1 + a_{22}x_2 + \cdots + a_{2n}x_n &= b_2, \\
\cdots\cdots\cdots\cdots\cdots\cdots\cdots & \cdots\cdots \\
a_{m1}x_1 + a_{m2}x_2 + \cdots + a_{mn}x_n &= b_m.
\end{aligned}
\tag{A1}
$$

The coefficients a_{ij} $(i = 1, \ldots, m; j = 1, \ldots, n)$ and the right-hand sides b_i $(i = 1, \ldots, m)$ are known numbers. The numbers x_j $(j = 1, \ldots, n)$ are unknown, and should be determined in such a way that Eqs. (A1) are satisfied simultaneously.

A *matrix* is a rectangular array of numbers arranged in rows and columns. The coefficients of the system (A1) form a matrix

$$
\mathbf{A} = \begin{bmatrix} a_{11} & a_{12} & \ldots & a_{1n} \\ a_{21} & a_{22} & \ldots & a_{2n} \\ \vdots & \vdots & \ddots & \vdots \\ a_{m1} & a_{m2} & \ldots & a_{mn} \end{bmatrix}. \tag{A2}
$$

Sometimes the matrix notation is brief, namely, $\mathbf{A} = \{a_{ij}\}$. The matrix $\mathbf{A}$ has m rows and n columns, or $\mathbf{A}$ is of order $m \times n$. The matrix element a_{ij} is located at the intersection of the ith row and the jth column of $\mathbf{A}$.

A matrix with only one column is called a column vector (a matrix with only one row is called a row vector). The right-hand side constants b_i $(i = 1, \ldots, m)$ and the unknowns

$x_j (j = 1, \ldots, n)$ form column vectors:

$$\mathbf{b} = \begin{bmatrix} b_1 \\ b_2 \\ \vdots \\ b_m \end{bmatrix}, \qquad \mathbf{x} = \begin{bmatrix} x_1 \\ x_2 \\ \vdots \\ x_n \end{bmatrix}. \tag{A3}$$

A row vector can be presented as a transpose to a column vector, for example:

$$\mathbf{b}^T = [b_1, b_2, \ldots, b_m], \tag{A4}$$

where $\mathbf{b}^T$ denotes the *transposed vector*.

A2 Matrix multiplication

According to (A1)–(A3), the matrix **A** combined in a certain way with the one-column matrix (vector **x**) should equal the one-column matrix (vector **b**). The process of combining matrices involved here is called *matrix multiplication* and is defined, in general, as follows: If the matrix $\mathbf{C} = \{c_{ij}\}$ of order $m \times p$ is the product of matrix $\mathbf{A} = \{a_{ij}\}$ of order $m \times n$ and $\mathbf{B} = \{b_{ij}\}$ of order $n \times p$ (or $\mathbf{C} = \mathbf{A}\,\mathbf{B}$), then the elements of the matrix C are represented as

$$c_{ij} = \sum_{k=1}^{n} a_{ik} b_{kj} \qquad \text{for } i = 1, \ldots, m; j = 1, \ldots, p. \tag{A5}$$

Note that it when repeated subscripts, in this case k, exist in an expression, it is normally assumed that summation occurs over that subscript or subscripts, as is done in this book. Therefore Eq. (A5) could also be written simply as $c_{ij} = a_{ik} b_{kj}$, with the summation over k assumed. With this definition of matrix product, the set of linear algebraic equations (A1) can be written in the following form:

$$\mathbf{A}\,\mathbf{x} = \mathbf{b}. \tag{A6}$$

Matrix multiplication is associative operation: If **A**, **B**, and **C** are matrices of order $m \times n$, $n \times p$, $p \times q$, respectively, then $(\mathbf{A}\,\mathbf{B})\,\mathbf{C} = \mathbf{A}\,(\mathbf{B}\,\mathbf{C})$. The matrix $\mathbf{A} = \{a_{ij}\}$ equals the matrix $\mathbf{B} = \{b_{ij}\}$ $(\mathbf{A} = \mathbf{B})$, if the matrices have the same order and $a_{ij} = b_{ij}$ for all i and j.

A3 Transposed, square, and symmetric matrices. Determinants

The *transposed matrix* $\mathbf{A}^T$ arises from the matrix **A** by interchanging the column vectors and the row vectors: $\{a_{ji}\}^T = \{a_{ij}\}$. If **A** is a $m \times n$ matrix, $\mathbf{A}^T$ is $n \times m$ matrix. If **A** is an $n \times n$ matrix, it is referred to as a *square matrix* of order n. Elements of a square matrix lying on the diagonal from top left a_{11} to bottom right a_{nn} are called *main diagonal elements* of matrix **A**. A square matrix that equals its transpose matrix, is called a *symmetric matrix*. A symmetric matrix is mirror-symmetric about its main diagonal elements, namely, $a_{ij} = a_{ji}$ for all i and j.

The *determinant* of a square matrix **A**, denoted det **A** (or |A|), refers to both the collection of the elements of the square matrix, enclosed in vertical lines, and the scalar value represented by that array. Thus,

$$\det \mathbf{A} = \begin{vmatrix} a_{11} & a_{12} & \dots & a_{1n} \\ a_{21} & a_{22} & \dots & a_{2n} \\ \vdots & \vdots & \ddots & \vdots \\ a_{n1} & a_{n2} & \dots & a_{nn} \end{vmatrix}. \tag{A7}$$

The determinant can be presented by the following formula:

$$\det \mathbf{A} = \sum_{j=1}^{n} (-1)^{i+j} a_{ij} |A_{ij}|,$$

where $|A_{ij}|$ is the ijth minor, that is, the determinant of the matrix **A** obtained by deleting the ith row and the jth column of the matrix. For example, the value of the determinant of a 2 × 2 (and 3 × 3) matrix is calculated as the sum of the products of the matrix elements:

$$\det \begin{pmatrix} a_{11} & a_{12} \\ a_{21} & a_{22} \end{pmatrix} = (-1)^{1+1} a_{11} \det a_{22} + (-1)^{1+2} a_{12} \det a_{21} = a_{11}a_{22} - a_{12}a_{21},$$

$$\begin{aligned} \det \begin{pmatrix} a_{11} & a_{12} & a_{13} \\ a_{21} & a_{22} & a_{23} \\ a_{31} & a_{32} & a_{33} \end{pmatrix} &= a_{11} \det \begin{pmatrix} a_{22} & a_{23} \\ a_{32} & a_{33} \end{pmatrix} - a_{12} \det \begin{pmatrix} a_{21} & a_{23} \\ a_{31} & a_{33} \end{pmatrix} + a_{13} \det \begin{pmatrix} a_{21} & a_{22} \\ a_{31} & a_{32} \end{pmatrix} \\ &= a_{11}(a_{22}a_{33} - a_{23}a_{32}) - a_{12}(a_{21}a_{33} - a_{23}a_{31}) \\ &\quad + a_{13}(a_{21}a_{32} - a_{22}a_{31}). \end{aligned} \tag{A8}$$

A4 Upper and lower triangular matrices

An *upper triangular matrix* **U** is a matrix in which all elements below the main diagonal are equal to zero: $a_{ij} = 0$ for all $i > j$. A *lower triangular matrix* **L** is a matrix in which all elements above the main diagonal are equal to zero: $a_{ij} = 0$ for all $i < j$. We can denote the matrices as:

$$\mathbf{U} = \begin{bmatrix} a_{11} & a_{12} & \dots & a_{1n} \\ 0 & a_{22} & \dots & a_{21} \\ \vdots & \vdots & \ddots & \vdots \\ 0 & 0 & \dots & a_{nn} \end{bmatrix}, \quad \mathbf{L} = \begin{bmatrix} a_{11} & 0 & \dots & 0 \\ a_{21} & a_{22} & \dots & 0 \\ \vdots & \vdots & \ddots & \vdots \\ a_{n1} & a_{n2} & \dots & a_{nn} \end{bmatrix}. \tag{A9}$$

The transpose of an upper triangular matrix is a lower triangular matrix and vice versa: $\mathbf{U}^T = \mathbf{L}$, and $\mathbf{L}^T = \mathbf{U}$.

A square matrix in which all elements outside the main diagonal are zero is called a *diagonal matrix*. A diagonal matrix has the following properties: (i) the matrix is a right and left triangular matrix at the same time; (ii) the matrix is symmetric; and (iii) the transpose of a diagonal matrix is the same matrix. A square matrix is called a *tridiagonal matrix* if the matrix has non-zero elements only on the main diagonal a_{ij}, $i = j$ and on the subdiagonal immediately above and below the main diagonal a_{ij}, $i = j \pm 1$. A square matrix is called a *band matrix* if the matrix has non-zero elements on the main diagonal and on the diagonals lying between the main diagonal and k above the main diagonal and the diagonals lying between the main diagonal and l below the main diagonal. Elements of a band matrix outside the diagonal band are zero, namely, $a_{ij} = 0$ for $i - j > l, j - i > k, l < n$.

A5 Identity matrix. Inverse matrix. M-matrix

A square matrix is called an *identity* (or a *unit*) *matrix* if the elements of its main diagonal equal 1 and all other matrix elements equal 0:

$$\mathbf{I} = \begin{bmatrix} 1 & 0 & \dots & 0 \\ 0 & 1 & \dots & 0 \\ \vdots & \vdots & \ddots & \vdots \\ 0 & 0 & \dots & 1 \end{bmatrix}. \tag{A10}$$

The square matrix **A** of order n is invertible provided there is a square matrix **B** of order n such that

$$\mathbf{A\,B} = \mathbf{I} = \mathbf{B\,A}. \tag{A11}$$

The matrix **B** is referred to as the *inverse* of matrix **A** and is denoted by $\mathbf{A}^{-1}$. If two matrices **A** and **B** are invertible square matrices of the same order, then their product is also invertible: $(\mathbf{AB})^{-1} = \mathbf{B}^{-1}\mathbf{A}^{-1}$.

M-*matrix* is a matrix with non-positive off-diagonal entries and with the positive real parts of the matrix eigenvalues. **M**-matrix has to be non-singular and its inverse matrix non-negative. **M**-matrices are important in the consideration of rates of convergence of iterative methods for solving systems of linear algebraic equations. Note that if **A** is a $k \times k$ square matrix and there is a k-component vector $\mathbf{X} \neq 0$ such that $\mathbf{AX} = \lambda\mathbf{X}$ for some scalar λ, then the scalar λ is called the *eigenvalue* of matrix **A** with corresponding *eigenvector* **X**.

A6 Matrix addition and scalar multiplication

It is possible to multiply a matrix by a scalar (or a number) and to add two matrices of the same order in a reasonable way. Let $\mathbf{A} = \{a_{ij}\}$ and $\mathbf{B} = \{b_{ij}\}$ be matrices of the same order and d is a number. We say that **B** is the product of d with **A**, or $\mathbf{B} = d\mathbf{A}$, if $b_{ij} = da_{ij}$ for all

i and j. Further, let $\mathbf{C} = \{c_{ij}\}$ be a matrix of the same order as $\mathbf{A}$ and $\mathbf{B}$. We say that $\mathbf{C}$ is the sum of $\mathbf{A}$ and $\mathbf{B}$, or $\mathbf{C} = \mathbf{A} + \mathbf{B}$, if $c_{ij} = a_{ij} + b_{ij}$ for all i and j. Hence multiplication of a matrix by a number and addition of matrices is done entry by entry.

The following rules regarding these operations, and also matrix multiplication, are easily verified. Assume that $\mathbf{A}$, $\mathbf{B}$, and $\mathbf{C}$ are matrices such that all the sums and products mentioned below are defined, and let d and f be some numbers. Then

- $\mathbf{A} + \mathbf{B} = \mathbf{B} + \mathbf{A}$,
- $(\mathbf{A} + \mathbf{B}) + \mathbf{C} = \mathbf{A} + (\mathbf{B} + \mathbf{C})$,
- $d(\mathbf{A} + \mathbf{B}) = d\mathbf{A} + d\mathbf{B}$,
- $(d + f)\mathbf{A} = d\mathbf{A} + f\mathbf{A}$,
- $(\mathbf{A} + \mathbf{B})\mathbf{C} = \mathbf{AC} + \mathbf{BC}$,
- $\mathbf{A}(\mathbf{B} + \mathbf{C}) = \mathbf{AB} + \mathbf{AC}$,
- $d(\mathbf{AB}) = (d\mathbf{A})\mathbf{B} = \mathbf{A}(d\mathbf{B})$, and
- if $d \neq 0$ and $\mathbf{A}$ is invertible, then $a\mathbf{A}$ is invertible and $(a\mathbf{A})^{-1} = \frac{1}{a}\mathbf{A}^{-1}$.

A7 Linear combination of vectors

The above definitions (sums of matrices and products of numbers with matrices) allow the summation of vectors of n components and multiplication of these vectors by numbers. If $\mathbf{x}^{(1)}, \mathbf{x}^{(2)}, \ldots, \mathbf{x}^{(k)}$ are k n-component vectors and $b_1, b_2, \ldots, b_k$ are k numbers, then the weighted sum

$$b_1\mathbf{x}^{(1)} + b_2\mathbf{x}^{(2)} + \cdots + b_k\mathbf{x}^{(k)} \tag{A12}$$

is also an n-component vector, called the linear combination of $\mathbf{x}^{(1)}, \mathbf{x}^{(2)}, \ldots, \mathbf{x}^{(k)}$ with weights, or coefficients, $b_1, b_2, \ldots, b_k$.

A8 Vector and matrix norm

When a vector $\mathbf{v} = (v_1, v_2, v_3)$ belonging to the three-dimensional Euclidean R^3 space is interpreted geometrically, its length is given by $\|\mathbf{v}\| = (v_1^2 + v_2^2 + v_3^2)^{1/2}$. The well-known properties of vector lengths include:

(a) $\|\mathbf{v}\| > 0$ if $\mathbf{v} \neq 0$,
(b) $\|a\mathbf{v}\| = |a|\,\|\mathbf{v}\|$ for any scalar a and any vector $\mathbf{v}$,
(c) $\|\mathbf{v} + \mathbf{w}\| \leq \|\mathbf{v}\| + \|\mathbf{w}\|$ for any vectors $\mathbf{v}$ and $\mathbf{w}$.

We can extend the vector norm for the n-dimensional Euclidean R^n space: for any vector $\mathbf{v} = (v_1, v_2, \ldots, v_n)$, its length is given by $\|\mathbf{v}\| = (v_1^2 + v_2^2 + \cdots + v_n^2)^{1/2}$. In general, several different notions of the length (norm) of a vector may also be useful. The most commonly used *vector norms* in R^n are the following:

- Euclidean vector norm: $\|\mathbf{v}\|_2 = \left(\sum_{t-1}^{n} |v_i|^2\right)^{1/2}$,
- vector max norm: $\|\mathbf{v}\|_m = \max_i |v_i|$,
- vector sum norm: $\|\mathbf{v}\|_1 = \sum_{t-1}^{n} |v_i|$.

All of these expressions satisfies the conditions (a)–(c), and hence are true vector norms.

Analogous to vector norms is the *matrix norm* $\|\mathbf{A}\|$, which is a scalar measure of square matrix $\mathbf{A}$. A matrix norm must satisfy the following conditions:

(A) $\|\mathbf{A}\| > 0$ if $\mathbf{A} \neq 0$,
(B) $\|a\mathbf{A}\| = |a|\,\|\mathbf{A}\|$ for any scalar a and any matrix $\mathbf{A}$,
(C) $\|\mathbf{A}+\mathbf{B}\| \leq \|\mathbf{A}\| + \|\mathbf{B}\|$ for any matrices $\mathbf{A}$ and $\mathbf{B}$,
(D) $\|\mathbf{AB}\| \leq \|\mathbf{A}\|\;\|\mathbf{B}\|$ for any matrices $\mathbf{A}$ and $\mathbf{B}$.

Although there are many ways to define matrix norms, it is especially useful to use matrix norms that are connected to existing vector norms. Namely, the following two norms are used frequently in matrix analysis:

- Euclidean matrix norm: $\|\mathbf{A}\|_2 = \left(\sum_{i,j=1}^{n} |a_{ij}|^2\right)^{1/2}$,
- matrix sum norm: $\|\mathbf{A}\|_1 = \sum_{i,j=1}^{n} |a_n|$.

A9 Divergence, gradient, curl, and Laplacian operators

We define initially the scalar and vector products.

The multiplication of two n-vectors $\mathbf{a}$ and $\mathbf{b}$ resulting in a scalar (real number) is called the *scalar product* (denoted by $\cdot$): $\mathbf{a} \cdot \mathbf{b} = |\mathbf{a}||\mathbf{b}|\cos\alpha$, where $|\mathbf{a}|$ and $|\mathbf{b}|$ are the vector magnitudes, and α is the angle between the vectors. We can represent this multiplication by components in the following form:

$$\mathbf{a}^T \cdot \mathbf{b} = (a_1, a_2, \ldots, a_n) \cdot \begin{pmatrix} b_1 \\ b_2 \\ \vdots \\ b_n \end{pmatrix} = a_1b_1 + a_2b_2 + \cdots + a_nb_n. \qquad \text{(A13)}$$

The vector product (defined only in 3-D space) of two vectors $\mathbf{a}$ and $\mathbf{b}$ is a vector $\mathbf{c}$ and is denoted by $\times$. The magnitude of $\mathbf{c} = \mathbf{a} \times \mathbf{b}$ is the area of the parallelogram spanned by vectors $\mathbf{a}$ and $\mathbf{b}$: $|\mathbf{a} \times \mathbf{a}| = |\mathbf{a}| \cdot |\mathbf{b}| \cdot |\sin\alpha|$. The direction of the vector $\mathbf{c} = \mathbf{a} \times \mathbf{b}$ is perpendicular to the plane spanned by vectors **a** and **b**. We can represent the vector product by vector components:

$$\mathbf{a} \times \mathbf{b} = (a_1\mathbf{i} + a_2\mathbf{j} + a_3\mathbf{k}) \times (b_1\mathbf{i} + b_2\mathbf{j} + b_3\mathbf{k})$$
$$= (a_2b_3 - a_3b_2)\mathbf{i} + (a_3b_1 - a_1b_3)\mathbf{j} + (a_1b_2 - a_2b_1)\mathbf{k} = \det\begin{pmatrix} \mathbf{i} & a_1 & b_1 \\ \mathbf{j} & a_2 & b_2 \\ \mathbf{k} & a_3 & b_3 \end{pmatrix}, \tag{A14}$$

where **i**, **j** and **k** are unit vectors in the three coordinate directions.

The following properties of the scalar and vector products are important:

(1) $(\mathbf{a} \times \mathbf{b}) \cdot \mathbf{c} = (\mathbf{b} \times \mathbf{c}) \cdot \mathbf{a} = (\mathbf{c} \times \mathbf{a}) \cdot \mathbf{b}$,
(2) $(\mathbf{a} \times \mathbf{b}) \times \mathbf{c} = (\mathbf{c} \cdot \mathbf{a})\mathbf{b} - (\mathbf{b} \cdot \mathbf{c})\mathbf{a}$, and
(3) $\mathbf{a} \times (\mathbf{b} \times \mathbf{c}) = (\mathbf{c} \cdot \mathbf{a})\mathbf{b} - (\mathbf{a} \cdot \mathbf{b})\mathbf{c}$.

Let a scalar P or a vector **a** be defined at any point of a domain with volume V bounded by surface S. Let **n** be the unit normal on this surface, directed outwards. Assume that P or **a** are continuous functions of the coordinates with finite first derivatives. Then a scalar function div **a**, and vector functions ∇P (or grad P) and curl **a** can be defined in the following way:

$$\iiint_V \operatorname{div} \mathbf{a}\, dV = \iint_S \mathbf{a} \cdot \mathbf{n}\, dS, \tag{A15}$$

$$\iiint_V \nabla P\, dV = \iint_S P\mathbf{n}\, dS, \tag{A16}$$

and

$$\iiint_V \operatorname{curl} \mathbf{a}\, dV = \iint_S \mathbf{n} \times \mathbf{a}\, dS. \tag{A17}$$

In Cartesian coordinates (x_1, x_2, x_3), we can express the divergence, gradient, and curl via components of the vector $\mathbf{a} = (a_1, a_2, a_3)$ and scalar P:

$$\operatorname{div} \mathbf{a} = \nabla \cdot \mathbf{a} = \partial a_1/\partial x_1 + \partial a_2/\partial x_2 + \partial a_3/\partial x_3, \tag{A18}$$

$$\nabla P \text{ has components } \partial P/\partial x_1, \partial P/\partial x_2, \partial P/\partial x_3, \text{ and} \tag{A19}$$

$$\operatorname{curl} \mathbf{a} = \nabla \times \mathbf{a} \text{ has components } \frac{\partial a_3}{\partial x_2} - \frac{\partial a_2}{\partial x_3}, \frac{\partial a_1}{\partial x_3} - \frac{\partial a_3}{\partial x_1}, \frac{\partial a_2}{\partial x_1} - \frac{\partial a_1}{\partial x_2}. \tag{A20}$$

We present below the identities, which are important in vector analysis:

$$\operatorname{div}(\operatorname{curl}\mathbf{a}) = \nabla\cdot(\nabla\times\mathbf{a}) = 0, \tag{A21}$$

$$\operatorname{curl}(\nabla P) = \nabla\times(\nabla P) = 0, \tag{A22}$$

$$\operatorname{div}(P\mathbf{a}) = \nabla\cdot(P\mathbf{a}) = P\nabla\cdot\mathbf{a} + \mathbf{a}\cdot\nabla P, \tag{A23}$$

$$\operatorname{curl}(P\mathbf{a}) = \nabla\times(P\mathbf{a}) = P\nabla\times\mathbf{a} + \nabla P\times\mathbf{a}, \text{ and} \tag{A24}$$

$$\operatorname{div}(\mathbf{a}\times\mathbf{b}) = \nabla\cdot(\mathbf{a}\times\mathbf{b}) = (\nabla\times\mathbf{a})\cdot\mathbf{b} - \mathbf{a}\cdot(\nabla\times\mathbf{b}). \tag{A25}$$

The divergence of a gradient is the Laplacian:

$$\text{div grad P} = \Delta P = (\nabla\cdot\nabla)P = \nabla^2 P = \frac{\partial^2 P}{\partial x_1^2} + \frac{\partial^2 P}{\partial x_2^2} + \frac{\partial^2 P}{\partial x_3^2}. \tag{A26}$$

The biharmonic operator (the Laplacian of the Laplacian) may be written as

$$\nabla^4 = (\nabla\cdot\nabla)(\nabla\cdot\nabla) = \frac{\partial^4}{\partial x_1^4} + \frac{\partial^4}{\partial x_2^4} + \frac{\partial^4}{\partial x_3^4} + 2\frac{\partial^4}{\partial x_2^2\partial_3^2} + 2\frac{\partial^4}{\partial x_3^2\partial x_1^2} + 2\frac{\partial^4}{\partial x_1^2\partial x_2^2}. \tag{A27}$$

B Appendix B
Spherical coordinates

B1 Spherical grids

The main problem with modelling a sphere is that a simple (longitude, latitude) grid contains a singularity at the poles, where the grid lines of longitude converge. While this does not matter when the equations are solved and time-stepped in spectral space by the spectral transform method (Chapter 5), when using a spatial discretisation this grid convergence gives severely non-uniform resolution, can lead to convergence problems with iterative solvers and forces a small time step if using explicit time-stepping. Although some codes have successfully used such a grid (see Zebib *et al.*, 1980; Kageyama and Sato, 1995; Iwase and Honda, 1997; Trubitsyn and Rykov, 2000), it is far from optimal. An optimal grid would have approximately uniform resolution over the spherical surface, and if it is desired to use a finite difference (FD) or straightforward finite volume (FV) approach, the grid lines should be orthogonal. Various grids have been used and are reviewed here and illustrated in Fig. B1.

Several grids arise from projecting Platonic solids onto a sphere then subdividing each face. One the earliest such grids is the *isocahedral grid* (Fig. B1a,b), as utilised in the FE code TERRA (Baumgardner, 1985; Baumgardner and Frederickson, 1985; Baumgardner, 1988). The resulting cells or elements are triangular on the 2-D sphere; TERRA uses elements that are triangular prisms in 3-D.

Projecting a cube onto a sphere then subdividing each of the six patches using quadrilateral cells results in the *cubed sphere grid* (Ronchi *et al.*, 1996), shown in Fig. B1e. The patches mesh cleanly, although there is an abrupt change in the grid direction at the boundaries. In the basic version of the cubed sphere grid, the grid lines subdividing each patch are great circles and the grid lines are not orthogonal. This is no problem if using a finite element (FE) or spectral element discretisation, as done for global seismic wave propagation (Komatitsch and Tromp, 2002) by extending the mesh to the centre of the Earth and by inserting a cube at the centre. If an FD discretisation is being used, however, the non-orthogonality of the mesh results in many additional terms in the discretised equations when expressed in terms of local coordinates. Nevertheless, successful FD implementations exist for constant viscosity (Hernlund and Tackley, 2003) and for variable viscosity (Choblet, 2005). Stemmer *et al.* (2006) found a modified cubed sphere grid in which the grid lines are almost orthogonal and the cross terms can be neglected. Their implementation uses a collocated grid, with all variables defined at the same nodes, rather than the more conventional staggered grid. Use of a collocated grid is more convenient provided care is taken with the treatment of pressure.

An FE mesh that can be regarded as subdividing a projected *octahedron* was used by Suzuki *et al.* (1999), Tabata and Suzuki (2000), and Tabata (2006) (Fig. B1c). Each face of the octahedron is divided into six triangles, which are further subdivided using used tetrahedral elements.

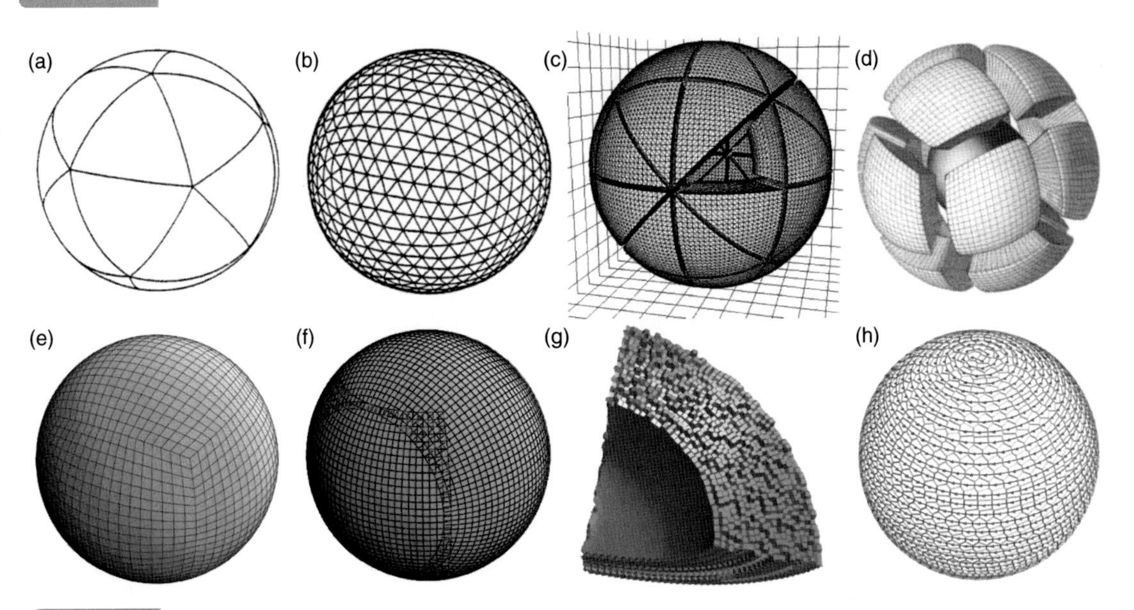

Fig. B1. Spherical meshes. (a) An isocahedron projected onto a sphere, and (b) subdivided eight ways in each direction (reprinted from Baumgardner, 1985). (c) The mesh of Tabata and Suzuki (2000), which can be regarded as subdividing a projected octahedron, first into six triangular blocks for each face then further using tetrahedral finite elements (the figure taken from Tabata, 2006). (d) The mesh used by CitComS, which can be regarded as subdividing each face of a projected tetrahedron into three rhombohedral blocks that are then further subdivided (the figure reprinted from Zhong *et al.*, 2000). (e) The cubed sphere grid (figure reprinted from Hernlund and Tackley, 2003). (f) The Yin-Yang grid (Kageyama and Sato, 2004). (g)(h) The spiral grid (from Huettig and Stemmer, 2008b).

Another mesh is utilised in the CitComS code developed by Zhong *et al.* (2000) from the original Cartesian CitCom (Moresi and Solomatov, 1995). This mesh can be regarded as a subdivided projected *tetrahedron*. Each of the four faces of the projected tetrahedron is divided into three rhomboidal blocks, resulting in twelve rhomboidal blocks, which are then further subdivided using rhomboidal finite elements (Fig. B1d).

A relatively new spherical grid that is unrelated to a Platonic solid is the *Yin-Yang grid* (Kageyama and Sato, 2004), which consists of two (longitude, latitude) patches, each covering $\pm 45°$ in latitude and $270°$ in longitude, meshed together like the two halves of a tennis ball (Fig. B1f). Unlike the grids discussed above, the two patches do not mesh cleanly, but rather overlap by 6% of the total surface area for the basic version, although it is possible to minimise this overlap by deleting grid points in the overlapping region (Kageyama and Sato, 2004). Each patch provides the boundary condition for the other patch, with interpolation being required to obtain the new edge values after each update. The major advantage of the Yin-Yang grid over the grids discussed above is that the grid lines are orthogonal, so that a simple FD or FV approximation with the equations written in spherical polar coordinates (Section B2) can be used. At the time of writing this grid has been successfully implemented in three different mantle convection codes (Yoshida and Kageyama, 2004; Kameyama *et al.*, 2008; Tackley, 2008), all of which use a staggered-grid finite difference/finite volume discretisation, with the latter two using multigrid solvers.

A novel grid, *the spiral grid*, has been proposed and implemented by Huettig and Stemmer (2008b) and is shown in Fig. B1g,h. This grid has arbitrarily selectable lateral and radial resolution owing to the complete removal of symmetries, and can be refined in some areas as required. After constructing the grid points using a spiral, cells are constructed using a Voronoi diagram. Huettig and Stemmer (2008a) show how a variable-viscosity FV scheme can be implemented with arbitarily shaped finite volumes; this approach is used by Huettig and Stemmer (2008b) to solve the variable viscosity Stokes flow using the spiral grid in the spherical shell.

A final approach to modelling a sphere or a spherical shell is to *embed a sphere in a Cartesian mesh*. This approach has the advantages of simplicity and extending to the centre of the planet without a singularity, but the disadvantages that a large fraction of the grid cells are not used, and that the surface of the sphere cuts through grid cells, requiring some special treatment. This approach has been applied to model convection inside giant planets both in 2-D (Evonuk and Glatzmaier, 2006), for which the effective geometry of the planet is cylindrical, and in 3-D (Evonuk, 2006), to model 3-D convection in the Earth's mantle (Kotelkin and Lobkovsky, 2007), and to model core–mantle differentiation (core formation) both in 2-D (Golabek *et al.*, 2009; Lin *et al.*, 2009) and 3-D (Honda *et al.*, 1993), and is implemented in the 2-D and 3-D codes of Gerya and Yuen (2007). For the problems of core formation and giant planet convection, it is useful that the grid extends naturally to the centre of the planet.

In the radial direction, most geodynamic codes using a spherical mesh simply repeat the same azimuthal mesh at each radius, which means that the azimuthal grid spacing becomes larger with increasing radius. To compensate for this, one method is to halve the angular grid spacing at some radius, as done in the finite element discretisations of Tabata and Suzuki (2000), Davies (2008) and Wolstencroft *et al.* (2009) (based on TERRA), and the spectral element discretisation of Komatitsch and Tromp (2002). The spiral grid naturally avoids this problem, as does the Cartesian mesh-embedded sphere.

B2 Vector operators, strain rates and stress equations in spherical polar coordinates

While some spherical codes use equations expressed in Cartesian coordinates, others use equations expressed in the orthogonal curvilinear spherical coordinates of radius r, latitude θ, and longitude ϕ. To derive these equations it is useful to know the form of the gradient, divergence and curl operators:

$$\nabla f = \frac{\partial f}{\partial r}\mathbf{e}_r + \frac{1}{r}\frac{\partial f}{\partial \theta}\mathbf{e}_\theta + \frac{1}{r\sin\theta}\frac{\partial f}{\partial \phi}\mathbf{e}_\phi$$

$$\nabla \cdot \mathbf{u} = \frac{1}{r^2}\frac{\partial}{\partial r}(r^2 u_r) + \frac{1}{r\sin\theta}\frac{\partial}{\partial \theta}(u_\theta \sin\theta) + \frac{1}{r\sin\theta}\frac{\partial}{\partial \phi}u_\phi \qquad \text{(B1)}$$

$$\nabla \times \mathbf{u} = \mathbf{e}_r \frac{1}{r \sin\theta}\left[\frac{\partial}{\partial\theta}(u_\phi \sin\theta) - \frac{\partial u_\theta}{\partial\phi}\right] + \mathbf{e}_\theta \frac{1}{r}\left[\frac{1}{\sin\theta}\frac{\partial u_r}{\partial\phi} - \frac{\partial}{\partial r}(r u_\phi)\right]$$
$$+ \mathbf{e}_\phi \frac{1}{r}\left[\frac{\partial}{\partial r}(r u_\theta) - \frac{\partial u_r}{\partial\theta}\right],$$

where f is a scalar field, $\mathbf{e}_r$, $\mathbf{e}_\theta$, $\mathbf{e}_\phi$ are the unit vectors in the respective directions, and $\mathbf{u}$ is a vector field. This leads to an incompressible continuity equation:

$$\nabla \cdot \mathbf{v} = \frac{1}{r^2}\frac{\partial}{\partial r}(r^2 v_r) + \frac{1}{r\sin\theta}\frac{\partial}{\partial\theta}(\sin\theta v_\theta) + \frac{1}{r\sin\theta}\frac{\partial}{\partial\phi}v_\phi = 0, \tag{B2}$$

where v is velocity. In the momentum (Stokes) equation, the three components of the divergence of total stress become

$$(\nabla\cdot\sigma)_r = -\frac{\partial p}{\partial r} + \frac{1}{r^2}\frac{\partial}{\partial r}(r^2\tau_{rr}) + \frac{1}{r\sin\theta}\frac{\partial}{\partial\theta}(\tau_{r\theta}\sin\theta) + \frac{1}{r\sin\theta}\frac{\partial\tau_{r\phi}}{\partial\phi} - \frac{\tau_{\theta\theta}+\tau_{\phi\phi}}{r}$$
$$(\nabla\cdot\sigma)_\theta = -\frac{1}{r}\frac{\partial p}{\partial\theta} + \frac{1}{r^2}\frac{\partial}{\partial r}(r^2\tau_{r\theta}) + \frac{1}{r\sin\theta}\frac{\partial}{\partial\theta}(\tau_{\theta\theta}\sin\theta)$$
$$+ \frac{1}{r\sin\theta}\frac{\partial\tau_{\phi\theta}}{\partial\phi} + \frac{1}{r}(\tau_{r\theta} - \tau_{\phi\phi}\cot\theta) \tag{B3}$$
$$(\nabla\cdot\sigma)_\phi = -\frac{1}{r\sin\theta}\frac{\partial p}{\partial\phi} + \frac{1}{r^2}\frac{\partial}{\partial r}(r^2\tau_{r\phi}) + \frac{1}{r\sin\theta}\frac{\partial}{\partial\theta}(\tau_{\theta\phi}\sin\theta)$$
$$+ \frac{1}{r\sin\theta}\frac{\partial\tau_{\phi\phi}}{\partial\phi} + \frac{1}{r}(\tau_{r\phi} + \tau_{\theta\phi}\cot\theta),$$

where σ is the total stress tensor, and τ is the deviatoric stress tensor. The components of deviatoric stress tensor for incompressible deformation are given by:

$$\tau_{rr} = 2\eta\frac{\partial v_r}{\partial r},\ \tau_{\theta\theta} = 2\eta\left(\frac{1}{r}\frac{\partial u_\theta}{\partial\theta} + \frac{v_r}{r}\right),\ \tau_{\phi\phi} = 2\eta\left(\frac{1}{r\sin\theta}\frac{\partial v_\phi}{\partial\phi} + \frac{v_r}{r} + \frac{v_\theta\cot\theta}{r}\right),$$
$$\tau_{r\theta} = \eta\left(r\frac{\partial}{\partial r}\left(\frac{v_\theta}{r}\right) + \frac{1}{r}\frac{\partial v_r}{\partial\theta}\right),\ \tau_{r\phi} = \eta\left(\frac{1}{r\sin\theta}\frac{\partial v_r}{\partial\phi} + r\frac{\partial}{\partial r}\left(\frac{v_\phi}{r}\right)\right), \tag{B4}$$
$$\tau_{\theta\phi} = \eta\left(\frac{\sin\theta}{r}\frac{\partial}{\partial\theta}\left(\frac{v_\phi}{\sin\theta}\right) + \frac{1}{r\sin\theta}\frac{\partial v_\theta}{\partial\phi}\right).$$

The appropriate terms in the energy equation can be straightforwardly derived; for example the diffusion term is given by

$$\nabla\cdot(\kappa\nabla T) = \frac{1}{r^2}\left[\frac{\partial}{\partial r}\left(r^2\kappa\frac{\partial T}{\partial r}\right) + \frac{1}{\sin\theta}\frac{\partial}{\partial\theta}\left(\kappa\sin\theta\frac{\partial T}{\partial\theta}\right) + \frac{1}{\sin^2\theta}\frac{\partial}{\partial\phi}\left(\kappa\frac{\partial T}{\partial\phi}\right)\right], \tag{B5}$$

where κ is the coefficient of thermal diffusivity and T is temperature.

B3 Two-dimensional approximations of spherical geometry

Because of the high computational expense of 3-D spherical calculations, 2-D calculations are often attractive. It is desirable that the influences of curvature, such as the relative surface areas of the outer and inner boundaries (for the mantle, the surface and core–mantle boundary) and the surface to volume ratio, are correctly accounted for in 2-D calculations.

Axisymmetric sphere. The oldest and probably most frequently used 2-D approximation of 3-D spherical geometry (dating back to at least Zebib *et al.* (1983) in the geodynamics community) is to assume symmetry around the polar axis, i.e. all variables are assumed to be constant in the ϕ direction so all terms involving ϕ are dropped. This naturally accounts for geometrical and curvature effects, but has the disadvantage that when projected in three dimensions, the planform of upwellings and downwellings depends on their latitude, which may force some unnatural behaviour. Specifically, upwellings/downwellings at the north or south pole are plume-like, thereas those away from the poles are conical sheets.

Rescaled cylinder. Cylindrical geometry with the actual inner and outer boundary radii gives an incorrect ratio of the surface areas of the inner and outer boundaries. This can be corrected by reducing the radii of both the inner and outer boundaries while keeping the depth of the domain constant, as proposed by (van Keken, 2001) and used in several subsequent papers (see, for example, van Keken *et al.*, 2001; Nakagawa and Tackley, 2004).

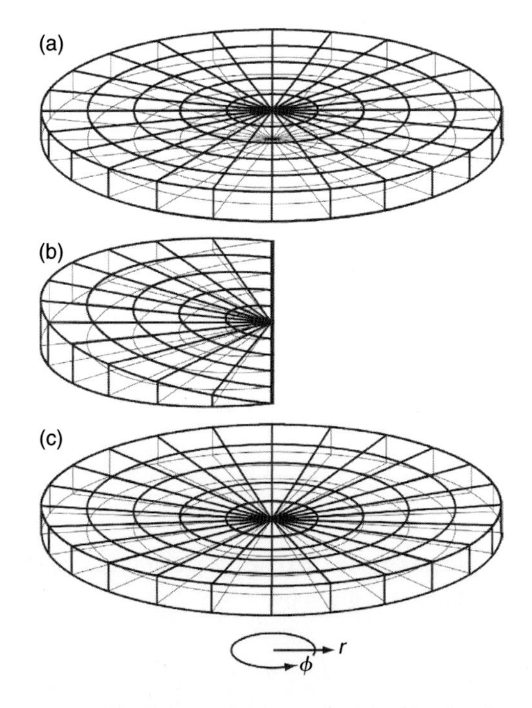

Fig. B2. A comparison of the effective cell or element thickness implied by the three 2-D approximations of 3-D spherical geometry: (a) cylindrical, (b) axisymmetric sphere, (c) spherical annulus (modified from Hernlund and Tackley, 2008).

This rescaling also gives approximately the correct ratio of mantle volume to surface area. A disadvantage is that the inner boundary is smaller than realistic.

Spherical annulus. Hernlund and Tackley (2008) proposed modelling convection using a spherical annulus, i.e. a 2-D slice around the equator, with the thickness of the slice proportional to radius. Practically, this involves using the equations expressed in spherical polar coordinates with all θ-derivatives and θ-velocities dropped and $\theta = \pi/2$. Like the axisymmetric sphere, this approach obtains the correct surface areas and geometry.

These three approaches are compared in Fig. B2, which shows the variation in effective thickness of cells or elements. For cylindrical geometry (Fig. B2a) the cells have the same thickness everywhere, while for an axisymmetric sphere (Fig. B2b) they become thinner towards the poles (like slices of an orange), whereas for a spherical annulus (Fig. B2c) they become thinner towards the centre. Different pros and cons are associated with these different approximations. Hernlund and Tackley (2008) discuss this, and present example convection calculations showing that the behaviour of convection in the spherical annulus compares favourably to calculations performed in rescaled cylindrical and axisymmetric spherical geometries when measured relative to calculations in a fully 3-D spherical shell.

Appendix C
Freely available geodynamic modelling codes

To an increasing extent, and mainly because of various large initiatives or organisations like CIG (Computational Infrastructure in Geodynamics), funded by the US National Science Foundation, the Earth Simulator Project, funded by the Japanese government, and AuScope, funded by the Australian government, sophisticated geodynamic codes are becoming freely available for online download, under various licences. Here we give a list of the codes available at the time of writing, including where they are available from. This list is a snapshot of the situation in October 2009; as this changes with time, no guarantee is made about its future validity. It does not include commercial codes that are used in the geodynamic community, nor those that are not openly distributed. Descriptions are taken from the online documentation, where available. The titles of the codes are listed here in an alphabetic order.

CitComCU. A finite element parallel code capable of modelling thermo-chemical convection in a 3-D domain appropriate for convection within the Earth's mantle. Developed from CitCom (Moresi and Solomatov, 1995; Moresi *et al.*, 1996). Available from CIG (http://www.geodynamics.org).

CitComS. A finite element code designed to solve thermal convection problems relevant to Earth's mantle in 3-D spherical geometry, developed from CitCom by Shijie Zhong (Zhong *et al.*, 2000). Available from CIG (http://www.geodynamics.org).

ConMan. A finite element program for the solution of the equations of incompressible, infinite-Prandtl number convection in two dimensions, originally written by Scott King, Arthur Raefsky and Brad Hager (King *et al.*, 1990). Available from CIG (http://www.geodynamics.org).

Ellipsis3D. A 3-D particle-in-cell finite element solid modelling code for visco-elastoplastic materials, as described in O'Neill *et al.* (2006). Available from CIG (http://www.geodynamics.org).

Gale. An Arbitrary Lagrangian Eulerian (ALE) code that solves problems related to orogenesis, rifting, and subduction with coupling to surface erosion models. This is an application of the Underworld platform listed below. Available from CIG (http://www.geodynamics.org).

GeoFEM. Developed as part of the Earth Simulator Project in Japan, GeoFEM has two standpoints: GeoFEM/Tiger, which is multi-purpose parallel finite-element software that can be applied to various fields in engineering and sciences, and GeoFEM/Snake, a software system optimised for the Earth Simulator and specialised for the simulation of solid Earth phenomena such as mantle–core convection, plate tectonics, seismic wave propagation and their coupled phenomena. See Okuda *et al.* (2003). Available from http://geofem.tokyo.rist.or.jp.

HC. A global mantle circulation solver following Hager and O'Connell (1981) which can compute velocities, tractions, and geoid for simple density distributions and plate velocities. Available from CIG (http://www.geodynamics.org).

LAYER. A finite element code developed over the years by Maria Zuber of the Massachusetts Institute of Technology and her students and postdocs. It models a 2-D rectangular cross-section through the lithosphere and the basic setup is for modelling shortening or extension. The rheology is visco-plastic with strain rate weakening (Neumann and Zuber, 1995) and strain weakening (Delescluse *et al.*, 2008). A version maintained by Laurent Montesi at the University of Maryland is available at: http://www.geol.umd.edu/~montesi/Geodynamics/LAYER.shtml.

MAG. A serial version of a rotating spherical convection/magnetoconvection/dynamo code that solves the non-dimensional Boussinesq equations for time-dependent thermal convection in a rotating spherical shell filled with an electrically conducting fluid, developed by Gary Glatzmaier and modified by Uli Christensen and Peter Olson. Available from CIG (http://www.geodynamics.org).

MILAMIN. A finite element method implementation in (almost) native MATLAB that is capable of modelling viscous flow with one million degrees of freedom per minute on a normal computer. This includes pre-processing, solving and post-processing. For details see Dabrowski *et al.* (2008). Available from http://milamin.org/.

PyLith (formerly known as **LithMop**). A finite element code for the solution of visco-elastic/plastic deformation that was designed for lithospheric modeling problems. Available from CIG (http://www.geodynamics.org).

SNAC. SNAC (StGermaiN Analysis of Continua) is an updated Lagrangian explicit finite difference code for modelling a finitely deforming elasto-visco-plastic solid in 3-D. Available from CIG (http://www.geodynamics.org).

Underworld. A 3-D-parallel geodynamic modelling framework capable of deriving viscous/visco-plastic thermal, chemical and thermo-chemical models consistent with tectonic processes, such as mantle convection and lithospheric deformation over long time scales. It utilises a Lagrangian particle-in-cell finite element scheme (the prototype of which is the Ellipsis code). The core set of phenonema modelled and distributed with Underworld include lithospheric deformation, slab subduction and mantle convection. Underworld is (at the time of writing) under collaborative development by Monash University and the Victoria Partnership for Advanced Computing (VPAC), as part of the Victorian node of the NCRIS AuScope 'Simulation, Analysis, Modelling' capability. Available from http://www.underworldproject.org/.

In addition to the complete modelling programs or packages listed above, we also list below some useful solvers that can be used in the construction of one's own code. There are many free general-purpose linear algebra packages available; those listed below are selected because they seem to have the greatest use in the geodynamics community at the time of writing.

Cigma. The CIG Model Analyzer (Cigma) is a suite of tools that facilitates the comparison of numerical models, and performs error analysis, benchmarking, and code verification. Available from CIG (http://www.geodynamics.org).

FEniCS. FEniCS is free software for automated solution of differential equations. It provides software tools for working with computational meshes, finite element variational formulations of PDEs, ODE solvers and linear algebra. Available from http://www.fenics.org/wiki/FEniCS_Project.

Geodynamics AMR Suite. The Geodynamics AMR Suite (deal.II) is a C++ program library targeted at the computational solution of partial differential equations using adaptive finite elements. Available from CIG (http://www.geodynamics.org).

Pardiso. A thread-safe, high-performance, robust, memory efficient and easy to use software for solving large sparse symmetric and unsymmetric linear systems of equations on shared memory multiprocessors. Available from http://www.pardiso-project.org/index.html.

PetSc. The Portable, Extensible Toolkit for Scientific Computation is a suite of data structures and routines for the scalable (parallel) solution of scientific applications modelled by partial differential equations. Available from (http://www.mcs.anl.gov/petsc).

References

Adams, R. A. (1975). *Sobolev Spaces*. Pure and Applied Mathematics Series, vol. 65. New York: Academic Press.

Aharonov, E., Spiegelman, M. and Kelemen, P. (1997). Three-dimensional flow and reaction in porous-media – implications for the Earth's mantle and sedimentary basins. *Journal of Geophysical Research*, **102**, 14 821–14 833.

Albers, M. (2000). A local mesh refinement multigrid method for 3-D convection problems with strongly variable viscosity. *Journal of Computational Physics*, **160**, 126–150.

Albers, M. and Christensen, U. R. (1996). The excess temperature of plumes rising from the core-mantle boundary. *Geophysical Research Letters,* **23**, 3567–3570.

Alekseev, A. K. and Navon, I. M. (2001). The analysis of an ill-posed problem using multiscale resolution and second order adjoint techniques. *Computational Methods in Applied Mechanics and Engineering,* **190**, 1937–1953.

Allgower, E. and Georg, K. (eds.) (1990). *Computational Solution of Nonlinear Systems of Equations*. Providence, RI: American Mathematical Society.

Ames, W. F. (1965). *Nonlinear Partial Differential Equations in Engineering*, vol. 1. New York: Academic Press.

Ames, W. F. (1972). *Nonlinear Partial Differential Equations in Engineering*, vol. 2. New York: Academic Press.

Auth, C. and Harder, H. (1999). Multigrid solution of convection problems with strongly variable viscosity. *Geophysical Journal International*, **137**, 793–804.

Auth, C., Bercovici, D. and Christensen, U. R. (2003). Two-dimensional convection with a self-lubricating, simple-damage rheology. *Geophysical Journal International*, **154**, 783–800.

Axelsson, O. (1996). *Iterative Solution Methods*. Cambridge: Cambridge University Press.

Babeyko, A. Y., Sobolev, S. V., Trumbull, R. B., Oncken, O. and Lavier, L. L. (2002). Numerical models of crustal-scale convection and partial melting beneath the Altiplano-Puna plateau. *Earth and Planetary Science Letters*, **199**, 373–388.

Babuska, I. and Rheinboldt, C. (1978). A-posteriori error estimates for the finite element method. *International Journal for Numerical Methods in Engineering*, **12**, 1597–1615.

Babuska, I. and Rheinboldt, C. (1979). Adaptive approaches and reliability estimations in finite element analysis. *Computer Methods in Applied Mechanics and Engineering*, **17/18**, 519–540.

Babuska, I., Chandra, J. and Flaherty, J. E. (eds.) (1983). *Adaptive Computational Methods for Partial Differential Equations*. Philadelphia: SIAM.

Babuska, I., Zienkiewicz, O. C., Gago, J. *et al.* (eds.) (1986). *Accuracy Estimates and Adaptive Refinements in Finite Element Computations*. Chichester: John Wiley & Sons.

Babuska, I., Flaherty, J. E., Henshaw, W. D. *et al.* (eds.) (1995). *Modeling, Mesh Generation, and Adaptive Numerical Methods for Partial Differential Equations*. The IMA Volumes in Mathematics and its Applications, vol. 75. New York: Springer.

Badro, J., Rueff, J.-P., Vanko, G. *et al.* (2004). Electronic transitions in perovskite: Possible nonconvecting layers in the lower mantle. *Science,* **305**, 383–386.

Balachandar, S. and Yuen, D. A. (1994). 3-Dimensional fully spectral numerical-method for mantle convection with depth-dependent properties. *Journal of Computational Physics*, **113**, 62–74.

Balachandar, S., Yuen, D. A. and Reuteler, D. (1993). Viscous and adiabatic heating effects in 3-dimensional compressible convection at infinite Prandtl number. *Physics of Fluids, a-Fluid Dynamics*, **5**, 2938–2945.

Balachandar, S., Yuen, D. A. and Reuteler, D. M. (1995). Localization of toroidal motion and shear heating in 3-D high Rayleigh number convection with temperature-dependent viscosity. *Geophysical Research Letters*, **22**, 477–480.

Balla, Z. (1987). Tertiary paleomagnetic data for the Carpatho-Pannonian region in the light of Miocene rotation kinematics. *Tectonophysics*, **139**, 67–98.

Barney, B. (2009) *Introduction to Parallel Computing*. Livermore: Lawrence Livermore National Laboratory (https://computing.llnl.gov/tutorials/parallel_comp/, accessed on 1 March 2009).

Bashforth, F. and Adams, J. D. (1883). *An Attempt to Test the Theories of Capillary Action with an Explanation of the Method of Integration Employed*. Cambridge: Cambridge University Press.

Bathe, K.-J. (1996). *Finite Element Procedures*. New Jersey: Prentice Hall.

Batt, G. E. and Braun, J. (1997). On the thermomechanical evolution of compressional orogens. *Geophysical Journal International*, **128**, 364–382.

Baumgardner, J. R. (1985). Three dimensional treatment of convection flow in the Earth's mantle. *Journal of Statistical Physics*, **39**, 501–511.

Baumgardner, J. R. (1988). Application of supercomputers to 3-D mantle convection. In: Runcorn, S. K. (ed.) *The Physics of the Planets*. New York: Wiley, pp. 199–231.

Baumgardner, J. R. and Frederickson, P. O. (1985). Icosahedral discretization of the 2-sphere. *SIAM Journal on Numerical Analysis*, **22**, 1107–1115.

Bayliss, A., and Turkel, E. (1992). Mapping and accuracy for Chebyshev pseudo-spectral approximations. *Journal of Computational Physics*, **101**, 349–359.

Beaumont, C., Munoz, J. A., Hamilton, J. and Fullsack, P. (2000). Factors controlling the Alpine evolution of the central Pyrenees inferred from a comparison of observations and geodynamical models. *Journal of Geophysical Research*, **105**, 8121–8145.

Bennett, A. F. (1992). *Inverse Methods in Physical Oceanography*. Cambridge: Cambridge University Press.

Bercovici, D. (1993). A simple model of plate generation from mantle flow. *Geophysical Journal International*, **114**, 635–650.

Bercovici, D. (1995). A source-sink model of the generation of plate tectonics from non-Newtonian mantle flow. *Journal of Geophysical Research*, **100**, 2013–2030.

Bercovici, D., Schubert, G. and Glatzmaier, G. A. (1989). 3-dimensional spherical-models of convection in the Earth's mantle. *Science*, **244**, 950–955.

Bercovici, D., Schubert, G. and Glatzmaier, G. A. (1992). Three-dimensional convection of an infinite-Prandtl-number compressible fluid in a basally heated spherical-shell. *Journal of Fluid Mechanics*, **239**, 683–719.

Bercovici, D., Ricard, Y. and Schubert, G. (2001). A two-phase model for compaction and damage 1. General Theory. *Journal of Geophysical Research*, **106**, 8887–8906.

Bern, M. W., Flaherty, J. E. and Luskin, M., (eds.) (1999). *Grid Generation and Adaptive Algorithms. The IMA Volumes in Mathematics and its Applications*, vol. 113. New York: Springer.

Blankenbach, B., Busse, F., Christensen, U. *et al.* (1989). A benchmark comparison for mantle convection codes. *Geophysical Journal International*, **98**, 23–38.

Boussinesq, J. (1903). *Theorie Analytique de la Chaleur, Vol. 2.* Paris: Gauthier-Villars.

Brandt, A. (1982). Guide to multigrid development. *Lecture Notes In Mathematics*, **960**, 220–312.

Braun, J. (1993). Three-dimensional numerical modeling of compressional orogenies: Thrust geometry and oblique convergence. *Geology*, **21**, 153–156.

Braun, J. (1994). Three dimensional numerical simulations of crustal-scale wrenching using a non-linear failure criterion. *Journal of Structural Geology*, **16**, 1173–1186.

Braun, J. and Beaumont, C. (1995). Three-dimensional numerical experiments of strain partitioning at oblique plate boundaries: Implications for contrasting tectonic styles in the southern Coast Ranges, California, and central South Island, New Zealand. *Journal of Geophysical Research*, **100**, 18059–18074.

Braun, J. and Sambridge, M. (1994). Dynamical Lagrangian Remeshing (DLR): A new algorithm for solving large strain deformation problems and its application to fault-propagation folding. *Earth and Planetary Science Letters*, **124**, 211–220.

Braun, J., Thieulot, C., Fullsack, P. *et al.* (2008). DOUAR: a new three-dimensional creeping flow model for the solution of geological problems. *Physics of the Earth and Planetary Interiors*, **171**, 76–91.

Bubnov, I. G. (1913). *Sbornik Instituta Inzhenerov Putei Soobshcheniya, vol. 81*. St. Petersburg.

Bubnov, V. A. (1976). Wave concepts in the theory of heat. *International Journal of Heat and Mass Transfer*, **19**, 175–184.

Bubnov, V. A. (1981). Remarks on wave solutions of the nonlinear heat conduction equation. *Journal of Engineering Physics and Thermophysics*, **40**, 907–913.

Buiter, S. J. H., Babeyko, A. Y., Ellis, S. *et al.* (2006). The numerical sandbox: Comparison of model results for a shortening and an extension experiment. In: Buiter, S. J. H. and Schreurs, G. (eds.) *Analogue and numerical modelling of crustal-scale processes, Special publications*. London: Geological Society.

Bunge, H.-P., Hagelberg, C. R. and Travis, B.J. (2003). Mantle circulation models with variational data assimilation: Inferring past mantle flow and structure from plate motion histories and seismic tomography. *Geophysical Journal International*, **152**, 280–301.

Bunge, H. P., Richards, M. A. and Baumgardner, J. R. (1996). Effect of depth-dependent viscosity on the planform of mantle convection. *Nature*, **379**, 436–438.

Bunge, H. P., Richards, M. A. and Baumgardner, J. R. (1997). A sensitivity study of 3-dimensional spherical mantle convection at 108 Rayleigh number – Effects of depth-dependent viscosity, heating mode, and an endothermic phase change. *Journal of Geophysical Research*, **102**, 11 991–12 007.

Bunge, H.-P., Richards, M. A., Lithgow-Bertelloni, C. *et al.* (1998). Time scales and heterogeneous structure in geodynamic earth models. *Science*, **280**, 91–95.

Bunge, H.-P., Richards, M. A. and Baumgardner, J. R. (2002). Mantle circulation models with sequential data-assimilation: Inferring present-day mantle structure from plate motion histories. *Philosophical Transactions of the Royal Society A*, **360**, 2545–2567.

Burov, E., Jolivet, L., Le Pourheit, L. and Poliakov, A. (2001). A thermomechanical model of exhumation of high pressure (HP) and ultra-high pressure (UHP) metamorphic rocks in Alpine-type collision belts. *Tectonophysics*, **342**, 113–136.

Burstedde, C., Ghattas, O., Gurnis, M. *et al.* (2008). Scalable adaptive mantle convection simulation on petascale supercomputers. In *Proceedings of the ACM/IEEE Conference on High Performance Computing, SC 2008, November 15–21, 2008*. Austin, Texas, USA. 15 pp.

Busse, F. H. (1978). A model of time-periodic mantle flow. *Geophysical Journal of the Royal Astronomical Society*, **52**, 1–12.

Busse, F. H., Christensen, U., Clever, R. *et al.* (1993). 3D convection at infinite Prandtl number in Cartesian geometry – a benchmark comparison. *Geophysical and Astrophysical Fluid Dynamics*, **75**, 39–59.

Cacuci, D. G. (2003). *Sensitivity and Uncertainty Analysis. Volume I: Theory*. Boca Raton: Chapman & Hall.

Cacuci, D. G., Ionescu-Bujor, M. and Navon, I. M. (2005). *Sensitivity and Uncertainty Analysis. Volume II: Applications to Large-Scale Systems.* Boca Raton: Chapman & Hall.

Cadek, O. and Fleitout, L. (2003). Effect of lateral viscosity variations in the top 300 km on the geoid and dynamic topography. *Geophysical Journal International*, **152**, 566–580.

Cadek, O. and Matyska, C. (1992). Variational approach to modeling present-time mantle convection. *Studia Geophysica et Geodaetica*, **36**, 215–229.

Cahouet, J. and Chabard, J.-P. (1988). Some fast 3D finite element solvers for the generalized Stokes problem. *International Journal of Numerical Methods in Fluids*, **8**, 869–895.

Carey, G.F. (1997). *Computational Grids: Generation, Adaptation, and Solution Strategies*. Series in Computational and Physical Processes in Mechanics and Thermal Science. New York: Taylor and Francis.

Castro, A. and Gerya, T. V. (2007). Magmatic implications of mantle wedge plumes: Experimental study. *Lithos*, **103**, 138–148.

Cattaneo, C. (1958). Sur une forme de l'equation de la chaleur elinant le paradox d'une propagation instantance. *Comptes Rendus*, **247**, 431–433 (in French).

Chan, K. H., Zhang, K. K., Zou, J. and Schubert, G. (2001). A non-linear, 3-D spherical alpha^2 dynamo using a finite element method. *Physics of the Earth and Planetary Interiors*, **128**, 35–50.

Chan, K. H., Zhang, K., Li, L. and Liao, X. (2007). A new generation of convection-driven spherical dynamos using EBE finite element method. *Physics of the Earth and Planetary Interiors*, **163**, 251–265.

Chandrasekhar, S. (1961). *Hydrodynamic and Hydromagnetic Stability*. New York: Oxford University Press.

Chery, J., Vilotte, J.-P. and Daignieres, M. (1991). Thermomechanical evolution of a thinned continental lithosphere under compression: implications for the Pyrenees. *Journal of Geophysical Research*, **96**, 4385–4412.

Choblet, G. (2005). Modelling thermal convection with large viscosity gradients in one block of the 'cubed sphere'. *Journal of Computational Physics*, **205**, 269–291.

Choblet, G. and Sotin, C. (2000). 3D thermal convection with variable viscosity: can transient cooling be described by a quasi-static scaling law? *Physics of the Earth and Planetary Interiors*, **119**, 321–36.

Choi, E., Lavier, L. and Gurnis, M. (2008). Thermomechanics of mid-ocean ridge segmentation. *Physics of the Earth and Planetary Interiors*, **171**, 374–386.

Chopelas, A. and Boehler, R. (1989). Thermal expansion measurements at very high pressure, systematics and a case for a chemically homogeneous mantle. *Geophysical Research Letters*, **16**, 1347–1350.

Christensen, U. (1981). Numerical experiments on convection in a chemically layered mantle. *Journal of Geophysics*, **49**, 82–84.

Christensen, U. (1983). Convection in a variable-viscosity fluid – Newtonian versus power-law rheology. *Earth and Planetary Science Letters*, **64**, 153–162.

Christensen, U. R. (1984). Convection with pressure- and temperature-dependent non-Newtonian rheology. *Geophysical Journal of the Royal Astronomical Society*, **77**, 343–384.

Christensen, U. (1989). Mixing by time-dependent convection. *Earth and Planetary Science Letters*, **95**, 382–394.

Christensen, U. R. (1992). An Eulerian technique for thermomechanical modelling of lithospheric extension. *Journal of Geophysical Research*, **97**, 2015–2036.

Christensen, U. and Harder, H. (1991). 3-D convection with variable viscosity. *Geophysical Journal International*, **104**, 213–226.

Christensen, U. R. and Hofmann, A. W. (1994). Segregation of subducted oceanic crust in the convecting mantle. *Journal of Geophysical Research*, **99**, 19 867–19 884.

Christensen, U. R. and Wicht, J. (2007). Numerical dynamo simulations. In: Olson, P. and Schubert, G. (eds.) *Treatise of Geophysics Volume 8: Core dynamics*. Elsevier.

Christensen, U. R. and Yuen, D. A. (1984). The interaction of a subducting lithospheric slab with a chemical or phase-boundary. *Journal of Geophysical Research*, **89**, 4389–4402.

Christensen, U. R. and Yuen, D. A. (1985). Layered convection induced by phase transitions. *Journal of Geophysical Research*, **90**, 10 291–10 300.

Christensen, U., Olson, P. and Glatzmaier, G. A. (1999). Numerical modelling of the geodynamo: a systematic parameter study. *Geophysical Journal International*, **138**, 393–409.

Christensen, U. R., Aubert, J., Cardin, P. *et al.* (2001). A numerical dynamo benchmark. *Physics of the Earth and Planetary Interiors*, **128**, 25–34.

Christie, I., Griffiths, D. F., Mitchell, A. R. and Zienkiewicz, O. C. (1976). Finite element methods for second order differential equations with significant first derivatives. *International Journal of Numerical Methods in Engineering*, **10**, 1389–1396.

Cloetingh, S. A. P. L., Burov, E., Matenco, L. *et al.* (2004). Thermo-mechanical controls on the model of continental collision in the SE Carpathians (Romania). *Earth and Planetary Science Lettters*, **218**, 57–76.

Cloetingh, S. A. P. L., Ziegler, P. A., Bogaard, P. J. F. *et al.* (2007). TOPO-EUROPE: The geoscience of coupled deep Earth – surface processes. *Global and Planetary Change*, **58**, 1–118.

Connolly, J. A. D. (1990). Multivariable phase diagrams: an algorithm based on generalized thermodynamics. *American Journal of Science*, **290**, 666–718.

Connolly, J. A. D. (2005). Computation of phase equilibria by linear programming: a tool for geodynamic modeling and an application to subduction zone decarbonation. *Earth and Planetary Science Letters*, **236**, 524–541.

Connolly, J. A. D. and Petrini, K. (2002). An automated strategy for calculation of phase diagram sections and retrieval of rock properties as a function of physical conditions. *Journal of Metamorphic Geology*, **20**, 697–708.

Connolly, J. A. D. and Podladchikov, Y. Y. (1998). Compaction-driven fluid flow in viscoelastic rock. *Geodinamica Acta*, **11**, 55–84.

Connolly, J. A. D. and Podladchikov, Y. Y. (2007). Decompaction weakening and channeling instability in ductile porous media: Implications for asthenospheric melt segregation. *Journal of Geophysical Research*, **112**, doi:10.1029/2005JB004213.

Conrad, C.P. and Gurnis, M. (2003). Seismic tomography, surface uplift, and the breakup of Gondwanaland: Integrating mantle convection backwards in time. *Geochemistry, Geophysics, Geosystems*, **4*(3)***, doi:10.1029/2001GC000299.

Cooper, C. M., Lenardic, A. and Moresi, L. (2004). The thermal structure of stable continental lithosphere within a dynamic mantle. *Earth and Planetary Science Letters*, **222**, 807–817.

Coumou, D., Driesner, T., Weis, P. and Heinrich, C. A. (2009). Phase separation, brine formation, and salinity variation at Black Smoker hydrothermal systems. *Journal of Geophysical Research*, **114**, doi:10.1029/2008JB005764.

Courant, R. (1943). Variational methods for the solution of problems of equilibrium and vibrations. *Bulletin of the American Mathematical Society*, **49**, 1–23.

Courant, R., Friedrichs, K. and Lewy, H. (1928). Uber die partiellen Differenzengleichungen der mathematischen Physik. *Mathematische Annalen*, **100**, 32–74 (in German).

Courant, R., Isaacson, E. and Rees, M. (1952). On the solution of nonlinear hyperbolic differential equations by finite differences. *Communications on Pure and Applied Mathematics*, **5**, 243–255.

Crichlow, J. M. (1988). *Introduction to Distributed and Parallel Computing*. Englewood Cliffs: Prentice Hall.

Cserepes, L., Rabinowicz, M. and Rosembergborot, C. (1988). Three-dimensional infinite Prandtl number convection in one and two layers with implications for the Earth's gravity-field. *Journal of Geophysical Research*, **93**, 12 009–12 025.

Csontos, L., Nagymarosy, A., Horvath, F. *et al.* (1992). Tertiary evolution of the intra-Carpathian area; a model. *Tectonophysics*, **208**, 221–241.

Cundall, P. A. (1989). Numerical experiments on localization in frictional materials. *Ingenieur–Archiv*, **58**, 148–159.

Cundall, P. A., and Strack, O. D. L. (1979). A discrete numerical model for granular assemblies. *Geotechnique*, **29**, 47–65.

Dabrowski, M., Krotkiewski, M. and Schmid, D. W. (2008). MILAMIN: MATLAB-based finite element method solver for large problems. *Geochemisty, Geophysics, Geosystems*, **9**, doi:10.1029/2007GC001719.

Daescu, D. N. and Navon, I. M. (2003). An analysis of a hybrid optimization method for variational data assimilation. *International Journal of Computational Fluid Dynamics*, **17**, 299–306.

Daudre, B. and Cloetingh, S. (1994). Numerical modelling of salt diapirism: influence of the tectonic regime. *Tectonophysics*, **240**, 59–79.

Davaille, A. and Vatteville, J. (2005). On the transient nature of mantle plumes. *Geophysical Research Letters*, **32**, L14309, doi:10.1029/2005GL023029.

Davies, D. R. (2008). Applying multi-resolution numerical methods to geodynamics. PhD thesis, Cardiff University (available at the web-site: http://amcg.ese.ic.ac.uk/~rhodrid/papers/Thesis.pdf).

Davies, D. R., Davies, J. H., Hassan, O., Morgan, K. and Nithiarasu, P. (2007). Investigations into the applicability of adaptive finite element methods to two-dimensional infinite Prandtl number thermal and thermo-chemical convection. *Geochemistry, Geophysics, Geosystems*, **8**, doi:10.1029/2006GC001470.

Delescluse, M., Montési, L. G. J. and Chamot-Rooke, N. (2008). Fault reactivation and selective abandonment in the oceanic lithosphere. *Geophysical Research Letters*, **35**, L16312, DOI: 10.1029/2008GL035066.

Demetrescu, C., Nielsen, S. B., Ene, M. *et al.* (2001). Lithosphere thermal structure and evolution of the Transylvanian Depression – insight from new geothermal measurements and modelling results. *Physics of the Earth and Planetary Interiors*, **126**, 249–267.

Dennis, J. E. and Schnabel, R. B. (1983). *Numerical Methods of Unconstrained Optimization and Nonlinear Equations*. Englewood Cliffs, NJ: Prentice Hall.

Deubelbeiss, Y. and Kaus, B. J. P. (2008). A comparison of Eulerian and Lagrangian numerical techniques for the Stokes equation in the presence of strongly varying viscosity. *Physics of the Earth and Planetary Interiors*, **171**, 92–111.

Donea, J. and Huerta, A. (2003). *Finite Element Methods for Flow Problems*. West Sussex: John Wiley & Sons.

Donea, J., Roig, B. and Huerta. A. (2000). High-order accurate time-stepping schemes for convection–diffusion problems. *Computer Methods in Applied Mechanics and Engineering*, **182**, 249–275.

Dormy, E., Cardin, P. and Jault, D. (1998). MHD flow in a slightly differentially rotating spherical shell, with conducting inner core, in a dipolar magnetic field. *Earth and Planetary Science Letters*, **160**, 15–30.

Egholm, D. L. (2007). A new strategy for discrete element numerical models: 1. Theory. *Journal of Geophysical Research*, **112**, B05203, doi:10.1029/ 2006JB004557.

Euler, L. (1768). *Institutiones Calculi Integralis*, Vol. 1. Petersburg: Académie del Science (in French).

Evonuk, M. (2006). *Numerical modeling of convection in the interiors of giant planets*. PhD thesis, University of California, Santa Cruz.

Evonuk, M. and Glatzmaier, G. A. (2006). A 2D study of the effects of the size of a solid core on the equatorial flow in giant planets. *Icarus*, **181**, 458–464.

Ewing, R. E. and Wang, H. (2001). A summary of numerical methods for time-dependent advection-dominated partial differential equations. *Journal of Computational and Applied Mathematics*, **128**, 423–445.

Faccenda, M., Minelli, G. and Gerya, T. (2009). Coupled and decoupled regimes of continental collision: Numerical modeling. *Earth and Planetary Science Letters*, **278**, 337–349.

Faddeev, D. K. and Faddeeva, V. N. (1963). *Computational Methods of Linear Algebra*. London: Freeman.

Falcone, M. and Ferretti, R. (1998). Convergence analysis for a class of high-order semi-Lagrangian advection schemes. *SIAM Journal of Numerical Analysis*, **35**, 909–940.

Ferziger, J. H. and Peric, M. (2002). *Computational Methods for Fluid Dynamics*. Berlin: Springer.

Finlayson, B. A. (1972). *The Method of Weighted Residuals and Variational Principles*. New York: Academic Press.

Flaherty, J. E., Paslow, P. J., Shephard, M. S. *et al.* (eds.) (1989). *Adaptive Methods for Partial Differential Equations*. Philadelphia: SIAM.

Foley, B. F. and Becker, T. W. (2009). Generation of plate-like behavior and mantle heterogeneity from a spherical, viscoplastic convection model. *Geochemistry, Geophysics, Geosystems*, **10**, Q08001, doi:10.1029/2009GC002378.

Fornberg, B. (1990). High order finite differences and the pseudospectral method on staggered grids. *SIAM Journal on Numerical Analysis*, **27**, 904–918.

Fornberg, B. (1995). *A Practical Guide to Pseudospectal Methods.* Cambridge: Cambridge University Press.

Forte, A. M. and Mitrovica, J. X. (2001). Deep-mantle high-viscosity flow and thermochemical structure inferred from seismic and geodynamic data. *Nature*, **410**, 1049–1056.

Forte, A. M. and Peltier, W. R. (1991). Viscous flow models of global geophysical observables 1. Forward problems. *Journal of Geophysical Research*, **96**, 20 131–20 159.

Forte, A. M. and Peltier, W. R. (1994). The kinematics and dynamics of poloidal-toroidal coupling in mantle flow: The importance of surface plates and lateral viscosity variations. *Advances in Geophysics*, **36**, 1–119.

Forte, A. M., Peltier, W. R. and Dziewonski, A. M. (1991). Inferences of mantle viscosity from tectonic plate velocities. *Geophysical Research Letters*, **18**, 1747–1750.

Foster, I. (1995). *Designing and Building Parallel Programs*. Reading: Addison-Wesley.

Fountain, T. J. (1994). *Parallel Computing: Principles and Practice*. Cambridge: Cambridge University Press.

Fournier, A., Bunge, H. P., Hollerbach, R. and Vilotte, J.-P. (2005). A Fourier-spectral element algorithm for thermal convection in rotating axisymmetric containers. *Journal of Computational Physics*, **204**, 462–489.

Fox, G. C., Johnson, M. A., Lyzenga, G. A. *et al.* (1988). *Solving Problems on Concurrent Processors.* Englewood Cliffs: Prentice Hall.

Frazer, R. A., Jones, W. P. and Skan, S. W. (1937). Approximations to functions and to the solutions of differential equations. *Great Britain Air Ministry, Aero Research Communication, Technical Report*, **1**, 517.

Frehner, M. and Schmalholz, S. M. (2006). Numerical simulations of parasitic folding in multilayers. *Journal of Structural Geology*, **28**, 1647–1657.

Frick, H., Busse, F. H. and Clever, R. M. (1983). Steady 3-dimensional convection at high Prandtl numbers. *Journal of Fluid Mechanics*, **127**, 141–153.

Frigo, M. and Johnson, S. G. (2005). The design and implementation of FFTW3. *Proceedings of the IEEE*, **93**, 216–231.

Fuchs, K., Bonjer, K., Bock, G. *et al.* (1979). The Romanian earthquake of March 4, 1977. II. Aftershocks and migration of seismic activity. *Tectonophysics*, **53**, 225–247.

Fuller, C. W., Willett, S. D. and Brandon, M. T. (2006). Formation of forearc basins and their influence on subduction zone earthquakes. *Geology*, **34**, 65–68.

Fullsack, P. (1995). An arbitrary Lagrangian–Eulerian formulation for creeping flows and its application in tectonic models. *Geophysical Journal International*, **120**, 1–23.

Furuichi, M., Kameyama, M. and Kageyama, A. (2008). Three-dimensional Eulerian method for large deformation of viscoelastic fluid: Toward plate-mantle simulation. *Journal of Computational Physics*, doi:10.1016/j.jcp.2008.01.052.

Fügenschuh, B. and Schmid, S.M. (2005). Age and significance of core complex formation in a very curved orogen: evidence from fission track studies in the South Carpathians (Romania). *Tectonophysics*, **404**, 33–53.

Gable, C. W., O'Connell, R. J. and Travis, B. J. (1991). Convection in 3 dimensions with surface plates – Generation of toroidal flow. *Journal of Geophysical Research*, **96**, 8391–8405.

Galerkin, B. G. (1915). *Vestnik Inzhenernoy Techniki, vol. 19*. St. Petersburg (in Russian).

Galpin, P. F. and Raithby, G. D. (1986). Numerical solution of problems in incompressible fluid flow: treatment of the temperature–velocity coupling. *Numerical Heat Transfer*, **10**, 105–129.

Geiger, S., Driesner, T., Heinrich, C. A. and Matthai, S. K. (2006). Multiphase thermohaline convection in the Earth's crust: I. A new finite element – finite volume solution technique combined with a new equation of state for NaCl–H_2O. *Transport in Porous Media*, **63**, 399–434.

Gerya, T. V. and Yuen, D. A. (2003). Characteristics-based marker-in-cell method with conservative finite-differences schemes for modeling geological flows with strongly variable transport properties. *Physics of the Earth and Planetary Interiors*, **140**, 293–318.

Gerya, T. V. and Yuen, D. A. (2007). Robust characteristics method for modelling multiphase visco-elasto-plastic thermo-mechanical problems. *Physics of the Earth and Planetary Interiors*, **163**, 83–105.

Gerya, T. V., Perchuk, L. L., Maresch, W. V. and Willner, A. P. (2004). Inherent gravitational instability of hot continental crust: Implication for doming and diapirism in granulite facies terrains. In: Whitney, D., Teyssier, C. and Siddoway, C. S. (eds.) *Gneiss Domes in Orogeny*. GSA.

Gerya, T. V., Connolly, J. A. D., Yuen, D. A., Gorczyk, W. and Capel, A. M. (2006). Seismic implications of mantle wedge plumes. *Physics of the Earth and Planetary Interiors*, **156**, 59–74.

Gerya, T. V., Perchuk, L. L. and Burg, J.-P. (2007). Transient hot channels: Perpetrating and regurgitating ultrahigh-pressure, high-temperature crust-mantle associations in collision belts. *Lithos*, **103**, 236–256.

Gerya, T. V., Connolly, J. A. D. and Yuen, D. A. (2008). Why is terrestrial subduction one-sided? *Geology*, **36**, 43–46.

Gibbons, A. and Rytter, W. (1988). *Efficient Parallel Algorithms*. Cambridge: Cambridge University Press.

Girbacea, R. and Frisch, W. (1998). Slab in the wrong place: Lower lithospheric mantle delamination in the last stage of the Eastern Carpathian subduction retreat. *Geology*, **26**, 611–614.

Glatzmaier, G. A. (1988). Numerical simulations of mantle convection – time-dependent, 3-dimensional, compressible, spherical-shell. *Geophysical and Astrophysical Fluid Dynamics*, **43**, 223–264.

Glatzmaier, G. A. and Roberts, P. H. (1995a). A 3-dimensional convective dynamo solution with rotating and finitely conducting inner-core and mantle. *Physics of the Earth and Planetary Interiors*, **91**, 63–75.

Glatzmaier, G. A. and Roberts, P. H. (1995b). A 3-dimensional self-consistent computer-simulation of a geomagnetic-field reversal. *Nature*, **377**, 203–209.

Glatzmaier, G. A. and Schubert, G. (1993). 3-dimensional spherical models of layered and whole mantle convection. *Journal of Geophysical Research*, **98**, 21 969–21 976.

Glatzmaier, G. A., Schubert, G. and Bercovici, D. (1990). Chaotic, subduction-like downflows in a spherical model of convection in the Earth's mantle. *Nature*, **347**, 274–277.

Godard, V., Cattin, R. and Lavé, J. (2004). Numerical modeling of mountain building: Interplay between erosion law and crustal rheology. *Geophysical Research Letters*, **31**, doi:10.1029/2004GL021006.

Godard, V., Lave, J. and Cattin, R. (2006). Geological Society Special Publications: Numerical modelling of erosion processes in the Hamalaya of Nepal: Effects of spatial variations of rock strength and precipitation. In: Buiter, S. J. H. and Shreurs, G. (eds.) *Analogue and Numerical Modelling of Crustal-Scale Processes*. London: Geological Society of London.

Golabek, G., Gerya, T. V., Ziethe, R., Kaus, B. J. P. and Tackley, P. J. (2009). Rheological controls on the terrestrial core formation mechanism. *Geochemistry, Geophysics, Geosystems*, **10** Q11007, doi: 10.1029/2009GC002552.

Golub, G. H. and Ortega, J. M. (1992). *Scientific Computing and Differential Equations: An Introduction to Numerical Methods*. New York: Academic Press.

Golub, G. H. and Van Loan, C. F. (1989). *Matrix Computations*, 2nd edn. Baltimore: Johns Hopkins University Press.

Gorczyk, W., Gerya, T. V., Connolly, J. A. D., Yuen, D. A. and Rudolph, M. (2006). Large-scale rigid-body rotation in the mantle wedge and its implications for seismic tomography. *Geochemistry, Geophysics, Geosystems*, **7**, doi:10.1029/2005GC001075.

Gorcyzk, W., Gerya, T. V., Connolly, J. A. D. and Yuen, D. A. (2007a). Growth and mixing dynamics of mantle wedge plumes. *Geology*, **35**, 587–590.

Gorcyzk, W., Guillot, S., Gerya, T. V. and Hattori, K. (2007b). Asthenospheric upwelling, oceanic slab retreat, and exhumation of UHP mantle rocks: Insights from Greater Antilles. *Geophysical Research Letters*, **34**, doi:10.1029/2007GL031059.

Gorcyzk, W., Willner, A. P., Gerya, T. V., Connolly, J. A. D. and Burg, J. P. (2007c). Physical controls of magmatic productivity at Pacific-type convergent margins: Numerical modelling. *Physics of the Earth and Planetary Interiors*, **163**, 209–232.

Griffths, D. F. and Mitchell, A. R.(1979). On generating upwind finite element methods. In: Hughes, T. J. R. (ed.) *Finite Element Methods for Convection Dominated Flows*, AMD-Vol. 34, New York: ASME, pp. 91–104.

Grigne, C., Labrosse, S. and Tackley, P. J. (2007a). Convection under a lid of finite conductivity in wide aspect ratio models: Effect of continents on the wavelength of mantle flow. *Journal of Geophysical Research*, 112, doi:10.1029/2006JB004297.

Grigne, C., Labrosse, S. and Tackley, P. J. (2007b). Convection under a lid of finite conductivity: heat flux scaling and application to continents. *Journal of Geophysical Research*, **112**, doi:10.1029/2005JB004192.

Gropp, W., Lusk, E. and Skjellum, A. (1994). *Using MPI: Portable Parallel Programming with the Message-Passing Interface*. Cambridge: MIT Press.

Gurnis, M. (1988). Large-scale mantle convection and the aggregation and dispersal of supercontinents. *Nature*, **332**, 695–699.

Gurnis, M. (1989). A reassessment of the heat-transport by variable viscosity convection with plates and lids. *Geophysical Research Letters*, **16**, 179–182.

Gurnis, M. and Zhong, S. (1991). Generation of long wavelength heterogeneity in the mantle by the dynamic interaction between plates and convection. *Geophysical Research Letters*, 18, 581–584.

Gvirtzman, Z. (2002). Partial detachment of a lithospheric root under the southeast Carpathians: toward a better definition of the detachment concept. *Geology*, **30**, 51–54.

Hadamard, J. (1902). Sur les problèmes aux dérivées partielles et leur signification physique. *Princeton University Bulletin*, pp. 749–752 (in French).

Hageman, L. A. and Young, D. M. (1981). *Applied Iterative Methods*. New York: Academic Press.

Hager, B. H. and O'Connell, R. J. (1978). Subduction zone dip and flow driven by plate motions. *Tectonophysics*, **50**, 111–133.

Hager, B. H. and O'Connell, R. J. (1981). A simple global model of plate dynamics and mantle convection. *Journal of Geophysical Research*, **86**, 4843–4867.

Hall, T. (1970). *Carl Friedrich Gauss*. Cambridge, MA: M.I.T. Press.

Hamilton, R. S. (1982). Three manifolds with positive Ricci curvature. *Journal of Differential Geometry*, **17**, 255–306.

Hansen, D. L. (2003). A meshless formulation for geodynamic modeling. *Journal of Geophysical Research*, **108**, B11, 2549, doi:10.1029/2003JB002460.

Hansen, U. and Ebel, A. (1984). Experiments with a numerical model related to mantle convection: boundary layer behaviour of small and large-scale flows. *Physics of the Earth and Planetary Interiors*, **36**, 374–390.

Hansen, U., Yuen, D. A. and Kroening, S. E. (1990). Transition to hard turbulence in thermal convection at infinite Prandtl number. *Physics of Fluids*, **A2(12)**, 2157–2163.

Hansen, U., Yuen, D. A. and Kroening, S. E. (1991). Effects of depth-dependent thermal expansivity on mantle circulations and lateral thermal anomalies. *Geophysical Research Letters*, **18**, 1261–1264.

Hansen, U., Yuen, D. A., Kroening, S. E. *et al.* (1993). Dynamical consequences of depth-dependent thermal expansivity and viscosity on mantle circulations and thermal structure. *Physics of the Earth and Planetary Interiors*, **77**, 205–223.

Harder, H. (1991). Numerical-simulation of thermal-convection with Maxwellian viscoelasticity. *Journal of Non-Newtonian Fluid Mechanics*, **39**, 67–88.

Harder, H. (1998). Phase transitions and the three-dimensional planform of thermal convection in the Martian mantle. *Journal of Geophysical Research*, **103**, 16 775–16 797.

Harder, H. and Christensen, U. R. (1996). A one-plume model of Martian mantle convection. *Nature,* **380**, 507–509.

Harder, H. and Hansen, U. (2005). A finite-volume solution method fof thermal convection and dynamo problems in spherical shells. *Geophysical Journal International*, **161**, 522–532.

Harlow, F. H. and Welch, J. E. (1965). Numerical calculation of time-dependent viscous incompressible flow of fluid with free surface. *Physics of Fluids A*, **8**, 2182–2189.

Harten, A. (1983). High resolution schemes for hyperbolic conservation laws. *Journal of Computational Physics*, **49**, 357–393.

Healy, D. M., Kostelec, P. J. and Rockmore, D. (2004). Towards safe and effective high-order Legendre transforms with applications to FFTs for the 2-sphere. *Advances in Computational Mathematics*, **21**, 59–105.

Hebert, L. B., Antoschechkina, P., Asimow, P. and Gurnis, M. (2009). Emergence of a low-viscosity channel in subduction zones through the coupling of mantle flow and thermodynamics. *Earth and Planetary Science Letters*, **278**, 243–256.

Heinrich, J. C. and Zienkiewicz, O. C. (1979). The finite element method and "upwinding" techniques in the numerical solution of convection dominated flow problems. In: Hughes, T. J. R. (ed.) *Finite Element Methods for Convection Dominated Flows*, AMD-Vol. 34, New York: ASME, pp. 105–136.

Heinrich, J. C., Huyakorn, P. S., Zienkiewicz, C. and Mitchell A. R. (1977). An "upwind" finite element scheme for two-dimensional convective transport equation. *International Journal of Numerical Methods in Engineering*, **11**, 134–143.

Hejda, P. and Reshatnyak, M. (2003). Control volume method for the dynamo problem in a sphere with the free rotating inner core. *Studia Geophysica et Geodaetica*, **47**, 147–159.

Hernlund, J. W. and Tackley, P. J. (2003). Three-dimensional spherical shell convection at infinite Prandtl number using the 'cubed sphere' method. In: Bathe, K. J. (ed.) *Computational Fluid and Solid Mechanics*. Amsterdam: Elsevier Science, pp. 931–933.

Hernlund, J. W. and Tackley, P. (2008). Modeling mantle convection in the spherical annulus. *Physics of the Earth and Planetary Interiors*, **171**, 48–54.

Hernlund, J. W., Tackley, P. J. and Stevenson, D. J. (2008). Buoyant melting instabilities beneath extending lithosphere, 1. Numerical models. *Journal of Geophysical Research*, **113**, doi:10.1029/2006JB004862.

Hewitt, J. M., McKenzie, D. P. and Weiss, N. O. (1975). Dissipative heating in convective flows. *Journal of Fluid Mechanics*, **68**, 721–738.

Hier-Majumder, C. A., Belanger, E., DeRosier *et al.* (2005). Data assimilation for plume models. *Nonlinear Process in Geophysics*, **12**, 257–267.

Hier-Majumder, C. A., Travis, B. J., Belanger, E. *et al.* (2006). Efficient sensitivity analysis for flow and transport in the Earth's crust and mantle. *Geophysical Journal International,* doi: 10.1111/j.1365–246X.2006.02926.x.

Hirt, C. W., Amsden, A. A. and Cook, J. L. (1974). An arbitrary Lagrangian–Eulerian computing method for all flow speeds. *Journal of Computational Physics*, **14**, 227–253.

Hofmeister, A. M. (1999). Mantle values of thermal conductivity and the geotherm from photon lifetimes. *Science*, **283**, 1699–1706.

Honda, R., Mizutani, H. and Yamamoto, T. (1993). Numerical simulation of Earths core formation. *Journal of Geophysical Research*, **98**, 2075–2089.

Honda, S. (1996). Applicability of adaptive grid inversion to imaging thermal anomalies caused by convection. *Geophysical Research Letters*, **23**, 2733–2736.

Honda, S. (1997). A possible role of weak zone at plate margin on secular mantle cooling. *Geophysical Research Letters*, **24**, 2861–2864.

Honda, S., Balachandar, S., Yuen, D. A. and Reuteler, D. (1993a). Three-dimensional mantle dynamics with an endothermic phase transition. *Geophysical Research Letters*, **20**, 221–224.

Honda, S., Yuen, D. A., Balachandar, S. and Reuteler, D. (1993b). Three-dimensional instabilities of mantle convection with multiple phase transitions. *Science*, **259**, 1308–1311.

Honda, S., Morishige, M. and Orihashi, Y. (2007). Sinking hot anomaly trapped at the 410 km discontinuity near the Honshu subduction zone, Japan. *Earth and Planetary Science Letters*, **261**, 565–577.

Hord, R. M. (1999). *Understanding Parallel Supercomputing*. New York: IEEE Press.

Householder, A. S. (1970). *The Numerical Treatment of a Single Nonlinear Equation.* New York: McGraw-Hill.

Houseman, G. (1988). The dependence of convection planform on mode of heating. *Nature*, **332**, 346–349.

Howard, L. N. (1966). Convection at high Rayleigh number. In: Goertler, H. and Sorger, P. (eds.) *Applied Mechanics, Proc. of the 11th Intl Congress of Applied Mechanics, Munich, Germany 1964.* New York: Springer-Verlag, pp. 1109–1115.

Hrennikoff, A. (1941). Solutions of problems in elasticity by the frame work method. *Journal of Applied Mechanics*, **8**, 169–175.

Huettig, C. and Stemmer, K. (2008a). Finite volume discretization for dynamic viscosities on Voronoi grids. *Physics of the Earth and Planetary Interiors*, **171**, 137–146.

Huettig, C. and Stemmer, K. (2008b). The spiral grid: A new approach to discretize the sphere and its application to mantle convection. *Geochemistry, Geophysics, Geosystems*, **9**, doi:10.1029/2007GC001581.

Hughes, T. J. R. (1978). A simple scheme for developing "upwind" finite elements. *International Journal of Numerical Mehods in Engineering*, **12**, 1359–1365.

Hughes, T. J. R. and Brooks, A. (1979). A multidimensional upwind scheme with no crosswind diffusion. In: Hughes, T. J. R. (ed.) *Finite Element Methods for Convection Dominated Flows*, AMD-Vol. 34, New York: ASME, pp. 19–35.

Hughes, T. J. R. and Brooks, A. (1982). A theoretical framework for Petrov-Galerkin methods with discontinuous weighting functions: application to the streamline-upwind procedure, In: Gallagher, R. H. *et al.* (eds.) *Finite Element in Fluids*, Vol. 4, New York: John Wiley & Sons, pp. 47–65.

Hughes, T. J. R., Liu, W. K. and Zimmermann, T. K. (1981). Lagrangian–Eulerian finite element formulation for incompressible viscous flows. *Computational Methods in Applied Mechanics and Engineering*, **29**, 329–349.

Ismail-Zadeh, A. T., Korotkii, A. I., Naimark, B. M., Suetov, A. P. and Tsepelev, I. A. (1998). Implementation of a three-dimensional hydrodynamic model for evolution of sedimentary basins. *Computational Mathematics and Mathematical Physics*, **38**, 1138–1151.

Ismail-Zadeh, A. T., Panza, G. F. and Naimark B. M. (2000a). Stress in the descending relic slab beneath the Vrancea region, Romania. *Pure and Applied Geophysics*, **157**, 111–130.

Ismail-Zadeh, A. T., Tsepelev, I. A., Talbot, C. J. and Oster, P. (2000b). A numerical method and parallel algorithm for three-dimensional modeling of salt diapirism. In: Keilis-Borok, V. I. Molchan and G. M. (eds.) *Problems in Dynamics and Seismicity of the Earth*. Moscow: GEOS, pp. **62**–76.

Ismail-Zadeh, A. T., Korotkii, A. I., Naimark, B. M. and Tsepelev, I. A. (2001a). Numerical simulation of three-dimensional viscous flow with gravitational and thermal effects. *Computational Mathematics and Mathematical Physics*, **41**, 1331–1345.

Ismail-Zadeh, A. T., Talbot, C. J. and Volozh, Y. A. (2001b). Dynamic restoration of profiles across diapiric salt structures: numerical approach and its applications. *Tectonophysics*, **337**, 21–36.

Ismail-Zadeh, A.T., Korotkii, A. I. and Tsepelev, I.A. (2003a). Numerical approach to solving problems of slow viscous flow backwards in time. In: Bathe, K. J. (ed.) *Computational Fluid and Solid Mechanics*, Amsterdam: Elsevier Science, pp. 938–941.

Ismail-Zadeh, A. T., Korotkii, A. I., Naimark, B. M. and Tsepelev, I. A. (2003b). Three-dimensional numerical simulation of the inverse problem of thermal convection. *Computational Mathematics and Mathematical Physics*, **43**, 587–599.

Ismail-Zadeh, A., Schubert, G., Tsepelev, I. and Korotkii, A. (2004a). Inverse problem of thermal convection: Numerical approach and application to mantle plume restoration. *Physics of the Earth and Planetary Interiors*, **145**, 99–114.

Ismail-Zadeh, A. T., Tsepelev, I. A., Talbot, C. J. and Volozh, Y. A. (2004b). Three-dimensional forward and backward modelling of diapirism: Numerical approach and its applicability to the evolution of salt structures in the Pricaspian basin. *Tectonophysics*, **387**, 81–103.

Ismail-Zadeh, A., Mueller, B. and Schubert, G. (2005a). Three-dimensional modeling of present-day tectonic stress beneath the earthquake-prone southeastern Carpathians based on integrated analysis of seismic, heat flow, and gravity observations. *Physics of the Earth and Planetary Interiors*, **149**, 81–98.

Ismail-Zadeh, A., Mueller, B. and Wenzel, F. (2005b). Modelling of descending slab evolution beneath the SE-Carpathians: Implications for seismicity. In: Wenzel, F. (ed.) *Perspectives in Modern Seismology, Lecture Notes in Earth Sciences*, Volume 105, Heidelberg: Springer-Verlag, pp. 205–226.

Ismail-Zadeh, A., Schubert, G., Tsepelev, I. and Korotkii, A. (2006). Three-dimensional forward and backward numerical modeling of mantle plume evolution: Effects of thermal diffusion. *Journal of Geophysical Research*, **111**, B06401, doi:10.1029/2005JB003782.

Ismail-Zadeh, A., Korotkii, A., Schubert, G. and Tsepelev, I. (2007). Quasi-reversibility method for data assimilation in models of mantle dynamics. *Geophysical Journal International*, **170**, 1381–1398.

Ismail-Zadeh, A., Schubert, G., Tsepelev, I. and Korotkii, A. (2008). Thermal evolution and geometry of the descending lithosphere beneath the SE-Carpathians: an insight from the past. *Earth and Planetary Science Letters*, **273**, 68–79.

Ismail-Zadeh, A., Korotkii, A., Schubert, G. and Tsepelev, I. (2009). Numerical techniques for solving the inverse retrospective problem of thermal evolution of the Earth interior. *Computers & Structures*, **87**, 802–811.

Iwase, Y. and Honda, S. (1997). An interpretation of the Nusselt–Rayleigh number relationship for convection in a spherical shell. *Geophysical Journal International*, **130**, 801–804.

Jackson, M. P. A., Vendeville, B. C. and Schultz-Ela, D. D. (1994), Structural dynamics of salt systems. *Annual Review of Earth and Planetary Sciences*, **22**, 93–117.

Jacoby, W. R. and Schmeling, H. (1982). On the effects of the lithosphere on mantle convection and evolution. *Physics of the Earth and Planetary Interiors*, **29**, 305–319.

Jamieson, R. A., Beaumont, C., Hamilton, J. and Fullsack, P. (1996). Tectonic assembly of inverted metamorphic sequences. *Geology*, **24**, 839–842.

Jarvis, G. T. and McKenzie, D. P. (1980). Convection in a compressible fluid with infinite Prandtl number. *Journal of Fluid Mechanics*, **96**, 515–583.

Jarvis, G. T. and Peltier, W. R. (1981). Effects of lithospheric rigidity on ocean floor bathymetry and heat flow. *Geophysical Research Letters*, **8**, 857–860.

Jiricek, R. (1979). Tectonic development of the Carpathian arc in the Oligocene and Neogene. In: Mahel, M. (ed.), *Tectonic Profiles Through the Western Carpathians*, Geol. Inst., Dionyz Stur, pp. 205–214.

Johnson, C. (1987). *Numerical Solution of Partial Differential Equations by the Finite Element Method*. Cambridge: Cambridge University Press.

Jones, C. A. and Roberts, P. H. (2000). Convection-driven dynamos in a rotating plane layer. *Journal of Fluid Mechanics*, **404**, 311–43.

Jones, M. N. (1985). *Spherical Harmonics and Tensors for Classical Field Theory*. New York: Wiley.

Kageyama, A. and Sato, T. (1995). Computer simulation of a magnetohydrodynamic dynamo. II. *Physics of Plasmas*, **2**, 1421–1431.

Kageyama, A. and Sato, T. (1997). Generation mechanism of a dipole field by a magnetohydrodynamic dynamo. *Physical Review E*, **55**, 4617–4626.

Kageyama, A. and Sato, T. (2004). The "Yin-Yang grid": An overset grid in spherical geometry. *Geochemistry, Geophysics, Geosystems*, **5**, doi:10.1029/2004GC000734.

Kageyama, A. and Yoshida, M. (2005). Geodynamo and mantle convection simulations on the Earth Simulator using the Yin-Yang grid. *Journal of Physics: Conference Series*, **16**, 325–338.

Kageyama, A., Miyagoshi, T. and Sato, T. (2008). Formation of current coils in geodynamo simulations. *Nature*, **454**, 1106–1109.

Kalnay, E. (2003). *Atmospheric Modeling, Data Assimilation and Predictability*. Cambridge: Cambridge University Press.

Kameyama, M. and Yuen, D. A. (2006). 3-D convection studies on the thermal state in the lower mantle with post-perovskite phase transition. *Geophysical Research Letters*, **33**, doi:10.1029/2006GL025744.

Kameyama, M., Kageyama, A. and Sato, T. (2005). Multigrid iterative algorithm using pseudo-compressibility for three-dimensional mantle convection with strongly variable viscosity. *Journal of Computational Physics*, **206**, 162–181.

Kameyama, M., Kageyama, A. and Sato, T. (2008). Multigrid-based simulation code for mantle convection in spherical shell using yin-yang grid. *Physics of the Earth and Planetary Interiors*, **171**, 19–32.

Karato, S. (1993). Importance of anelasticity in the interpretation of seismic tomography. *Geophysical Research Letters*, **20**, 1623–1626.

Karato, S. (2008). *Deformation of Earth Materials: An Introduction to the Rheology of Solid Earth*. Cambridge: Cambridge University Press.

Katayama, J. S., Matsushima, M. and Honkura, Y. (1999). Sine characteristics of magnetic field behavior in a model of MHD dynamo thermally driven in a rotating spherical shell. *Physics of the Earth and Planetary Interiors*, **111**, 141–159.

Katsura, T., Yamada, H., Nishikawa, O. *et al.* (2004). Olivine–wadsleyite transition in the system $(Mg, Fe)_2SiO_4$. *Journal of Geophysical Research*, **109**, B02209, doi:10.1029/2003JB002438.

Katz, R. F., Knepley, M. G., Smith, B., Spiegelman, M. and Coon, E. T. (2007). Numerical simulation of geodynamic processes with the Portable Extensible Toolkit for Scientific Computation. *Physics of the Earth and Planetary Interiors*, **163**, 52–68.

Kaus, B. J. P. (2005). *Modelling approaches to geodynamic processes*, Ph.D. Thesis, ETH Zurich.

Kaus, B. J. P. (2010). Factors that control the angle of shear bands in geodynamic numerical models of brittle deformation. *Tectonophysics*, **484**, 36–47.

Kaus, B. J. P., and Podladchikov, Y. Y. (2001). Forward and reverse modeling of the three-dimensional viscous Rayleigh–Taylor instability. *Geophysical Research Letters*, **28**(6), 1095–1098.

Kaus, B. J. P., Connolly, J. A. D., Podladchikov, Y. Y. and Schmalholz, S. M. (2005). Effect of mineral phase transitions on sedimentary basin subsidence and uplift. *Earth and Planetary Science Letters*, **233**, 213–228.

Kellogg, L. H. and Turcotte, D. L. (1990). Mixing and the distribution of heterogeneities in a chaotically convecting mantle. *Journal Of Geophysical Research*, **95**, 421–432.

Kikuchi, N. (1986). *Finite Element Methods in Mechanics*. Cambridge: Cambridge University Press.

King, S. D. and Hager, B. H. (1990). The relationship between plate velocity and trench viscosity in Newtonian and power-law subduction calculations. *Geophysical Research Letters*, **17**, 2409–2412.

King, S. D., Raefsky, A. and Hager, B. H. (1990). ConMan: Vectorizing a finite element code for incompressible two-dimensional convection in the Earth's mantle. *Physics of the Earth and Planetary Interiors*, **59**, 195–207.

King, S. D., Gable, C. W. and Weinstein, S. A. (1992). Models of convection-driven tectonic plates – a comparison of methods and results. *Geophysical Journal International*, **109**, 481–487.

King, S. D., Lee, C., van Keken, P. E. *et al.* (2009). Community benchmark for 2D Cartesian compressible convection in the Earth's mantle. *Geophysical Journal International*, **179**, 1–11.

Kirsch, A. (1996). *An Introduction to the Mathematical Theory of Inverse Problems*. New York: Springer-Verlag.

Kohlstedt, D. L., Evans, B. and Mackwell, S. J. (1995). Strength of the lithosphere–Constraints imposed by laboratory experiments. *Journal of Geophysical Research*, **100**, 17 587–17 602.

Komatitsch, D. and Tromp, J. (2002). Spectral-element somulations of global seismic wave propagation, Part I: Validation. *Geophysical Journal International*, **149**, 390–412.

Korotkii, A. I. and Tsepelev, I. A. (2003). Solution of a retrospective inverse problem for one nonlinear evolutionary model. *Proc. Steklov Institute of Mathematics*, **2**, 80–94.

Korotkii, A. I., Tsepelev, I. A., Ismail-Zadeh, A. T. and Naimark, B.M. (2002). Three-dimensional backward modeling in problems of Rayleigh-Taylor instability. *Proc. Ural State University*, **22**(4), 96–104.

Kopitzke, U. (1979). Finite element convection models: Comparison of shallow and deep mantle convection, and temperatures in the mantle. *Journal of Geophysics*, **46**, 97–121.

Kotelkin, V. D. and Lobkovsky, L. I. (2007). The Myasnikov global theory of the evolution of planets and the modern thermochemical model of the Earth's evolution. *Izvestiya, Physics of the Solid Earth*, **43**, 24–41.

Koyi, H. (1996). Salt flow by aggrading and prograding overburdens. In: Alsop, I., Blundell, D., Davison, I. (eds.) *Salt Tectonics*. Geological Society Special Publication 100, London, pp. 243–258.

Kuang, W. and Bloxham, J. (1999). Numerical modeling of magnetohydrodynamic convection in a rapidly rotating spherical shell: Weak and strong field dynamo action. *Journal of Computational Physics*, **153**, 51–81.

Landuyt, W., Bercovici, D. and Ricard, Y. (2008). Plate generation and two-phase damage theory in a model of mantle convection. *Geophysical Journal International*, **174**, 1065–1080.

Larsen, T. B., Yuen, D. A., Moser, J. and Fornberg, B. (1997). A high-order finite-difference method applied to large Rayleigh number mantle convection. *Geophysical and Astrophysical Fluid Dynamics*, **84**, 53–83.

Lattes, R. and Lions, J. L. (1969). *The Method of Quasi-Reversibility: Applications to Partial Differential Equations*. New York: Elsevier.

Lavier, L. L. and Buck, W. R. (2002). Half graben versus large-offset low-angle normal fault: Importance of keeping cool during normal faulting. *Journal of Geophysical Research*, **107**, doi:10.1029/2001JB000513.

Lavier, L. L., Buck, W. R. and Poliakov, A. N. B. (2000). Factors controlling normal fault offset in an ideal brittle layer. *Journal of Geophysical Research*, **105**, 23 431–23 442.

Lax, P. D. (1954). Weak solutions of nonlinear hyperbolic equations and their numerical computation. *Communications on Pure and Applied Mathematics*, **2**, 159–193.

Le Dimet, F.-X., Navon, I. M. and Daescu, D. N. (2002). Second-order information in data assimilation. *Monthly Weather Reviews*, **130**, 629–648.

Leitch, A. M., Steinbach, V. and Yuen, D. A. (1996). Centerline temperature of mantle plumes in various geometries: Incompressible flow. *Journal of Geophysical Research*, **101**, 21 829–21 846.

Lenardic, A. and Kaula, W. M. (1993). A numerical treatment of geodynamic viscous-flow problems involving the advection of material interfaces. *Journal of Geophysical Research*, **98**, 8243–8260.

Lenardic, A. and Kaula, W. M. (1994). Self-lubricated mantle convection – 2-dimensional models. *Geophysical Research Letters*, **21**, 1707–1710.

Lenardic, A. and Kaula, W. M. (1995a). Mantle dynamics and the heat-flow into the Earth's continents. *Nature*, **378**, 709–711.

Lenardic, A. and Kaula, W. M. (1995b). More thoughts on convergent crustal plateau formation and mantle dynamics with regard to Tibet. *Journal of Geophysical Research*, **100**, 15 193–15 203.

Lenardic, A. and Kaula, W. M. (1996). Near-surface thermal/chemical boundary-layer convection at infinite Prandtl number – 2-dimensional numerical experiments. *Geophysical Journal International*, **126**, 689–711.

Lenardic, A. and Moresi, L.-N. (1999). Some thoughts on the stability of cratonic lithosphere: Effects of buoyancy and viscosity. *Journal of Geophysical Research*, **104**, 12 747–12 758.

Lenardic, A., Kaula, W. M. and Bindschadler, D. L. (1995). Some effects of a dry crustal flow law on numerical simulations of coupled crustal deformation and mantle convection on Venus. *Journal of Geophysical Research*, **100**, 16 949–16 957.

Lenardic, A., Moresi, L.-N. and Muhlhaus, H. (2003). Longevity and stability of cratonic lithosphere: Insights from numerical simulations of coupled mantle convection and continental tectonics. *Journal of Geophysical Research*, **108**, doi:10.1029/2002JB001859.

Leonard, B. P. (1979). A stable and accurate convective modelling procedure based on quadratic upstream interpolation. *Computer Methods for Applied Mechanics and Engineering*, **19**, 59–98.

Lesur, V. and Gubbins, D. (1999). Evaluation of fast spherical transforms for geophysical applications. *Geophysical Journal International*, **139**, 547–555.

Li, J., Sato, T. and Kageyama, A. (2002). Repeated and sudden reversals of the dipole field generated by a spherical dynamo action. *Science*, **295**, 1887–90.

Lin, J., Gerya, T., Tackley, P. J., Yuen, D. A. and Golabek, G. (2009). Protocore destabilization during planetary accretion: Influence of a deforming planetary surface. *Icarus*, **204**, 732–748.

Lin, S.-C. and Van Keken, P. E. (2006). Dynamics of thermochemical plumes: 1. Plume formation and entrainment of a dense layer. *Geochemistry, Geophysics, Geosystems*, **7**, doi:10.1029/2005GC001071.

Linzer, H.-G. (1996). Kinematics of retreating subduction along the Carpathian arc, Romania. *Geology*, **24**, 167–170.

Lipovski, G. J. and Malek, M. (1987). *Parallel Computing: Theory and Comparisons*. New York: John Wiley & Sons.

Liu, D. C. and Nocedal, J. (1989). On the limited memory BFGS method for large scale optimization. *Mathematical Programming*, **45**, 503–528.

Liu, G. R. (2003). *Mesh Free Methods: Moving beyond the Finite Element Method*. Boca Raton: CRC Press.

Liu, L. and Gurnis, M. (2008). Simultaneous inversion of mantle properties and initial conditions using an adjoint of mantle convection. *Journal of Geophysical Research*, **113**, B08405, doi:10.1029/2008JB005594.

Liu, L., Spasojeviæ, S. and Gurnis, M. (2008). Reconstructing Farallon Plate subduction beneath North America back to the Late Cretaceous. *Science*, **322**, 934–938.

Liu, M., Yuen, D. A., Zhao, W. *et al.* (1991). Development of diapiric structures in the upper mantle due to phase transitions. *Science*, **252**, 1836–1839.

Lowman, J. P. and Jarvis, G. T. (1993). Mantle convection flow reversals due to continental collisions. *Geophysical Research Letters*, **20**, 2087–2090.

Lowman, J. P. and Jarvis, G. T. (1995). Mantle convection models of continental collision and breakup incorporating finite thickness plates. *Physics of the Earth and Planetary Interiors*, **88**, 53–68.

Lowman, J. P., King, S. D. and Gable, C. W. (2004). Steady plumes in a viscously stratified, vigorously convecting, three-dimensional numerical mantle convection models with mobile plates. *Geochemistry, Geophysics, Geosystems*, **5**, doi: 10.1029/2003GC000583.

Lyakhovsky, V., Podladchikov, Y. and Poliakov, A. (1993). A rheological model of a fractured solid. *Tectonophysics*, **226**, 187–198.

Machetel, P. and Yuen, D. A. (1989). Penetrative convective flows induced by internal heating and mantle compressibility. *Journal of Geophysical Research*, **94**, 10 609–10 626.

Machetel, P., Rabinowicz, M. and Bernadet, P. (1986). Three-dimensional convection in spherical shells. *Geophysical and Astrophysical Fluid Dynamics*, **37**, 57–84.

Malevsky, A. V. (1996). Spline-characteristic method for simulation of convective turbulence. *Journal of Computational Physics*, **123**, 466–475.

Malevsky, A. V. and Yuen, D. A. (1991). Characteristics-based methods applied to infinite Prandtl number thermal-convection in the hard turbulent regime. *Physics of Fluids A*, **3**, 2105–2115.

Malevsky, A. V. and Yuen, D. A. (1992). Strongly chaotic non-Newtonian mantle convection. *Geophysical and Astrophysical Fluid Dynamics*, **65**, 149–171.

Malevsky, A. V. and Yuen, D. A. (1993). Plume structures in the hard-turbulent regime of three-dimensional infinite Prandtl number convection. *Geophysical Research Letters*, **20**, 383–386.

Malevsky, A. V., Yuen, D. A. and Weyer, L. M. (1992). Viscosity and thermal fields associated with strongly chaotic non-Newtonian thermal convection. *Geophysical Research Letters*, **19**, 127–130.

Marchuk, G. I. (1958). *Numerical Methods for Solutions of Nuclear Reactors*. Moscow: Atomizdat (in Russian).

Marchuk, G. I. (1989). *Methods of Computational Mathematics*. Moscow: Nauka (in Russian).

Martin, M., Ritter, J. R. R. and the CALIXTO working group (2005). High-resolution teleseismic body-wave tomography beneath SE Romania – I. Implications for three-dimensional versus one-dimensional crustal correction strategies with a new crustal velocity model. *Geophysical Journal International*, **162**, 448–460.

Martin, M., Wenzel, F. and the CALIXTO working group (2006). High-resolution teleseismic body wave tomography beneath SE-Romania – II. Imaging of a slab detachment scenario. *Geophysical Journal International*, **164**, 579–595.

Martinec, Z., Matyska, C. Cadek, O. and Hrdina P. (1993). The Stokes problem with 3d Newtonian rheology in a spherical-shell. *Computer Physics Communications*, **76**, 63–79.

Matsui, H. and Buffett, B. A. (2005). Sub-grid scale model for convection-driven dynamos in a rotating plane layer. *Physics of the Earth and Planetary Interiors*, **153**, 108–123.

Matsui, H. and Okuda, H. (2004). Treatment of the magnetic field for geodynamo simulations using the finite element method. *Earth Planets Space*, **56**, 945–954.

May, D. A. and Moresi, L. (2008). Preconditioned iterative methods for Stokes flow problems arising in computational geodynamics. *Physics of the Earth and Planetary Interiors*, **171**, 33–47.

McDonald, A. and Bates, J. R. (1987). Improving the estimate of the departure point iosition in a two-time level semi-Lagrangian and semi-implicit scheme. *Monthly Weather Review*, **115**, 737–739.

McHenry, D. (1943). A lattice analogy for the solution of plane stress problems. *Journal of the Institute of Civil Engineers*, **21**, 59–82.

McKenzie, D. P. (1972). Active tectonics of the Mediterranean region. *Geophysical Journal of the Royal Astronomical Society*, **30**, 109–185.

McKenzie, D. (1979). Finite deformation during fluid flow. *Geophysical Journal of the Royal Astronomical Society*, **58**, 689–715.

McKenzie, D. (1984). The generation and compaction of partially molten rock. *Journal of Petrology*, **25**, 713–765.

McKenzie, D. P., Roberts, J. M. and Weiss, N. O. (1974). Convection in the Earth's mantle: towards a numerical simulation. *Journal of Fluid Mechanics*, **62**, 465–538.

McLaughlin D. (2002). An integrated approach to hydrologic data assimilation: Interpolation, smoothing, and forecasting. *Advances in Water Resources*, **25**, 1275–1286.

McNamara, A. K. and Zhong, S. (2004). Thermochemical structures within a spherical mantle: Superplumes or piles? *Journal of Geophysical Research*, **109**, doi:10.1029/2003JB00287.

McNamara, A. K. and Zhong, S. (2005). Thermochemical piles beneath Africa and the Pacific Ocean. *Nature*, **437**, 1136–1139.

Minear, J. W. and Toksöz, M. N. (1970). Thermal regime of a downgoing slab and new global tectonics. *Journal of Geophysical Research*, **75**, 1397–1419.

Mishin, Y. A., Gerya, T. V., Burg, J.-P. and Connolly, J. A. D. (2008). Dynamics of double subduction: Numerical modeling. *Physics of the Earth and Planetary Interiors*, **171**, 280–295.

Mitrovica, J. X. (1996). Haskell (1935) revisited. *Journal of Geophysical Research*, **101**, 555–569.

Mitrovica, J. X. and Forte, A. M. (2004). A new inference of mantle viscosity based upon joint inversion of convection and glacial isostatic adjustment data. *Earth and Planetary Science Letters*, **225**, 177–189.

Monnereau, M. and Quere, S. (2001). Spherical shell models of mantle convection with tectonic plates. *Earth and Planetary Science Letters*, **184**, 575–587.

Monnereau, M. and Yuen, D. A. (2007). Topology of the postperovskite phase transition and mantle dynamics. *Proceedings of the National Academy of Sciences*, **104**, 9156–9161.

Montelli, R., Nolet, G., Dahlen, F.A. *et al.* (2004). Finite-frequency tomography reveals a variety of plumes in the mantle. *Science*, **303**, 338–343.

Moore, W. B., Schubert, G. and Tackley, P. (1998). Three-dimensional simulations of plume–lithosphere interaction at the Hawaiian Swell. *Science*, **279**, 1008–1011.

Moresi, L. and Gurnis, M. (1996). Constraints on the lateral strength of slabs from 3-dimensional dynamic flow models. *Earth and Planetary Science Letters*, **138**, 15–28.

Moresi, L. N. and Lenardic, A. (1997). 3-Dimensional numerical simulations of crustal deformation and subcontinental mantle convection. *Earth and Planetary Science Letters*, **150**, 233–243.

Moresi, L. and Muhlhaus, H.-B. (2006). Anisotropic viscous models of large-deformation Mohr–Coulomb failure. *Philosophical Magazine*, **86**, 3287–3305.

Moresi, L. N. and Solomatov, V. S. (1995). Numerical investigation of 2d convection with extremely large viscosity variations. *Physics of Fluids*, **7**, 2154–2162.

Moresi, L. and Solomatov, V. (1998). Mantle convection with a brittle lithosphere – Thoughts on the global tectonic styles of the Earth and Venus. *Geophysical Journal International*, **133**, 669–682.

Moresi, L., Zhong, S. and Gurnis, M. (1996). The accuracy of finite-element solutions of Stokes-flow with strongly varying viscosity. *Physics of the Earth and Planetary Interiors*, **97**, 83–94.

Moresi, L. N., Dufour, F. and Muhlhaus, H.-B. (2003). A Lagrangian integration point finite element method for large deformation modeling of viscoelastic geomaterials. *Journal of Computational Physics*, **184**, 476–497.

Moresi, L., Quenette, S., Lemiale, V. *et al.* (2007). Computational approaches to studying non-linear dynamics of the crust and mantle. *Physics of the Earth and Planetary Interiors*, **163**, 69–82.

Morgan, W. J. (1972). Plate motions and deep convection. *Geological Society of America Memoirs*, **132**, 7–22.

Morley, C. K. (1996). Models for relative motion of crustal blocks within the Carpathian region, based on restorations of the outer Carpathian thrust sheets. *Tectonics*, **15**, 885–904.

Morra, G., Regenauer-Lieb, K. and Giardini, D. (2006). Curvature of oceanic arcs. *Geology*, **34**, 877–880.

Morse, P.M. and Feshbach, H. (1953). *Methods of Theoretical Physics*. New York: McGraw-Hill.

Morton, K. W. and Mayers, D. F. (2005). *Numerical Solution of Partial Differential Equations, An Introduction.* Cambridge: Cambridge University Press.

Moucha, R., Forte, A. M., Mitrovica, J. X. and Daradich, A. (2007). Lateral variations in mantle rheology: implications for convection related surface observables and inferred viscosity models. *Geophysical Journal International*, **169**, 113–135.

Muhlhaus, H.-B. and Regenauer-Lieb, K. (2005). Towards a self-consistent plate mantle model that includes elasticity: simple benchmarks and application to basic modes of convection. *Geophysical Journal International*, **163**, 788–800.

Naimark, B. M. and Ismail-Zadeh, A. T. (1995). Numerical models of subsidence mechanism in intracratonic basins: Application to North American basins. *Geophysical Journal International*, **123**, 149–160.

Naimark, B. M. and Malevsky, A. V. (1984). Economic method of bicubic spline interpolation. *Computational Seismology* (in Russian, translated to English by Allerton Press, N. Y.), **17**, 141–149.

Naimark, B. M. and Malevsky, A. V. (1988). Approximate method of solving the problem of gravitational and thermal stability and calculations of displacement and stress fields for upper mantle models. *Computational Seismology* (in Russian, translated to English by Allerton Press, N. Y.), **20**, 33–52.

Naimark, B. M., Ismail-Zadeh, A. T. and Jacoby, W. R. (1998). Numerical approach to problems of gravitational instability of geostructures with advected material boundaries. *Geophysical Journal International*, **134**, 473–483.

Nakagawa, T. and Tackley, P. J. (2004). Effects of thermo-chemical mantle convection on the thermal evolution of the Earth's core. *Earth and Planetary Science Letters*, **220**, 107–119.

Nakagawa, T. and Tackley, P. J. (2006). Three-dimensional structures and dynamics in the deep mantle: Effects of post-perovskite phase change and deep mantle layering. *Geophysical Research Letters*, **33**, doi:10.1029/2006GL025719.

Nakagawa, T. and Tackley, P. J. (2008). Lateral variations in CMB heat flux and deep mantle seismic velocity caused by a thermal-chemical-phase boundary layer in 3D spherical convection. *Earth and Planetary Science Letters*, **271**, 348–358.

Nakagawa, T., Tackley, P. J., Deschamps, F. and Connolly, J. A. D. (2009). Incorporating self-consistently calculated mineral physics into thermo-chemical mantle convection simulations in a 3D spherical shell and its influence on seismic anomalies in Earth's mantle. *Geochemistry, Geophysics, Geosystems*, **10**, doi:10.1029/2008GC002280.

Neumann, G. A. and Zuber, M. T. (1995). A continuum approach to the development of normal faults. In: Daemen, J. J. K. and Schultz, R. A. (eds.) *Rock Mechanics: Proceedings of the 35th U. S. Symposium on Rocks Mechanics*. Rotterdam: AA Balkema, pp. 191–198.

Nikolaeva, K., Gerya, T. V. and Connolly, J. A. D. (2008). Numerical modelling of crustal growth in intraoceanic volcanic arcs. *Physics of the Earth and Planetary Interiors*, **171**, 336–358.

Nitoi, E., Munteanu, M., Marincea, S. *et al.* (2002). Magma-enclave interactions in the East Carpathian Subvolcanic Zone, Romania: petrogenetic implications. *Journal of Volcanology and Geothermal Resources*, **118**, 229–259.

Nolet, G. and Dahlen, F. A. (2000). Wave front healing and the evolution of seismic delay times. *Journal of Geophysical Research*, **105**, 19 043–19 054.

Ogawa, M. (2003). Plate-like regime of a numerically modeled thermal convection in a fluid with temperature-, pressure-, and stress-dependent viscosity. *Journal of Geophysical Research*, **108**, doi:10.1029/2000JB000069.

Ogawa, M., Schubert, G. and Zebib, A. (1991). Numerical simulations of 3-dimensional thermal convection in a fluid with strongly temperature-dependent viscosity. *Journal of Fluid Mechanics*, **233**, 299–328.

Okuda, H., Yagawa, G. and Nakamura, H. (2003). Parallel FE solid Earth simulator Tiger and Snake, *SC2003 November 15–21, 2003. Phoenix, Arizona, U.S.A.*

Olson, P. and Singer, H. (1985). Creeping plumes. *Journal of Fluid Mechanics*, **158**, 511–531.

Oncescu, M. C. (1984). Deep structure of the Vrancea region, Romania, inferred from simultaneous inversion for hypocentres and 3-D velocity structure. *Annals Geophysicae*, **2**, 23–28.

Oncescu, M. C. and Bonjer, K. P. (1997). A note on the depth recurrence and strain release of large Vrancea earthquakes. *Tectonophysics*, **272**, 291–302.

O'Neill, C., Moresi, L., Mueller,D., Albert, R. and Dufour, F. (2006). Ellipsis 3D: A particle-in-cell finite-element hybrid code for modelling mantle convection and lithospheric deformation. *Computers & Geosciences*, **32**, 1769–1779.

O'Neill, C. J., Lenardic, A., Griffin, W. L. and O'Reilly, S. Y. (2008). Dynamics of cratons in an evolving mantle. *Lithos*, **102**, 12–24.

Ortega, J. M. (1988). *Introduction to Parallel and Vector Solution of Linear Systems*. New York: Plenum Press.

Ortega, J. and Rheinboldt, W. (1970). *Iterative Solution of Nonlinear Equations in Several Variables*. New York: Academic Press.

Osher, S. and Chakravarthy, S. (1984). High resolution schemes and the entropy condition. *SIAM Journal of Numerical Analysis*, **21**, 984–995.

Pacheco, P. S. (1997). *Programming Parallel with MPI*. San Francisco: Morgan Kaufmann.

Pana, D. and Erdmer, P. (1996). Kinematics of retreating subduction along the Carpathian arc, Romania: Comment. *Geology*, **24**, 862–863.

Pana, D. and Morris, G. A. (1999). Slab in the wrong place: Lower lithospheric mantle delamination in the last stage of the Eastern Carpathian subduction retreat: Comment. *Geology*, **27**, 665–666.

Parmentier, E. M. and Turcotte, D. L. (1974). An explanation of the pyroxene geotherm based on plume convection in the upper mantle. *Earth and Planetary Science Letters*, **24**, 209–212.

Parmentier, E. M., Turcotte, D. L. and Torrance, K. E. (1975). Numerical experiments on the structure of mantle plumes. *Journal of Geophysical Research*, **80**, 4417–4424.

Parmentier, E. M., Sotin, C. and Travis, B. J. (1994). Turbulent 3-D thermal convection in an infinite Prandtl number, volumetrically heated fluid – Implications for mantle dynamics. *Geophysical Journal International*, **116**, 241–251.

Parter, S. (ed.) (1979). *Numerical Methods for Partial Differential Equations*. New York: Academic Press.

Patankar, S. V. (1980). *Numerical Heat Transfer and Fluid Flow*. New York: McGraw-Hill.

Patankar, S. V. and Spalding, D. B. (1972). A calculation procedure for heat and mass transfer in three-dimensional parabolic flows. *International Journal of Heat and Mass Transfer*, **15**, 1787–1806.

Peltier, W. R. (1972). Penetrative convection in the planetary mantle. *Geophysical and Astrophysical Fluid Dynamics*, **5**, 47–88.

Peltier, W. R. and Solheim, L. P. (1992). Mantle phase-transitions and layered chaotic convection. *Geophysical Research Letters*, **19**, 321–324.

Perelman, G. (2002). The entropy formula for the Ricci flow and its geometric applications, http://arxiv.org/abs/math.DG/0211159.

Petersen, W. P. and Arbenz, P. (2004). *Introduction to Parallel Computing: A Practical Guide with Examples in C*. Oxford: Oxford University Press.

Petrini, K., Connolly, J. A. D. and Podladchikov, Y. Y. (2001). A coupled petrological-tectonic model for sedimentary basin evolution: the influence of metamorphic reactions on basin subsidence. *Terra Nova*, **13**, 354–359.

Petrunin, A. and Sobolev, S. V. (2006). What controls thickness of sediments and lithospheric deformation at a pull-apart basin? *Geology*, **34**, 389–392.

Petrunin, A. G. and Sobolev, S. V. (2008). Three-dimensional numerical models of the evolution of pull-apart basins. *Physics of the Earth and Planetary Interiors*, **171**, 387–399.

Phillips, B. R. and Bunge, H. P. (2005). Heterogeneity and time dependence in 3D spherical mantle convection models with continental drift. *Earth and Planetary Science Letters*, **233**, 121–135.

Phillips, B. R. and Bunge, H. P. (2007). Supercontinent cycles disrupted by strong mantle plumes. *Geology*, **35**, 847–850.

Phinney, R. A. and Burridge, R. (1973). Representation of the elastic-gravitational excitation of a spherical Earth model by generalized spherical harmonics. *Geophysical Journal of the Royal Astronomical Society*, **74**, 451–487.

Piazzoni, A., Steinle-Neumann, G., Bunge, H. P. and Dolejs, D. (2007). A mineralogical model for density and elasticity of the Earth's mantle. *Geochemistry, Geophysics, Geosystems*, **8**, doi: 10.1029/2007GC001697.

Poliakov, A. and Podladchikov, Y. (1992). Diapirism and topography. *Geophysical Journal International*, **109**, 553–564.

Poliakov, A. N. B., van Balen, R., Podladchikov, Yu. *et al.* (1993a). Numerical analysis of how sedimentation and redistribution of surficial sediments affects salt diapirism. *Tectonophysics*, **226**, 199–216.

Poliakov, A. N. B., Podladchikov, Y. and Talbot, C. (1993b). Initiation of salt diapirs with frictional overburdens: numerical experiments. *Tectonophysics,* **228**, 199–210.

Poliakov, A., Cundall, P., Podladchikov, Y. and Lyakhovsky, V. (1993c). An explicit inertial method for the simulation of viscoelastic flow: an evaluation of elastic effects on diapiric flow in two- and three- layers models. In: Stone, D. B. and Runcorn, S. K. (eds.) *Flow and Creep in the Solar System: Observations, Modeling and Theory*. Dortrecht: Kluwer, pp. 175–195.

Poliakov, A., Podladchikov, Y., Dawson, E. *et al.* (1996). Salt diapirism with simultaneous brittle faulting and viscous flow. In: Alsop, I., Blundell, D., Davison, I. (eds.) *Salt Tectonics*. Geological Society Special Publication No. 100, London, pp. 291–302.

Popa, M., Radulian, M., Grecu, B. *et al.* (2005). Attenuation in Southeastern Carpathians area: Result of upper mantle inhomogeneity. *Tectonophysics*, **410**, 235–249.

Popov, A. A. and Sobolev, A. V. (2008). SLIM3D: A tool for three-dimensional thermomechanical modeling of lithospheric deformation with elasto-visco-plastic rheology. *Physics of the Earth and Planetary Interiors*, **171**, 55–75.

Prager, W. and Hodge, P. (1951). *Theory of Perfectly Plastic Solids*. London: Chapman & Hall.

Press, W. H., Teukolsky, S. A., Vetterling, W. T. and Flannery, B. P. (2007). *Numerical Recipes –The Art of Scientific Computing*, 3rd edn. Cambridge: Cambridge University Press.

Quinn, M. J. (2003). *Parallel Programming in C with MPI and OpenMP.* New York: McGraw-Hill.

Ramberg, H. (1968). Instability of layered system in the field of gravity. II. *Physics of the Earth and Planetary Interiors*, **1**, 448–474.

Ranalli, G. (1995). *Rheology of the Earth*, 2nd edn. London: Chapman and Hall.

Reese, C. C., Solomatov, V. S., Baumgardner, J. R. and Yang, W.-S. (1999). Stagnant lid convection in a spherical shell. *Physics of the Earth and Planetary Interiors*, **116**, 1–7.

Reiner, M. (1964). The Deborah number. *Physics Today*, 17, 62.

Rhie, C. M. and Chow, W. L. (1983). A numerical study of the turbulent flow past an airfoil with trailing edge separation. *American Institute of Aeronautics and Astronautics Journal*, 21, 1525–1532.

Ribe, N. M. and Christensen, U. (1994). Three-dimensional modeling of plume–lithosphere interaction. *Journal of Geophysical Research*, **99**, 669–682.

Ricard, Y. and Vigny, C. (1989). Mantle dynamics with induced plate tectonics. *Journal of Geophysical Research*, **94**, 17 543–17 559.

Ricard, Y., Fleitout, L. and Froidevaux, C. (1984). Geoid heights and lithospheric stresses for a dynamic Earth. *Annales Geophysicae*, **2**, 267–286.

Ricard, Y., Richards, M., Lithgow-Bertelloni, C. and Le Stunff, Y. (1993). A geodynamic model of mantle density heterogeneity. *Journal of Geophysical Research*, **98**, 21 895–21 909.

Ricard, Y., Bercovici, D. and Schubert, G. (2001). A two-phase model for compaction and damage 2. Applications to compaction, deformation, and the role of interfacial surface tension. *Journal of Geophysical Research*, **106**, 8907–8924.

Ricard, Y., Mattern, E. and Matas, J. (2005). Synthetic tomographic images of slabs from mineral physics. In: van der Hilst, R. D., Bass, J. D., Matas, J., and Trampert, J. (eds.) *Earth's Deep Mantle: Structure, Composition, and Evolution*. Washington D.C.: American Geophysical Union, pp. 285–302.

Richards, M. A. and Hager, B. H. (1984). Geoid anomalies in a dynamic Earth. *Journal of Geophysical Research*, **89**, 5987–6002.

Richards, M. A., Duncan, R. A. and Courtillot, V. (1989). Flood basalts and hot spot tracks: Plume heads and tails. *Science*, **246**, 103–107.

Richards, M. A., Yang, W. S., Baumgardner, J. R. and Bunge, H. P. (2001). Role of a low-viscosity zone in stabilizing plate tectonics: Implications for comparative terrestrial planetology. *Geochemistry, Geophysics, Geosystems*, **2**, doi:10.1029/2000GC000115.

Richardson, L. F. (1910). The approximate arithmetical solution by finite differences of physical problems including differential equations, with an application to the stresses in a masonry dam, *Philosophical Transactions of the Royal Society of London, Series A*, **210**, 307–357.

Richter, F. M. (1973). Finite amplitude convection through a phase boundary. *Geophysical Journal of the Royal Astronomical Society*, **35**, 265–276.

Richtmayer, R. D. and Morton, K. W. (1967). *Difference Methods for Initial-Value Problems*, 2nd edn. New York: Interscience.

Roe, P. L. (1985). Some contributions to the modelling of discontinuous flows. *Lectures in Applied Mechanics*, **22**, 163–193.

Romanowicz, B. and Gung, Y. (2002). Superplumes from the core-mantle boundary to the lithosphere: Implications for heat flux. *Science*, **296**, 513–516.

Römer, M.-M. and Neugebauer, H. J. (1991). The salt dome problem: a multilayered approach. *Journal of Geophysical Research*, **96**, 2389–2396.

Ronchi, C., Iacono, R., Paolucci, P. S., Hertzberger, B. and Serazzi, G. (1996). The "cubed sphere": a new method for the solution of partial differential equations in spherical geometry. *Journal of Computational Physics*, **124**, 93–114.

Roosta, S. H. (2000). *Parallel Processing and Parallel Algorithms: Theory and Computation*. New York: Springer.

Rotvig, J. and Jones, C. A. (2002). Rotating convection-driven dynamos at low Ekman number. *Physical Review E. Statistical Physics, Plasmas, Fluids, & Related Interdisciplinary Topics*, **66**, 56 308–1-15.

Royden, L. H. (1988). Late Cenozoic tectonics of the Pannonian basin system. *American Association of Petroleum Geology Memoirs*, **45**, 27–48.

Ruepke, L., Phipps Morgan, J., Hort, M. and Connolly, J. A. D. (2004). Serpentine and the subduction zone water cycle. *Earth and Planetary Science Letters*, **223**, 17–34.

Runge, C. (1908). Über eine Methode, die partielle Differentialgleichung Δu = constant numerisch zu integrieren. *Zeitschrift für Mathematische Physik*, **56**, 225–232 (in German).

Rykov, V. V. and Trubitsyn, V. (2006). Numerical technique for calculation of three-dimensional mantle convection and tectonics of continental plates. *Computational Seismology and Geodynamics*, **3**, 17–22.

Saad, Y. (1996). *Iterative Methods for Sparse Linear Systems*. Boston: PWS Publishing.

Samarskii, A. A. (1967). On regularisation of finite difference schemes. *Computational Mathematics and Mathematical Physics*, **7**, 62–93.

Samarskii, A. A. (1977). *Theory of Finite-Difference Schemes*. Moscow: Nauka (in Russian).

Samarskii, A. A. and Vabishchevich, P. N. (1995). *Computational Heat Transfer. Vol. 1. Mathematical Modelling.* New York: John Wiley & Sons.

Samarskii, A. A. and Vabishchevich, P. N. (2004). *Numerical Methods for Solving Inverse Problems of Mathematical Physics*, Moscow: URSS (in Russian).

Samarskii, A. A., Vabishchevich, P. N. and Vasiliev, V. I. (1997). Iterative solution of a retrospective inverse problem of heat conduction. *Mathematical Modeling*, **9**(5), 119–127.

Sandulescu, M. (1988). Cenozoic tectonic history of the Carpathians. *American Association of Petroleum Geology Memoirs*, **45**, 17–25.

Schmalholz, S. M. (2006). Scaled amplification equation: a key to the folding history of buckled viscous single layers. *Tectonophysics*, **419**, 41–53.

Schmalholz, S. M., Podladchikov, Y. Y. and Schmid, D. W. (2001). A spectral/finite-difference method for simulating large deformations of heterogeneous, viscoelastic materials. *Geophysical Journal International*, **145**, 188–208.

Schmalholz, S. M., Kaus, B. J. P. and Burg, J. P. (2009). Stress–strength relationship in the lithosphere during continental collision. *Geology*, **37**, 775–778.

Schmalzl, J. and Hansen, U. (2000). A fully implicit model for simulating dynamo action in a Cartesian domain. *Physics of the Earth and Planetary Interiors*, **120**, 339–49.

Schmalzl, J. and Loddoch, A. (2003). Using subdivision surfaces and adaptive surface simplification algorithms for modeling chemical heterogeneities in geophysical flows. *Geochemistry, Geophysics, Geosystems*, **4**, doi:10.1029/2003GC000578.

Schmeling, H. (1987). On the relation between initial conditions and late stages of Rayleigh–Taylor instabilities. *Tectonophysics*, **133**, 65–80.

Schmeling, H. (1989). Compressible convection with constant and variable viscosity – the effect on slab formation, geoid, and topography. *Journal of Geophysical Research*, **94**, 12 463–12 481.

Schmeling, H. (2000). Partial melting and melt segregation in a convecting mantle. In: Bagdassarov, N., Laporte, D. and Thompson, A. B. (eds.) *Physics and Chemistry of Partially Molten Rocks*. Dordrecht: Kluwer, pp. 141–178.

Schmeling, H. and Jacoby, W. R. (1981). On modeling the lithosphere in mantle convection with non-linear rheology. *Journal of Geophysics*, **50**, 89–100.

Schmeling, H., Babeyko, A., Enns, A. *et al.* (2008). A benchmark comparison of spontaneous subduction models – towards a free surface. *Physics of the Earth and Planetary Interiors*, **171**, 198–223.

Schubert, G. (ed.) (2007). *Treatise on Geophysics*, 11-Volume Set, Elsevier.

Schubert, G. and Anderson, C. A. (1985). Finite-element calculations of very high Rayleigh number thermal-convection. *Geophysical Journal of the Royal Astronomical Society*, **80**, 575–601.

Schubert, G., Yuen, D. A. and Turcotte, D. L. (1975). Role of phase transitions in a dynamic mantle. *Geophysical Journal of the Royal Astronomical Society*, **42**, 705–735.

Schubert, G., Turcotte, D. L. and Olson, P. (2001). *Mantle Convection in the Earth and Planets*. Cambridge: Cambridge University Press.

Schuberth, B. S. A., Bunge, H. P., Steinle-Neumann, G., Moder, C. and Oeser, J. (2009). Thermal versus elastic heterogeneity in high-resolution mantle circulation models with pyrolite composition: High plume excess temperatures in the lowermost mantle. *Geochemistry, Geophysics, Geosystems*, **10**, doi: 10.1029/2008GC002235.

Scott, D. R. and Stevenson, D. J. (1989). A self-consistent model of melting, magma migration and buoyancy-driven circulation beneath mid-ocean ridges. *Journal of Geophysical Research*, **94**, 2973–88.

Seghedi, I., Downes, H., Vaselli, O. *et al.* (2004). Post-collisional Tertiary-Quaternary mafic alkalic magmatism in the Carpathian-Pannonian region: a review. *Tectonophysics*, **393**, 43–62.

Shu, C.-W. (1997). Essentially nonoscillatory (ENO) and weighted essentially nonoscillatory (WENO) schemes for hyperbolic conservation laws. In: Quarteroni, A. (ed.) *Advanced Numerical Approximation of Nonlinear Hyperbolic Equations*, Lecture Notes in Mathematics, vol. 1697. New York: Springer, pp. 325–432.

Simpson, G. (2004). Dynamic interactions between erosion, deposition, and three-dimensional deformation in compressional fold belt settings. *Journal of Geophysical Research*, **109**, doi:10.1029/2003JF000111.

Simpson, G. and Castelltort, S. (2006). Coupled model of surface water flow, sediment transport and morphological evolution. *Computers & Geosciences*, **32**, 1600–1614.

Sleep, N. H. (1990). Hotspots and mantle plumes: some phenomenology. *Journal of Geophysical Research*, **95**, 6715–6736.

Smolarkiewicz, P. K. (1984). A fully multidimensional positive definite advection transport algorithm with small implicit diffusion, *Journal of Computational Physics*, **54**, 325–362.

Snyder, L., and Lin, C. (2008). *Principles of Parallel Programming*. Boston: Addison-Wesley.

Sobolev, S. V., Petrunin, A., Garfunkel, Z. and Babeyko, A. Y. (2005). Thermo-mechanical model of the Dead Sea Transform. *Physics of the Earth and Planetary Interiors*, **238**, 78–95.

Solheim, L. P. and Peltier, W. R. (1990). Heat-transfer and the onset of chaos in a spherical, axisymmetrical, anelastic model of whole mantle convection. *Geophysical and Astrophysical Fluid Dynamics*, **53**, 205–255.

de Souza Neto, E. A., Peric, D. and Owen, D. R. J. (2008). *Computational Methods for Plasticity*. Chichester, UK: Wiley.

Sperner, B., Lorenz, F., Bonjer, K. *et al.* (2001). Slab break-off – abrupt cut or gradual detachment? New insights from the Vrancea region (SE Carpathians, Romania). *Terra Nova*, **13**, 172–179.

Sperner, B. and the CRC 461 Team (2005). Monitoring of slab detachment in the Carpathians. In: Wenzel, F. (ed.) *Challenges for Earth Sciences in the 21st Century*. Heidelberg: Springer, pp. 187–202.

Spiegelman, M. (1993). Flow in deformable porous-media.1. Simple analysis. *Journal of Fluid Mechanics*, **247**, 17–38.

Spotz, W. F. and Swarztrauber, P. N. (2001). A performance comparison of associated Legendre projections. *Journal of Computational Physics*, **168**, 339–355.

Sramek, O. (2007). Modèle d'écoulement biphasé en sciences de la Terre : fusion partielle, compaction et différenciation. Ecole Normale Superieure de Lyon, 166 p. (in French).

Sramek, O., Ricard, Y. and Bercovici, D. (2007). Simultaneous melting and compaction in deformable two-phase media. *Geophysical Journal International*, **168**, 964–982.

Stanica, D. and Stanica, M. (1993). An electrical resistivity lithospheric model in the Carpathian orogen from Romania. *Physics of the Earth and Planetary Interiors*, **81**, 99–105.

Staniforth, A. and Coté, J. (1991). Semi-Lagrangian integration schemes for atmospheric models – a review. *Monthly Weather Review*, **119**, 2206–2223.

Steinbach, V. and Yuen, D. A. (1992). The effects of multiple phase-transitions on Venusian mantle convection. *Geophysical Research Letters*, **19**, 2243–2246.

Steinberger, B., and O'Connell, R. J. (1998). Advection of plumes in mantle flow: implications for hotspot motion, mantle viscosity and plume distribution. *Geophysical Journal International*, **132**, 412–434.

Stellmach, S. and Hansen, U. (2004). Cartesian convection driven dynamos at low Ekman number. *Physical Review E*, **70**, DOI: 10.1103/PhysRevE.70.056312.

Stemmer, K., Harder, H. and Hansen, U. (2006). A new method to simulate convection with strongly temperature- and pressure-dependent viscosity in a spherical shell: Applications to the Earth's mantle. *Physics of the Earth and Planetary Interiors*, **157**, 223–249.

Stixrude, L. and Lithgow-Bertelloni, C. (2005a). Mineralogy and elasticity of the oceanic upper mantle: Origin of the low-velocity zone. *Journal of Geophysical Research*, **110**, doi:10.1029/2004JB002965.

Stixrude, L. and Lithgow-Bertelloni, C. (2005b). Thermodynamics of mantle minerals. Physical properties. *Geophysical Journal International*, **162**, 610–632.

Stixrude, L. and Lithgow-Bertelloni, C. (2007). Influence of phase transformations on lateral heterogeneity and dynamics in Earth's mantle. *Earth and Planetary Sciences Letters*, **263**, 45–55.

Suzuki, A., Tabata, M. and Honda, S. (1999). Numerical solution of an unsteady Earth's mantle convection problem by a stabilized finite element method. *Theoretical and Applied Mechanics*, **48**, 371–378.

Sweby, P. K. (1984). High resolution schemes using flux limiters for hyperbolic conservation laws. *SIAM Journal of Numerical Analysis*, **21**, 995–1011.

Szabó, C., Falus, G., Zajacz, Z. *et al.* (2004). Composition and evolution of lithosphere beneath the Carpathian-Pannonian Region: a review. *Tectonophysics*, **393**, 119–137.

Tabata, M. (2006). Finite element approximation to infinite Prandtl number Boussinesq equations with temperature dependent coefficients – thermal convection problems in a spherical shell. *Future Generation Computer Systems*, **22**, 521–531.

Tabata, M. and Suzuki, A. (2000). A stabilized finite element method for the Rayleigh–Benard equations with infinite Prandtl number in a spherical shell. *Computational Methods in Applied Mechanics and Engineering*, **190**, 387–402.

Tackley, P. J. (1993). Effects of strongly temperature-dependent viscosity on time-dependent, 3-dimensional models of mantle convection. *Geophysical Research Letters*, **20**, 2187–2190.

Tackley, P. J. (1994). *Three-dimensional models of mantle convection: Influence of phase transitions and temperature-dependent viscosity*, Ph.D. Thesis, California Institute of Technology 299 p.

Tackley, P. J. (1995). On the penetration of an endothermic phase transition by upwellings and downwellings. *Journal of Geophysical Research*, **100**, 15 477–15 488.

Tackley, P. J. (1996a). Effects of strongly variable viscosity on three-dimensional compressible convection in planetary mantles. *Journal of Geophysical Research*, **101**, 3311–3332.

Tackley, P. J. (1996b). On the ability of phase transitions and viscosity layering to induce long-wavelength heterogeneity in the mantle. *Geophysical Research Letters*, **23**, 1985–1988.

Tackley, P. J. (1998). Self-consistent generation of tectonic plates in three-dimensional mantle convection. *Earth and Planetary Sciences Letters*, **157**, 9–22.

Tackley, P. J. (2000a). Self-consistent generation of tectonic plates in time-dependent, three-dimensional mantle convection simulations Part 1: Pseudo-plastic yielding. *Geochemical Geophysics Geosystems*, **1**, Paper number 2000GC000036.

Tackley, P. J. (2000b). Self-consistent generation of tectonic plates in time-dependent, three-dimensional mantle convection simulations Part 2: Strain weakening and asthenosphere. *Geochemistry Geophysics Geosystems*, **1**, Paper number 2000GC000043.

Tackley, P. J. (2000c). The quest for self-consistent incorporation of plate tectonics in mantle convection. In: Richards, M. A., Gordon, R. and van der Hilst, R. (eds.) *History and Dynamics of Global Plate Motions*. Washington, D.C., American Geophysical Union, pp. 47–72.

Tackley, P. J. (2007). Mantle geochemical geodynamics. In: Bercovici, D. and Schubert, G. (eds.) *Treatise on Geophysics Volume 7: Mantle dynamics*. Amsterdam: Elsevier, pp. 437–505.

Tackley, P. J. (2008). Modelling compressible mantle convection with large viscosity contrasts in a three-dimensional spherical shell using the yin-yang grid. *Physics of the Earth and Planetary Interiors*, **171**, 7–18.

Tackley, P. J. and King, S. D. (2003). Testing the tracer ratio method for modeling active compositional fields in mantle convection simulations. *Geochemistry, Geophysics, Geosystems*, **4**, doi:10.1029/2001GC000214.

Tackley, P. J. and Stevenson, D. J. (1993). A mechanism for spontaneous self-perpetuating volcanism on the terrestrial planets. In: Stone, D. B. and Runcorn, S. K. (eds.) *Flow and Creep in the Solar System: Observations, Modeling and Theory*. Amsterdam: Kluwer, pp. 307–322.

Tackley, P. J., Stevenson, D. J., Glatzmaier, G. A. and Schubert, G. (1993). Effects of an endothermic phase transition at 670 km depth in a spherical model of convection in the Earth's mantle. *Nature*, **361**, 699–704.

Tackley, P. J., Stevenson, D. J., Glatzmaier, G. A. and Schubert, G. (1994). Effects of multiple phase transitions in a 3-dimensional spherical model of convection in Earth's mantle. *Journal of Geophysical Research*, **99**, 15 877–15 901.

Talbot, C. J. (1995). Molding of salt diapirs by stiff overburdens. In: Jackson, M. P. A., Roberts, D. G., Snelson, S. (eds.) *Salt Tectonics – A Global Perspective*. American Association of Petroleum Geologists. Memoir 65, Tulsa, pp. 61–75.

Tarapoanca, M., Carcia-Catellanos, D., Bertotti, G. *et al.* (2004). Role of the 3-D distributions of load and lithospheric strength in orogenic arcs: polystage subsidence in the Carpathians foredeep. *Earth and Planetary Science Letters*, **221**, 163–180.

Temam, R. (1977). *Navier–Stokes Equations*. Amsterdam: North-Holland.

Tikhonov, A. N. (1963). Solution of incorrectly formulated problems and the regularization method. *Doklady Akademii Nauk SSSR*, **151**, 501–504 (Engl. transl.: *Soviet Math. Dokl.*, **4**, 1035–1038, 1963).

Tikhonov, A. N. and Arsenin, V. Y. (1977). *Solution of Ill-Posed Problems*. Washington, DC: Winston.

Tikhonov, A. N., and Samarskii, A. A. (1990). *Equations of Mathematical Physics*. New York: Dover Publications.

Tilgner, A. (1999). Spectral methods for the simulation of incompressible flows in spherical shells. *International Journal of Numerical Methods in Fluids*, **30**, 713–724.

Tirone, M., Ganguly, J. and Morgan, J. P. (2009). Modeling petrological geodynamics in the Earth's mantle. *Geochemistry, Geophysics, Geosystems*, **10**, Q04012, doi:10.1029/2008GC002168.

Torrance, K. E. and Turcotte, D. L. (1971). Thermal convection with large viscosity variations. *Journal of Fluid Mechanics*, **47**, 113–125.

Travis, B., Olson, P. and Schubert, G. (1990a). The transition from 2-dimensional to 3-dimensional planforms in infinite-Prandtl-number thermal-convection. *Journal of Fluid Mechanics*, **216**, 71–91.

Travis, B. J., Anderson, C., Baumgardner, J. *et al.* (1990b). A benchmark comparison of numerical-methods for infinite prandtl number thermal-convection in 2-dimensional Cartesian geometry. *Geophysical and Astrophysical Fluid Dynamics*, **55**, 137–160.

Trompert, R. A. and Hansen, U. (1996). The application of a finite-volume multigrid method to 3-dimensional flow problems in a highly viscous fluid with a variable viscosity. *Geophysical and Astrophysical Fluid Dynamics*, **83**, 261–291.

Trompert, R. A., and Hansen, U. (1998a). On the Rayleigh number dependence of convection with a strongly temperature-dependent viscosity. *Physics of Fluids*, **10**, 351–360.

Trompert, R. and Hansen, U. (1998b). Mantle convection simulations with rheologies that generate plate-like behavior. *Nature*, **395**, 686–689.

Trubitsyn, V. P. and Rykov, V. V. (1995). A 3-D numerical-model of the Wilson cycle. *Journal of Geodynamics*, **20**, 63–75.

Trubitsyn, V. P. and Rykov, V. V. (2000). *3-D Spherical Models of Mantle Convection with Floating Continents*. Menlo Park: USGS.

Trubitsyn, V., Kaban, M., Mooney, W., Reigber, C. and Schwintzer, P. (2006). Simulation of active tectonics processes for a convecting mantle with moving continents. *Geophysical Journal International*, **164**, 611–623.

Trubitsyn, V., Kaban, M. K. and Rothacher, M. (2008). Mechanical and thermal effects of floating continents on the global mantle convection. *Physics of the Earth and Planetary Interiors*, **171**, 313–322.

Trushkov, V. V. (2002). An example of (3+1)-dimensional integrable system. *Acta Applied Mathematics*, **62**, 111–122.

Turcotte, D. L. and Schubert, G. (2002). *Geodynamics*, 2nd edn. Cambridge: Cambridge University Press.

Turcotte, D. L., Torrance, K. E. and Hsui, A. T. (1973). Convection in the Earth's mantle. In: Bolt, B. A. *et al.* (ed.) *Methods in Computational Physics: Geophysics*. New York: Academic Press, pp. 431–454.

Ueda, K., Gerya, T. and Sobolev, S. V. (2008). Subduction initiation by thermal-chemical plumes: Numerical studies. *Physics of the Earth and Planetary Interiors*, **171**, 296–312.

Vanka, S. P. (1986). Block implicit multigrid solution of Navier–Stokes equations with primitive variables. *Journal of Computational Physics*, **65**, 138–158.

van Heck, H. and Tackley, P. J. (2008). Planforms of self-consistently generated plate tectonics in 3-D spherical geometry. *Geophysical Research Letters*, **35**, doi:10.1029/2008GL035190.

van Keken, P. (2001). Cylindrical scaling for dynamical cooling models of the Earth. *Physics of the Earth and Planetary Interiors*, **124**, 119–30.

van Keken, P. E., Spiers, C. J., van den Berg, A. P. *et al.* (1993). The effective viscosity of rocksalt: implementation of steady-state creep laws in numerical models of salt diapirism. *Tectonophysics*, **225**, 457–476.

van Keken, P. E., King, S. D., Schmeling, H. *et al.* (1997). A comparison of methods for the modeling of thermochemical convection. *Journal of Geophysical Research*, **102**, 22 477–95.

van Keken, P. E., Ballentine, C. J. and Porcelli, D. (2001). A dynamical investigation of the heat and helium imbalance. *Earth and Planetary Science Letters*, **188,** 421–34.

van Leer, B. (1974). Towards the ultimate conservative difference scheme II. Monoticity and conservation combined in a second-order scheme. *Journal of Computational Physics*, **14**, 361–370.

Vasiliev, F. P. (2002). *Methody optimizatsii*. Moscow: Factorial Press (in Russian).

Vendeville, B. C. and Jackson, M. P. A. (1992). The rise of diapirs during thin-skinned extension. *Marine and Petroleum Geology*, **9**, 331–353.

Verfürth, R. (1996). *A Review of Posteriori Error Estimations and Adaptive Mesh-Refinement Techniques*. Stuttgart: Teubner-Wiley.

Vernotte, P. (1958). Les paradoxes de la theorie continue de l'equation de la chaleur. *Comptes Rendus*, **246**, 3154–3155 (in French).

Versteeg, H. K. and Malalasekera, W. (2007). *An Introduction to Computational Fluid Dynamics: The Finite Volume Method,* 2nd edn. Harlow: Pearson Education Limited.

Wang, Y. and Hutter, K. (2001). Comparison of numerical methods with respect to convectively dominated problems. *International Journal of Numerical Methods in Fluids*, **37**, 721–745.

Weidle, C., Wenzel, F. and Ismail-Zadeh, A. (2007). t* – an unsuitable parameter for anelastic attenuation in the Eastern Carpathians. *Geophysical Journal International*, **170**, 1139–1150.

Weinberg, R. F. and Schmeling, H. (1992). Polydiapirs – multiwavelength gravity structures. *Journal of Structural Geology*, **14**, 425–436.

Weinstein, S. A. (1993). Catastrophic overturn of the Earth's mantle driven by multiple phase-changes and internal heat-generation. *Geophysical Research Letters*, **20**, 101–104.

Weinstein, S. A. (1996). Thermal convection in a cylindrical annulus with a non-Newtonian outer surface. *Pure and Applied Geophysics*, **146**, 551–572.

Weinstein, S. A. and Olson, P. L. (1992). Thermal convection with non-Newtonian plates. *Geophysical Journal International*, **111**, 515–530.

Weiss, J., Maruszewski, J. P. and Smith, W. A. (1999). Implicit solution of preconditioned Navier–Stokes equations using algebraic multigrid. *American Institute of Aeronautics and Astronautics Journal*, **37**, 29–36.

Wesseling, P. (1992). *An Introduction to Multigrid Methods*. Chichester: John Wiley and Sons.

Wesseling, P. (2001). *Principles of Computational Fluid Dynamics*. Berlin: Springer.

Wicht, J. and Busse, F. H. (1997). Magnetohydrodynamic dynamos in rotating spherical shells. *Geophysical and Astrophysical Fluid Dynamics*, **86**, 103–129.

Wiggins, C. and Spiegelman, M. (1995). Magma migration and magmatic solitary waves in 3-D. *Geophysical Research Letters*, **22**, 1289–1292.

Willett, S., Beaumont, C. and Fullsack, P. (1993). Mechanical model for the tectonics of doubly vergent compressional orogens. *Geology*, **21**, 371–374.

Willett, S. D. (1999). Orogeny and orography: The effects of erosion on the structure of mountain belts. *Journal of Geophysical Research*, **104**, 28 957–28 981.

Wilson, J. T. (1966). Did the Atlantic close and then re-open? *Nature*, **211**, 676–681.

Woidt, W.-D. (1978). Finite element calculations applied to salt dome analysis. *Tectonophysics*, **50**, 369–386.

Wolstencroft, M., Davies, J. H. and Davies, D. R. (2009). Nusselt–Rayleigh number scaling for spherical shell Earth mantle simulations up to a Rayleigh number of 10^9. *Physics of the Earth and Planetary Interiors*, **176**, 132–141.

Wortel, M. J. R. and Spakman, W. (2000). Subduction and slab detachment in the Mediterranean–Carpathian region. *Science*, **290**, 1910–1917.

Xiao, F. and Yabe, T. (2001). Completely conservative and oscillationless semi-Lagrangian schemes for advection transportation. *Journal of Computational Physics*, **170**, 498–522.

Xie, S. and Tackley, P. J. (2004a). Evolution of helium and argon isotopes in a convecting mantle. *Physics of the Earth and Planetary Interiors*, **146**, 417–439.

Xie, S. and Tackley, P. J. (2004b). Evolution of U–Pb and Sm–Nd systems in numerical models of mantle convection. *Journal of Geophysical Research*, **109, B11204**, doi:10.1029/2004JB003176.

Xu, W., Lithgow-Bertelloni, C., Stixrude, L. and Ritsema, J. (2008). The effect of bulk composition on mantle seismic structure. *Earth and Planetary Science Letters*, **275**, 70–79.

Yabe, T., Ogata, Y., Takizawa, K. *et al.* (2002). The next generation CIP as a conservative semi-Lagrangian solver for solid, liquid and gas. *Journal of Computational and Applied Mathematics*, **149**, 267–77.

Yang, W.-S. and Baumgardner, J. R. (2000). A matrix-dependent transfer multigrid method for strongly variable viscosity infinite Prandtl number thermal convection. *Geophysical and Astrophysical Fluid Dynamics*, **92**, 151–195.

Yoshida, M. and Kageyama, A. (2004). Application of the Yin-Yang grid to a thermal convection of a Boussinesq fluid with infinite Prandtl number in a three-dimensional spherical shell. *Geophysical Research Letters*, **31**, doi:10.1029/2004GL019970.

Yoshida, M. and Ogawa, M. (2004). The role of hot uprising plumes in the initiation of plate-like regime of three-dimensional mantle convection. *Geophysical Research Letters*, **31**, 5607.

Yoshida, M., Iwase, Y. and Honda, S. (1999). Generation of plumes under a localized high viscosity lid in 3-D spherical shell convection. *Geophysical Research Letters*, **26**, 947–50.

Young, R. E. (1974). Finite-amplitude thermal convection in a spherical shell. *Journal of Fluid Mechanics*, **63**, 695–721.

Yu, N., Imatani, S. and Inoue, T. (2004). Characteristics of temperature field due to pulsed heat input calculated by non-Fourier heat conduction hypothesis. *JSME International Journal, Series A*, **47**(4), 574–580.

Yuen, D. A., Quareni, F. and Hong, H. J. (1987). Effects from equation of state and rheology in dissipative heating in compressible mantle convection. *Nature*, **326**, 67–69.

Zebib, A., Schubert, G. and Strauss, J. M. (1980). Infinite Prandtl number thermal convection in a spherical shell. *Journal of Fluid Mechanics*, **97**, 257–277.

Zebib, A., Schubert, G., Dein, J. L. and Paliwal, R. C. (1983). Character and stability of axisymmetric thermal convection in spheres and spherical shells. *Geophysical and Astrophysical Fluid Dynamics*, **23**, 1–42.

Zhang, S. and Christensen, U. (1993). Some effects of lateral viscosity variations on geoid and surface velocities induced by density anomalies in the mantle. *Geophysical Journal International*, **114**, 531–547.

Zhang, S. and Yuen, D. A. (1996a). Intense local toroidal motion generated by variable viscosity compressible convection in 3-D spherical-shell. *Geophysical Research Letters*, **23**, 3135–3138.

Zhang, S. and Yuen, D.A. (1996b). Various influences on plumes and dynamics in time-dependent, compressible, mantle convection in 3-D spherical shell. *Physics of the Earth and Planetary Interiors*, **94**, 241–267.

Zhong, S. (2005). Dynamics of thermal plumes in three-dimensional isoviscous thermal convection. *Geophysical Journal International*, **162**, 289–300.

Zhong, S. (2006). Constraints on thermochemical convection of the mantle from plume heat flux, plume excess temperature, and upper mantle temperature. *Journal of Geophysical Research*, **111**, doi:10.1029/2005JB003972.

Zhong, S. and Gurnis, M. (1992). Viscous-flow model of a subduction zone with a faulted lithosphere–long and short wavelength topography, gravity and geoid. *Geophysical Research Letters*, **19**, 1891–1894.

Zhong, S. and Gurnis, M. (1993). Dynamic feedback between a continent-like raft and thermal convection. *Journal of Geophysical Research*, **98**, 12 219–12 232.

Zhong, S. and Gurnis, M. (1995). Mantle convection with plates and mobile, faulted plate margins. *Science*, **267**, 838–843.

Zhong, S. and Gurnis, M. (1996). Interaction of weak faults and non-Newtonian rheology produces plate-tectonics in a 3d model of mantle flow. *Nature*, **383**, 245–247.

Zhong, S., Gurnis, M. and Moresi, L. (1996). Free-surface formulation of mantle convection. 1. Basic theory and application to plumes. *Geophysical Journal International*, **127**, 708–718.

Zhong, S., Zuber, M. T., Moresi, L. and Gurnis, M. (2000). Role of temperature-dependent viscosity and surface plates in spherical shell models of mantle convection. *Journal of Geophysical Research*, **105**, 11 063–82.

Zhong S., McNamara, A., Tan, E., Moresi, L. and Gurnis, M. (2008). A benchmark study on mantle convection in a 3-D spherical shell using CitcomS. *Geochemistry, Geophysics, Geosystems*, **9**, Q10017, doi:10.1029/2008GC002048.

Zienkiewicz, O. C. and Taylor, R. L. (2000). *Finite Element Method*, 5th edn. Oxford: Butterworth-Heinemann.

Zou, X., Navon, I. M., Berger, M. *et al.* (1993). Numerical experience with limited-memory quasi-Newton and truncated Newton methods. *SIAM Journal of Optimization*, **3**, 582–608.

Author index

Subject index

Printed in the United States
by Baker & Taylor Publisher Services